Manufacturing Assembly Handbook

Manufacturing Assembly Handbook

Bruno Lotter

Butterworths

London Boston Singapore Sydney Toronto Wellington

⬢ PART OF REED INTERNATIONAL P.L.C.

All rights reserved. No part of this publication may be reproduced or transmitted in any form or by any means (including photocopying and recording) without the written permission of the copyright holder except in accordance with the provisions of the Copyright Act 1956 (as amended) or under the terms of a licence issued by the Copyright Licensing Agency Ltd, 33–34 Alfred Place, London, England WC1E 7DP. The written permission of the copyright holder must also be obtained before any part of this publication is stored in a retrieval system of any nature. Applications for the copyright holder's written permission to reproduce, transmit or store in a retrieval system any part of this publication should be addressed to the Publishers.

Warning: The doing of an unauthorized act in relation to a copyright work may result in both a civil claim for damages and criminal prosecution.

This book is sold subject to the Standard Conditions of Sale of Net Books and may not be resold in the UK below the net price given by the Publishers in their current price list.

English edition first published 1989

Originally published as *Wirtschaftliche Montage – Ein Handbuch für Elektrogerätebau und Feinwerktechnik* by VDI-Verlag GmbH, Düsseldorf, West Germany
© **VDI-Verlag GmbH, Düsseldorf 1986**

British Library Cataloguing in Publication Data

Lotter, Bruno
    Manufacturing assembly handbook.
    1. Manufactured goods. Assembly
    I. Title II. Wirtschaftliche Montage. *English*
    670.42

ISBN 0-408-03561-7

Library of Congress Cataloguing in Publication Data applied for

Filmset by Mid-County Press, London SW15 2NW
Printed and bound by Hartnoll Ltd, Bodmin, Cornwall

Preface

The rational assembly of electrical and precision engineering products such as domestic appliances, electrical products, do-it-yourself equipment and car accessories has become a decisive competitive factor for quality improvements and cost reductions. The range of solutions offered extends from the ergonomically designed manual assembly station up to highly flexible robot-equipped and computer-controlled assembly cells and systems. In addition to product and operational knowledge, a whole range of methodic and technical expertise in the field of assembly technology is required to identify the most economic option in each and every case and which, by and large, has not previously been available in a practically relevant form for this product range.

It is therefore particularly gratifying when such a renowned expert as Bruno Lotter, the Technical Director of an electrical engineering company with production plants at home and overseas, combines his professional experience covering more than 20 years in a comprehensive book. The realization process for an assembly device from the user's point of view was taken as a guideline, the procedure in each phase being illustrated by numerous practical examples.

Precision and electrical engineering companies therefore have a guideline at their disposal for the planning and realization of assembly equipment which thoroughly examines all aspects of this modern branch of production technology and gives rise to a whole range of scientific ideas for independent problem solution. I wish the book rapid success and am convinced that it will occupy a permanent position amongst the literature of the relatively young science of assembly technology because it addresses itself to both the practising and student engineer alike.

Hanover, April 1986 *Hans-Peter Wiendahl*

Foreword

Today, virtually all companies are faced with the urgent necessity to either maintain or regain their competitiveness. Increased productivity is an important requirement in achieving this aim. Since, however, significant advances in productivity are no longer expected in the classical fields of production such as metal machining and forming, etc. increased attention is being focused on the rationalization of assembly processes, and particularly so since the proportion of the assembly costs related to the manufacturing costs of a product is continually increasing. The assembly process must therefore be highly optimized, mechanized and automated in companies which manufacture assembly-intensive products. Assembly can be rationalized from the manual single-station operation up to the point of automation by organizational and technical measures. The possible level of automation of a product is principally determined by two factors, namely the production rate and the product design.

The prime object of this book is to identify clearly the possibilities for the rationalization of assembly in relation to the above-mentioned factors. In this respect, an assembly-oriented product design forms the basis for both economic manual assembly and also the fundamental requirement for automated assembly. For this reason, particular emphasis is given to assembly-oriented product design. At low production rates or with complicated assembly processes, in the foreseeable future optimized manual assembly will remain the most economic. For this reason, in this book, assembly technology is not only viewed from the aspect of automation but covers the whole range of systems from entirely manual assembly to fully automated assembly systems.

This book is written based on practical experience for practical application and will give experts in the field of rationalization guidelines for the solution of rationalization problems. Furthermore, it will be of assistance to training establishments and, in addition, give students of production technology a general insight into the state of assembly technology, into systematic planning and realization procedures for assembly operations.

Thanks are extended to companies and institutions for the kind provision of material for the presentation of this book and also those who have assisted me by the sacrifice of their spare time. Special thanks are extended to Herr Professor Dr-Ing. Hans-Peter Wiendahl, Director of the Institut für Fabrikanlagen at Hanover University, for his many valued suggestions and the critical review of the manuscript. Herr Dr Klinger of Messrs Festo is also thanked for his assistance in the preparation of the chapter on controllers. Gratitude is likewise expressed to VDI-Verlag Düsseldorf for their close cooperation and production of the book.

Sulzfeld, May 1986 *Bruno Lotter*

Contents

Preface   v

Foreword   vii

1.   Introduction   1
1.1   Assembly   1
1.2   Status of assembly in the production operation   2

2.   Product design as a requirement for economic assembly   4
2.1   Product design   5
    2.1.1   Base part   6
    2.1.2   Number of parts   7
2.2   Assembly-extended ABC analysis   9
    2.2.1   Fundamental question 1: Price of individual parts and their manufacturing costs   9
    2.2.2   Fundamental question 2: Supply condition   11
    2.2.2.1   Bulk material   11
    2.2.2.2   Formatted packaging   11
    2.2.2.3   Magazining   12
    2.2.2.4   Strip parts, flow material, belted material   12
    2.2.3   Fundamental question 3: Ease of handling   13
    2.2.3.1   Ease of arrangement   14
    2.2.3.2   Material transfer   18
    2.2.4   Fundamental question 4: Assembly direction and ease of assembly   20
    2.2.4.1   Assembly direction   20
    2.2.4.2   Ease of assembly   21
    2.2.4.3   Assembly working spaces   22
    2.2.4.4   Stability   24
    2.2.5   Fundamental question 5: Assembly methods   26
    2.2.5.1   Screw connections   26
    2.2.5.2   Selection of assembly methods   27
    2.2.5.3   Examples   28
    2.2.6   Fundamental question 6: Quality   29
    2.2.6.1   Parts quality   30
    2.2.6.2   Cost penalties of poor parts quality   31

x Contents

	2.2.7	Fundamental question 7: Assembly costs 33
	2.2.8	Organizational implementation of the assembly-extended ABC analysis 33

3. Manual assembly 36

3.1 Introduction 36

3.2 Principles of work-point arrangement 37

3.3 Organizational forms of manual assembly 39
    3.3.1 Single-point assembly 40
    3.3.1.1 Arrangement examples for single-point assembly 40
    3.3.2 Line assembly 44
    3.3.2.1 Line assembly by manual transfer of the assembled part 44
    3.3.2.2 Line assembly by mechanical transfer of the assembled part in an unarranged form 45
    3.3.2.3 Line assembly by mechanical transfer of the assembled part in an arranged form 46
    3.3.2.3.1 Workpiece carriers 53
    3.3.2.3.2 Cycling 54
    3.3.2.3.3 Rating of line assembly by mechanical transfer of the assembled part in an arranged form 56

4. Primary–secondary analysis – an aid for the determination of the economic efficiency of assembly concepts 57

4.1 Introduction 57

4.2 Definition of the efficiency of assembly operations 57

4.3 Field of application 58
    4.3.1 Basic analysis 58
    4.3.2 Fine analysis of single assembly work points in terms of primary and secondary processes 61
    4.3.2.1 Reaching 62
    4.3.2.2 Gripping 64
    4.3.2.3 Collection 64
    4.3.2.4 Assembly 64
    4.3.2.5 Release 65

4.4 Application example of assembly analysis by primary and secondary activity 65
    4.4.1 Single assembly work point with provision of parts in manual parts dispensers 66
    4.4.2 Single assembly work point with parts provision by a parts paternoster 68
    4.4.3 Single assembly work point, parts provision partly by manual parts dispensers and partly by vibratory spiral conveyors 69
    4.4.4 Linking of three single assembly work points to form a line assembly with manual transfer of the assembled part 70

	4.4.5	Linking of three single assembly work points to form a line assembly with mechanical transfer of the assembled part in workpiece carriers 72
	4.4.6	Summary and efficiency consideration 74
	4.4.7	Primary–secondary fine analysis for the handling and assembly of a single part 77
4.5		Extended analysis in terms of primary and secondary requirements for the total sequence of an assembly operation 79
4.6		Practical examples 81
	4.6.1	Example 1: Switch assembly 81
	4.6.1.1	Initial basis 81
	4.6.1.2	Solution proposal 83
	4.6.1.3	Assembly cycle 83
	4.6.1.4	PAP–SAP analysis 87
	4.6.2	Example 2: Switch element 88
	4.6.2.1	Initial basis 88
	4.6.2.2	Solution proposal 89
	4.6.3	Example 3: Headlight assembly 93
	4.6.3.1	Initial basis 93
	4.6.3.2	Solution proposal 94

5. Modules for the automation of assembly processes   96

5.1		Introduction 96
	5.1.1	Handling 97
5.2		Feeder units 102
	5.2.1	Feeder units for parts with one arrangement feature 103
	5.2.1.1	Hopper with scoop segment 103
	5.2.1.2	Hopper with blade wheel 104
	5.2.1.3	Hopper with magnetic plate discharging system 105
	5.2.1.4	Inclined conveyors 107
	5.2.2	Feeder units for parts with several arrangement criteria 108
	5.2.2.1	Arrangement of parts in vibratory spiral conveyors 109
	5.2.2.2	Types of vibratory spiral conveyors 110
	5.2.2.3	Discharge rails 113
	5.2.2.4	Parts separation – removal – distribution 117
	5.2.2.5	Secondary sorting equipment 122
	5.2.2.6	Capacity of vibratory spiral conveyors 123
	5.2.3	Electronic position identification of parts 125
	5.2.4	The feed of interlocking parts 126
5.3		Handling equipment 129
	5.3.1	Positioning units 130
	5.3.1.1	Drives 130
	5.3.1.2	Kinematics 131
	5.3.1.3	Control 132
	5.3.1.4	Grippers 133
	5.3.1.5	Design of positioning units 134
	5.3.2	Industrial robots 135

xii Contents

        5.3.2.1   Kinematics, arm and gripper   136
        5.3.2.2   Control   139
        5.3.2.3   Drive   141
        5.3.2.4   Measurement system   142
        5.3.2.5   Sensors   143
        5.3.2.6   Types of industrial robots   144
        5.3.2.6.1 SCARA horizontal articulated-arm robots   144
        5.3.2.6.2 Robots with translational X-, Y- and Z-axes   147

5.4   Transfer equipment   150
        5.4.1   Cycled transfer equipment   150
        5.4.1.1   Circular cyclic transfer equipment   151
        5.4.1.1.1 Pneumatically driven circular cyclic units   152
        5.4.1.1.2 Circular cyclic units with a Maltese-cross drive   153
        5.4.1.1.3 Cam drives for circular cyclic units   155
        5.4.1.2   Cyclic longitudinal transfer equipment   157
        5.4.1.2.1 Overhead and underfloor longitudinal transfer systems   159
        5.4.1.2.2 Rotary longitudinal transfer systems   159
        5.4.1.2.3 Plate longitudinal transfer systems   159
        5.4.2   Non-cycled transfer equipment   159

5.5   Screw-inserting units   161
        5.5.1   Drop-tube-type automatic screw-insertion machines   164
        5.5.2   Automatic screw inserters with feed-rail feed   166

5.6   Riveting units   168
        5.6.1   Press-riveting   168
        5.6.2   Rotating-mandrel riveting   170

5.7   Welding units   171
        5.7.1   Resistance welding   173
        5.7.2   Laser welding equipment   174

5.8   Soldering equipment   179

5.9   Bonding   181

6.   Design of assembly machines   183

6.1   Introduction   183

6.2   Single-station assembly machines   184

6.3   Multi-station assembly machines   186
        6.3.1   Design of parts feed stations   189
        6.3.2   Checking stations   190
        6.3.3   Design of pneumatically operated multi-station assembly machines   192
        6.3.4   Design of electric-motor driven multi-station assembly machines   195
        6.3.4.1   Design of assembly machines with a vertical drive shaft arrangement   195
        6.3.4.2   Assembly machines with a horizontally arranged drive shaft   198

	6.3.4.3	Performance of several simultaneous principal movements by an oscillatory drive 200

        6.3.5   Assembly machine systems   205

6.4   Combining assembly machines to form assembly lines   210

6.5   Integration of manual work points in automated assembly lines   214
        6.5.1   Manual work points for parts provision   214
        6.5.2   Manual assembly work points   216

6.6   Uncycled assembly lines including manual work points   218

6.7   Availability of assembly systems   218
        6.7.1   Parameters of the operational characteristic   219
        6.7.2   Utilization   219
        6.7.3   Factors of influence on the availability of assembly systems   220
        6.7.3.1 Number of stations   220
        6.7.3.2 Time required for the rectification of breakdowns   221
        6.7.3.3 Cycle time   222
        6.7.3.4 Quality of individual parts   223
        6.7.4   Summary   224

7. Design of flexible-assembly systems   225

7.1   Introduction   225

7.2   Primary–secondary fine analysis with the application of assembly robots   226
        7.2.1   Reaching   227
        7.2.2   Gripping   228
        7.2.3   Collecting   228
        7.2.4   Assembly   228
        7.2.5   Release   229

7.3   Working space   229

7.4   Gripper   230

7.5   The design of flexible single-station assembly cells   230
        7.5.1   Semi-automatic flexible-assembly cells   231
        7.5.2   Automatic flexible-assembly cells   232

7.6   Assembly lines with flexible-assembly cells, interconnected by manual work points   235
        7.6.1   Solution examples   235
        7.6.1.1 Example 1   235
        7.6.1.2 Example 2   237
        7.6.2   Summary   238

8. Stored program controllers [46]   239

8.1   Introduction   239

8.2   Design of stored program controllers   239
        8.2.1   Input modules   239
        8.2.2   Signal processing modules   239

xiv Contents

        8.2.3   Output modules   240
        8.2.4   Network modules   240
        8.2.5   Program   240

8.3   Programming equipment   240

8.4   Modules   241

8.5   Operating system   242

8.6   Programming of SPCs   243

8.7   Ease of maintenance   246

8.8   Availability   247

8.9   Data exchange   247

9.   Practical examples   248

9.1   Assembly machines   248
        9.1.1   Example 1: Rocker   248
        9.1.1.1 Objective   248
        9.1.1.2 Operations to be performed   249
        9.1.1.3 Criteria for method selection   249
        9.1.1.4 Description of equipment   249
        9.1.2   Example 2: Valve plate   250
        9.1.2.1 Objective   250
        9.1.2.2 Operations to be performed   252
        9.1.2.3 Criteria for method selection   252
        9.1.2.4 Description of equipment   253
        9.1.3   Example 3: Spray nozzle–spray head   255
        9.1.3.1 Objective   255
        9.1.3.2 Operations to be performed   255
        9.1.3.3 Criteria for method selection   256
        9.1.3.4 Description of equipment   257
        9.1.4   Example 4: Terminal block   260
        9.1.4.1 Objective   260
        9.1.4.2 Operations to be performed   262
        9.1.4.3 Criteria for method selection   262
        9.1.4.4 Description of equipment   262
        9.1.5   Example 5: High-pressure nozzle   264
        9.1.5.1 Objective   264
        9.1.5.2 Operations to be performed   266
        9.1.5.3 Criteria for method selection   266
        9.1.5.4 Description of equipment   266
        9.1.6   Example 6: Audio cassettes   269
        9.1.6.1 Objective   269
        9.1.6.2 Operations to be performed   270
        9.1.6.3 Criteria for method selection   270
        9.1.6.4 Description of equipment   270
        9.1.7   Example 7: Car fan motor   272
        9.1.7.1 Objective   272
        9.1.7.2 Operations to be performed   272

	9.1.7.3	Criteria for method selection 273
	9.1.7.4	Description of equipment 273
9.2	Flexible-assembly systems 275
	9.2.1	Example 1: Switch block 275
	9.2.1.1	Objective 275
	9.2.1.2	Operations to be performed 275
	9.2.1.3	Criteria for method selection 276
	9.2.1.4	Description of equipment 277
	9.2.2	Example 2: Assembly of clips on car headlights 279
	9.2.2.1	Objective 279
	9.2.2.2	Operations to be performed 279
	9.2.2.3	Criteria for method selection 280
	9.2.2.4	Description of equipment 280
	9.2.3	Example 3: Domestic appliance drive 282
	9.2.3.1	Objective 282
	9.2.3.2	Operations to be performed 283
	9.2.3.3	Criteria for method selection 283
	9.2.3.4	Description of equipment 284
	9.2.4	Example 4: Equipping of printed circuit boards 287
	9.2.4.1	Objective 287
	9.2.4.2	Operations to be performed 287
	9.2.4.3	Criteria for method selection 287
	9.2.4.4	Description of equipment 287
	9.2.5	Example 5: Auxiliary contact block 288
	9.2.5.1	Objective 288
	9.2.5.2	Operations to be performed 288
	9.2.5.3	Criteria for method selection 289
	9.2.5.4	Description of equipment 289

10. The integration of parts manufacturing processes into assembly equipment or of assembly operations into parts production equipment   292

10.1 Introduction   292

10.2 Integrated parts production   292

10.3 The production machining of parts in assembly equipment   293

10.4 Practical example: Assembly system with integrated parts production   294

10.5 The integration of assembly processes into parts production processes   297

10.6 The integration of parts production into assembly equipment within the concept of just-in-time production   299

10.7 Limits for the integration of production processes   302

11. Planning and efficiency of automated assembly systems   303

11.1 Introduction   303

11.2 Requirement list   303

xvi Contents

11.3   Product analysis   306

11.4   Assembly sequence analysis   306
    11.4.1   Product design and assembly situation   306
    11.4.1.1 Example 1   307
    11.4.1.2 Example 2   307
    11.4.1.3 Example 3   308
    11.4.2   Assembly sequence   309

11.5   Workpiece carrier design   310
    11.5.1   Introduction   310
    11.5.2   Design examples of workpiece carriers   311
    11.5.2.1 Example 1   311
    11.5.2.2 Example 2   313
    11.5.2.3 Example 3   315

11.6   Function analysis   316

11.7   Determination of cycle time   317

11.8   Layout planning   317
    11.8.1   Principles of layout planning   317
    11.8.2   Layout examples   318
    11.8.2.1 Example 1   318
    11.8.2.2 Example 2   320

11.9   Determination of personnel requirement   321

11.10 Determination of availability   323
    11.10.1   Parts quality   323
    11.10.2   Number of stations   323
    11.10.3   Availability of individual stations   323
    11.10.4   System structure   324
    11.10.5   Initial operation characteristics   324
    11.10.6   Personnel qualifications   325

11.11 Assembly systems   325
    11.11.1   Cycle time   326
    11.11.2   System structure – integration of necessary manual operations   327
    11.11.3   Conditions on the periphery of automatic assembly   327
    11.11.4   Summary   327

11.12 Investment calculations   327

11.13 Evaluation and selection   329
    11.13.1   Machine hourly rate   330
    11.13.1.1 Estimated depreciation (C_D)   330
    11.13.1.2 Estimated interest (C_I)   330
    11.13.1.3 Space costs (C_S)   330
    11.13.1.4 Energy costs (C_E)   330
    11.13.1.5 Maintenance costs (C_M)   330
    11.13.1.6 Calculation of the machine hourly rate   331
    11.13.2   Personnel-related costs   331
    11.13.3   Work-point cost calculation   331

Contents xvii

11.14 Optimized overall solution 333

11.15 Computer-aided planning of automated assembly systems 333
    11.15.1 CAD layout planning 333
    11.15.2 Simulation technique 336

12. Practical example: Planning and realization of an automated assembly system 338

12.1 Introduction 338

12.2 Planning procedure 338
    12.2.1 Requirement list 338
    12.2.2 Product analysis 338
    12.2.3 Assembly sequence analysis 341
    12.2.3.1 Product design and assembly situation 341
    12.2.3.2 Assembly sequence 343
    12.2.3.3 Workpiece carrier design 345
    12.2.4 Function analysis 345
    12.2.5 Determination of cycle time 345
    12.2.6 Layout planning 345
    12.2.7 Determination of personnel requirement 347
    12.2.8 Determination of availability 348
    12.2.8.1 Parts quality 348
    12.2.8.2 Number of stations 348
    12.2.8.3 Individual availability 348
    12.2.8.4 System structure 348
    12.2.8.5 Initial operation characteristics 349
    12.2.8.6 Personnel qualifications 349

12.3 Detailed planning of assembly system 349
    12.3.1 Introduction 349
    12.3.2 Machine I 350
    12.3.3 Machine II 352
    12.3.4 Machine III 354
    12.3.5 Machine IV 355
    12.3.6 Machine V 357
    12.3.7 Machine VI 357
    12.3.8 Machine VII 359
    12.3.9 Machine VIII 361
    12.3.10 Machine IX 362

12.4 Investment calculations 364

12.5 Evaluation and selection/work point cost comparison 364

12.6 Investment risks 365

13. The operation of automated assembly systems 370

13.1 Prerequisites for initial running 370
    13.1.1 Parts quality 370
    13.1.2 Functional reliability of the system 371
    13.1.2.1 Rough analysis of the fault operation characteristic 371

        13.1.2.2 Derailed analysis of the fault operation characteristic   372
        13.1.3   Practical example of fault time determination by MANALYS on an automatic assembly system   375

13.2 Payment   376

13.3 Maintenance   377

13.4 Work safety   380

14. Outlook   381

15. References   384

16. Index   387

Chapter 1

Introduction

1.1 Assembly

As a general rule, industrially produced products are formed from a large number of single parts produced at different times and by various production processes. The object of assembly is to form a part of higher complexity with specified functions in a specific period of time from the individual parts [1].

As specified in DIN 3593, assembly operations are processes of assembling together and as defined by VDI-guidelines 3239, 3240 and 3244 are functions of workpiece manipulation. The principal activities during assembly are shown in Figure 1.1 including operations such as adjustment and inspection and also the performance of secondary functions such as cleaning and deburring, etc.

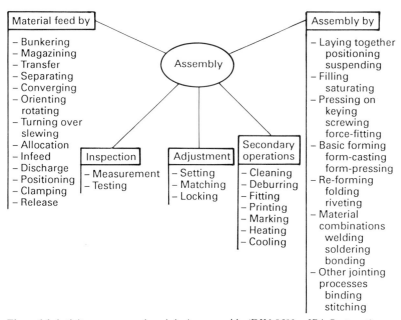

Figure 1.1 Activity groups employed during assembly (DIN 8593 – IPA Stuttgart)

1.2 Status of assembly in the production operation

In industrial production, the classical production technologies such as metal cutting, re-forming and basic forming, etc. have achieved a high level of automation. Parts are produced at low cost. On the other hand, assembly is largely a manual operation with the interplay of many different functions. In the fields of electrical and precision engineering technology, the wage cost components for assembly are between 25% and 70% of the total wage costs.

The wage costs included in the manufacturing costs of a product are segregated into the following components:

- metal-cutting production
- non-cutting production
- moulded parts production
- assembly
- testing.

It is clearly shown that the assembly operations are the most expensive production stage. Figure 1.2 shows an analysis of three selected products. The rationalization

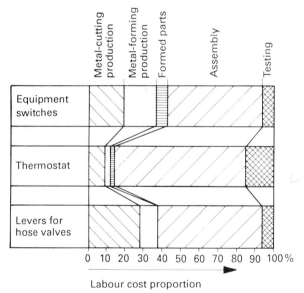

Figure 1.2 Composition of wage costs for selected precision engineering products

potential for assembly is quite evident [2]. However, comparable successes, for example in metal-cutting production technology, are yet to be seen. The principal obstacles to rapid rationalization achievements in assembly are as follows [3]:

- The product service life is decreasing, the batch sizes are smaller and the number of variants greater.
- Assembly is the last production stage and must continuously be adapted to changing market demands.

- All the errors in planning, product development and initial production directly affect assembly.
- The physical work components of assembly are widely varied and necessitate a high labour involvement to achieve higher flexibility.
- Product-specific assembly techniques and organizational forms such as individual work stations, line assembly, erection point assembly and automatic assembly obstruct the implementation of once established solutions to other products.
- Products are not designed in terms of assembly considerations.

The rationalization of assembly can only be successful if the existing product design is also brought into question. This aspect will therefore be the subject of closer consideration on the basis of practically considered fundamental questions.

Chapter 2

Product design as a requirement for economic assembly

The development of a product is characterized by the specified product function. The production processes are largely dictated by the design of the individual parts and the product assembly arrangement.

Figure 2.1 shows a correlation of cost responsibility and origination of the principal company areas of Design, Parts Manufacture and Assembly in accordance with Reference 4. It is clear that the Design Department only accounts for 12% of the manufacturing costs. However, it largely determines the production processes by the selection of material, tolerances and dimensions, etc. and is therefore responsible for 75% of the manufacturing costs. Since assembly is the most labour-intensive activity, particular importance is attached to assembly-oriented product design. An assembly-oriented product design is not only the fundamental requirement for automatic assembly, it also forms the basis for economic manual assembly. Its principal objectives are shown in Figure 2.2. In addition to simplification and the directly linked easier control of assembly processes, a higher repeat frequency of assembly processes is also achieved in spite of smaller batch sizes and a large number of product variants [5].

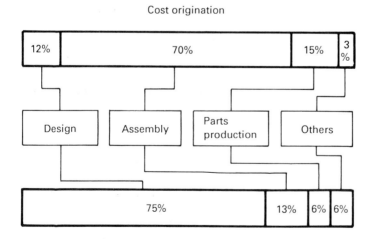

Figure 2.1 Cost responsibility – cost origination (Gairola)

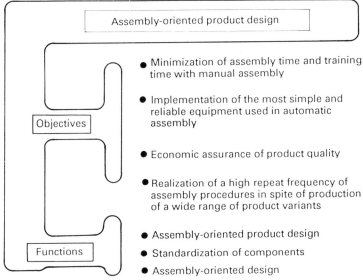

Figure 2.2 Objectives of assembly-oriented product design (Platos–Witte)

2.1 Product design

The degree of difficulty of assembly increases with the complexity of the product. Products with a small number of parts are easier to design to be assembly-oriented. As a general rule, resulting from the small number of parts or assembly operations, such products are finish-assembled at one assembly point. On the other hand, the assembly operations must be subdivided for products with a larger number of parts. A condition for this is that a product must be designed so that subassemblies can be pre-assembled. Figure 2.3 shows schematically a subdivision into subassemblies from pre-assembly to final assembly corresponding to assembly progress.

The following rules should be observed for the assembly-oriented design of subassemblies and final products:

- The whole assembly operation must be subdivided into clearly discernible stages by suitable subassembly formation.
- A subassembly must be completed as a unit so that it can continue to be handled and manipulated as a single part.
- It must be possible to test a subassembly separately.
- Every subassembly should have the minimum possible number of connections to other subassemblies.
- Variant-dependent subassemblies should not be included together with variant-neutral subassemblies.
- As far as possible, variant subassemblies should have an equal number of installation conditions.

To comply with these rules, a so-called base part is of prime importance for the individual subassemblies and the end product [5].

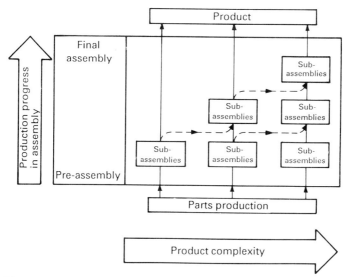

Figure 2.3 Assembly-oriented product subdivision into subassemblies

2.1.1 Base part

A base part is defined as the principal part (quite often the base plate or housing) to which further parts are attached in the course of assembly. This term is applicable to both subassemblies and end products. In the field of electronics, a classic example of a base part is the printed circuit board on which all other components are mounted.

In the assembly procedure, the base part need not be the first part to be handled; it is rather the part on to which other parts are mounted. It should be designed so that, by subdividing the assembly procedure, transfer from one assembly point to another is possible without any special device. This requirement cannot always be realized. However, with manual assembly in particular, which functions without a workpiece fixture, it should be a specific objective. With progressive assembly with circulating workpiece carriers or automated assembly with workpiece fixtures, to some degree the workpiece carrier or the assembly fixture can assume the function of the base part.

Base parts must be designed so that during assembly they are self-centring in the workpiece fixture. The requirement for clamping and therefore the provision of clamping faces on the part should be largely avoided, since clamping workpiece fixtures are more complicated and expensive than workpiece carriers without clamping equipment. With a subdivided assembly operation, i.e. with transfer from one assembly point to another, workpiece fixtures with clamping equipment represent a considerable financial outlay.

With a subdivided assembly, and particularly with mechanized assembly operations, the centring accuracy of the base part in the workpiece fixture is important. Base plates or housings as base parts must be designed so that with external centring their external tolerances (see Figure 2.4(a)) must be selected so that with regard to an automatic assembly procedure, the joint points inside the base part can be positioned with adequate accuracy. If this cannot be achieved by external centring, the base part must be equipped with centring holes as shown in Figure 2.4(b). With regard to their

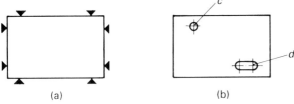

Figure 2.4 Centring of base parts: (a) external centring; (b) centring by a hole and slot

tolerances in relation to the function positions of the base part, the centring holes must be dimensioned so that the function points of the base part can be operated and manipulated via the centring holes. If the base part is located by centring pins, to avoid static reversibility, one centring hole is to be circular (c) and the second (d) in the form of a slot (see Figure 2.4(b)), [2,5].

Clamping workpiece carriers should function without an energy supply. As an example, Figure 2.5 shows the clamping of a right-angular base part with two reference faces *a* and *b*. Clamping is by a slide with an eccentric actuating mechanism.

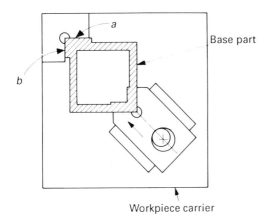

Figure 2.5 Clamping of a base part in a workpiece carrier

2.1.2 Number of parts

Minimization of the number of parts in a product is a highly important objective for assembly-oriented product design. Every part avoided need not be designed, progressed and assembled.

The use of modern materials and production methods facilitates the single-piece design form of parts which previously had to be manufactured from several parts. Figure 2.6 shows one such solution. Three parts manufactured from standardized parts means handling and assembly costs for three parts. The redesign of these three assembled parts in the form of a single part produced as a special part does indeed incur additional tooling costs but reduces the costs for handling and assembly to a single operation and is therefore more economic overall [6].

In the field of precision engineering and electrical equipment production, the use of high-performance plastics presents excellent design opportunities for parts reduction.

8   Product design as a requirement for economic assembly

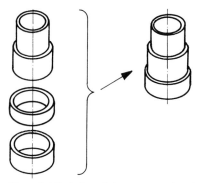

Figure 2.6 Redesign of three parts reduced to one part (Treer)

Figure 2.7 shows a further example. In the initial design of a pneumatic piston, the assembly consisted of seven parts. The problem in assembly was the need to hold the cover plate against the helical compression spring in order to fit the screws. A design revision within the context of assembly-oriented product form resulted in a cover design with an integral connection technique in the form of a snap connection between the cover and housing and an alteration of the piston to give location guidance of the helical compression spring. The number of individual parts was reduced from seven to four, and the expensive assembly operation 'insert screws' was also saved.

Figure 2.8 shows an example of a parts reduction possibility on a complex product. A switch element formed by 21 parts was difficult to assemble on account of the product design. Chiefly, a large proportion of the parts are unsuitable for automatic handling. A design revision along the lines for assembly-oriented product design resulted in a reduction to 16 parts. The proportion of parts which cannot be handled automatically was halved by the design revision [7].

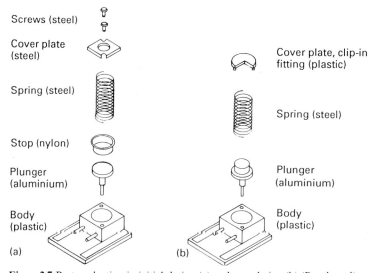

Figure 2.7 Parts reduction in initial design (a) and new design (b) (Boothroyd)

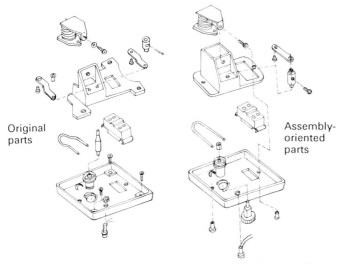

Original parts

Assembly-oriented parts

Figure 2.8 Parts reduction by assembly-oriented design (Boothroyd)

2.2 Assembly-extended ABC analysis

By reason of its specific requirements, every product has its own particular interrelationship in the product design and the resultant production methods. The greatest effect is achieved by the assembly-oriented form in the draft design phase of a product. A finally developed product can only be altered at considerable cost during its production period. The assembly-oriented product design is not only a technical but also an organizational and therefore a personnel problem. Simple and intelligible procedural directives must therefore be available to support the designers trained to think in unit cost terms in assembly-oriented product design.

The assembly-extended ABC analysis is one of these means for the assembly-oriented design of a product. This is derived from the generally known ABC analysis in the fields of company economics and value analysis.

The basic concept is expressed in the following question:

What does a part or subassembly cost up to the point where its required function is achieved following final assembly?

These seven fundamental questions — the basis of the assembly-extended ABC analysis — are illustrated in Figure 2.9. These fundamental questions require detailed concepts as answers which are formulated by a dialogue between product development and production planning. None of the seven fundamental questions can be answered in isolation. They all mutually affect each other [8,9].

2.2.1 Fundamental question 1: Price of individual parts and their manufacturing costs

The manufacturing costs of a part are principally determined by the following factors:

- Material selection
- Manufacturing process
- Batch size.

10 Product design as a requirement for economic assembly

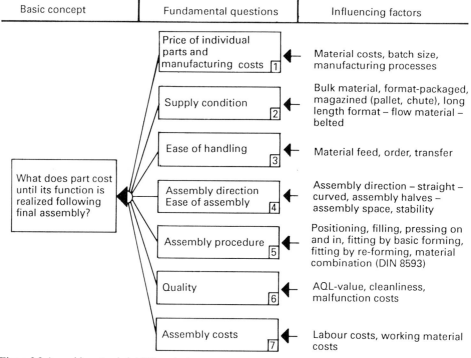

Figure 2.9 Assembly-extended ABC analysis

The specification of the material (raw material) affects the manufacturing process and consequently the supply condition of the individual part to assemble. If, for example, in the design of a punched-bent part, it is specified that it is to be galvanized to give surface protection, bulk material supply is perfectly satisfactory. The result of this production process is that, starting from the punching stage up to the point of supply to assembly, the individual parts are subjected to several intermediate handling procedures in the course of galvanizing by emptying transport bins into the galvanizing drums and then back into the bins. Consequently, the quality of the parts is not improved. Every intermediate handling process can result in deformations, the inclusion of foreign bodies or contamination of the parts. In turn, this results in disruptions in the assembly processes with considerable cost penalties. The possibility of part magazing is ruled out with such a production process.

In this case during the design of the part it should be considered whether the use of a higher-quality raw material could render surface treatment in the Galvanizing Department superfluous and, in so doing, to guarantee an assembly of suitable quality. Under certain circumstances, any such part manufacturing process in a production process lends itself to magazined production for the avoidance of bulk material (see Section 2.2.2).

The same applies to heat treatment. If heat-treatment processes are necessary on parts and they normally take place in a random order. Product Development should investigate if the process can be avoided by the use of a heat-treated raw material. High-form-stability, easily positionable parts are less disadvantageously affected by intermediate handling operations, galvanic or thermal processes than sensitive parts which are difficult to position.

The continuous standardization of single parts or the creation of part families with uniform production processes makes larger batch sizes possible in single-part production and consequently also a minimization of machine tooling time. In addition, the use of standardized parts leads to a higher proportion of repeat activities for assembly.

2.2.2 Fundamental question 2: Supply condition

To a large degree, together with fundamental questions 3 (ease of handling) and 4 (assembly direction and ease of assembly), the supply condition of single parts determines the economic form of assembly and its suitability for automation. The planned supply condition of single parts also affects the manufacturing process of these parts.

The most frequent supply conditions are:

- bulk material
- formatted packaging
- magazine (pallet, chute, etc.)
- bundled
- long length format.

The prime objective of part production must be that a once-achieved, correctly arranged condition is maintained in order to avoid repeated rearrangement.

2.2.2.1 Bulk material

Bulk material supply of single parts to assembly means that they must be arranged in the correct position for fitting. Manual arrangement incurs a high work involvement, and automatic arrangement a high level of capital investment. With bulk material supply, the size and weight of the bins containing the single parts is important. Large bins necessitate refilling into hand trays or automatic positioning and feed systems, for example vibratory spiral conveyors. If, during assembly, a part is taken directly from a large bin, the use of lifting equipment is necessary for ergonomic material availability. Every material transfer operation has the danger of part deformation, contamination and the inclusion of foreign bodies. The provision of bulk material in small containers without transfer for direct use is preferable to provision by large bins. Figure 2.10 shows the basic concept in the form of a schematic comparison of handling with large and small bins [2, 43].

2.2.2.2 Formatted packaging

As a general rule, complex, sensitive parts or subassemblies are not packed as bulk material, and hence unformatted, but rather formatted and, with further manual processing, facilitate an orderly presentation of the parts or subassemblies from the packaging at the assembly point. With automative further processing in assembly, the positional accuracy of the parts in the packaging is not adequate. If a single part or subassembly has to be 100% inspected before packaging, formatted packaging does not represent any additional expense and should, therefore, without question be used with further manual processing.

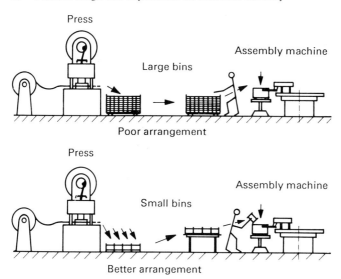
Figure 2.10 Bulk material handling between part production and assembly

2.2.2.3 Magazining
Magazined supply of parts has the advantage that they can be fed in an arranged form to the assembly operation. An investigation should be undertaken within Production to establish if magazining can be linked to a part-production process which covers the requirement for magazining. The ordered condition achieved can then be retained in order to avoid part-arrangement operations in assembly. Transportable magazines are a basic requirement.

In general, magazines can be subdivided into flat (pallettes) and chute magazines. Three different types of magazine are shown in Figure 2.11 independently of the part or subassembly form. Example A shows a flat magazine for pre-assembled micro-switch elements which holds 200 arranged elements in four rows. The flat magazine is a single-piece thermoplastic injection moulding and is interstackable. Example B shows a chute magazine in zig-zag form for the magazining of symmetrical cylindrical parts. The zig-zag arrangement relieves the sorting device of the magazine outlet point from the static pressure of the magazined parts. This type of magazine can either be filled directly on a lathe or during subsequent operations, e.g. inspection. Example C shows a chute magazine for plate parts. These magazines can be loaded directly during the punching operation and have a high capacity, particularly with the magazining of thin parts.

2.2.2.4 Strip parts, flow material, belted material
The most reliable form of feed known from the field of punching is long length feed, i.e. in the form of a strip. With reference to this technique, with the feed of pre-punched parts but with residual interconnecting webs in a strip, reference is made to flow material feed. The advantage of this type of feed is the retention of the arranged condition of the parts and therefore the realization of a high material availability in the material feed. Figure 2.12 shows the simplified solution by an example. The single part is fed as flow material and is separated from the strip immediately before fitting. With punched-bent parts, reattaching to strips is difficult or quite often even impossible. In this case, the bending must be included in the separating operation.

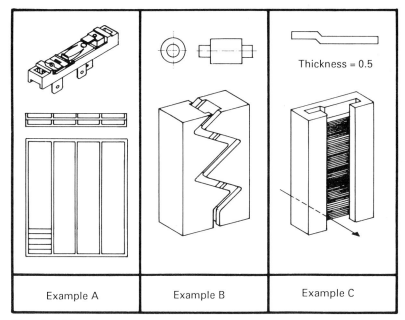

Figure 2.11 Magazine types

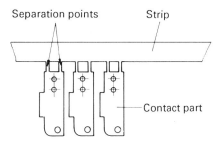

Figure 2.12 Flow material

Electronic components are preferably supplied in belt form. For fitting, the components are separated from the belt and, at the same time, the connection wires bent to the required hole spacing. The principle is illustrated in Figure 2.13.

Plastic injection-moulded parts can also be produced in belt form by injecting a virtually endless wire. An example is shown in Figure 2.14. The plastic parts are cut from the wire for fitting. The residual wire remains in the injection-moulded parts.

2.2.3 Fundamental question 3: Ease of handling

Details of the parameters and characteristics which defined the attitude of a single part at rest, in motion and in assembly are necessary for a systematic examination of single parts with regard to their suitable assembly design in order to identify their effect on automated handling. Single parts supplied as bulk material must have features for

14 Product design as a requirement for economic assembly

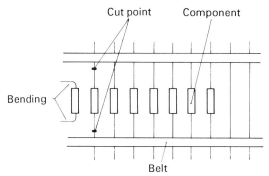

Figure 2.13 The belting of components

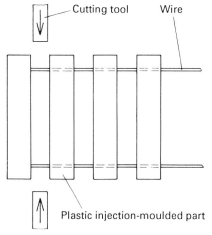

Figure 2.14 The belting of plastic parts

automatic arrangement in the required fitting position and also features for arranged transfer in the course of assembly.

2.2.3.1 Ease of arrangement
The performance of an automated feeder unit depends upon by how many features a part can be arranged. With manual feed also, the number of arrangement features is a factor in the performance. In accordance with MTM (Method Time Measurement, a system of pre-determined times) [44], the basic movement to grasp is differentiated in terms of its involvement as to whether or not a part can be grasped without eye contact and brought under control or if an additional action, for example regripping, reaching over, selecting, etc. is necessary.

Figure 2.15 demonstrates how large the number of theoretical workpiece positions can be for right-angular-bent sheet-metal parts with different flask lengths. By reason of their geometry, single parts have a preferred position. Respective examples are shown in Figure 2.16. Depending upon the part geometry, with many parts, the frequency of occurrence of a preferred position is not adequate in order to achieve a high performance of an automatic feed unit. This is demonstrated in Figure 2.17 by a

Assembly-extended ABC analysis 15

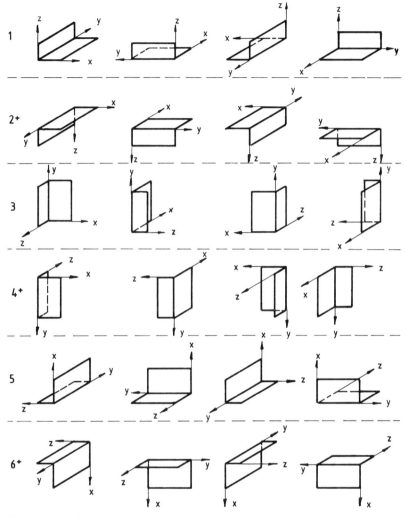

* realistically non-existent

Figure 2.15 Theoretically possible workpiece positions of a simple workpiece (Dolezalek)

sheet-metal part. By reason of its geometry, the flat position is the preferred position without any arrangement in terms of the asymmetrical outer contours. Automatic arrangement is made difficult in that the differentiating feature of the outer contour is very small and difficult to detect. If the design of the part can, for example, be modified by cutting and bending two tongues, an additional arrangement feature is formed (see Figure 2.17), and with automatic feed gives the possibility of arrangement by an arrangement feature.

The term 'arrangement difficulty' is of some assistance in forming an estimation of the arranging involvement. It is based on the question of how many angular rotations a part must describe from an arbitrary preferred position to be brought into a specified

16 Product design as a requirement for economic assembly

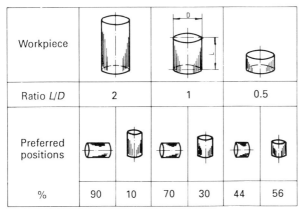

Figure 2.16 Preferred positions of cylindrical workpieces (IPA Stuttgart)

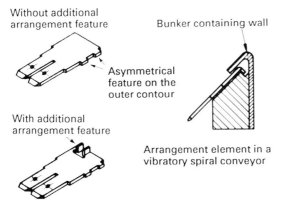

Figure 2.17 Addition of an arrangement feature on an asymmetrical part for improvement of the arrangement unit

position which, for example, is the required grasping position immediately before assembly.

The arrangement difficulty (D) can then be calculated by the following formula from the number of required angular rotations (R) and the number of opposite, asymmetrical sides (S):

$$D = R + S$$

Figure 2.18 illustrates the definition of arrangement difficulty by examples and Figure 2.19 some simple examples for the calculation of the arrangement difficulty.

Figure 2.20 shows six examples for the correct design of single parts. By a symmetrical design of example (a), the single part can be arranged in its preferred position and does not need to be sorted in relation to its end faces. Cylindrical parts as shown in (b), whose diameter to length ratio is not 1:1, can be fed with a higher work involvement. Open annular parts as shown in (c) have a tendency to interlock. Overlapping of the ends prevents interlocking, and the single parts can then be fed. The arrangement of cylindrical flat parts as shown in (d) is only possible by testing on the internal contours. Testing can only be avoided by providing an external chamfer on

Assembly-extended ABC analysis 17

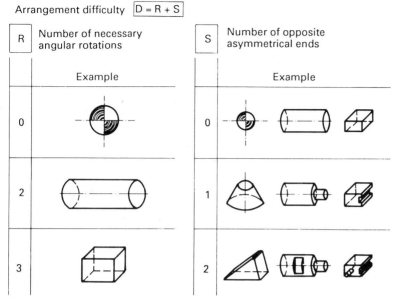

Figure 2.18 Arrangement difficulty (IPA Stuttgart)

one outer face of the part. The forming of a hook on the sheet-metal part shown in (e) makes it possible to arrange it by design feature (see also Figure 2.17). Particular difficulty is experienced with feeding coiled parts (coiled springs, hooks, spring washers, etc.). Example (f) shows a coil spring with and without closed ends. Coil springs with non-closed ends have a tendency to interlock.

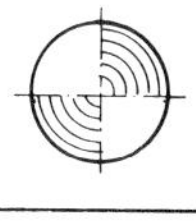

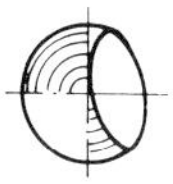

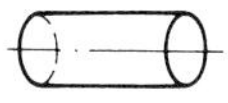

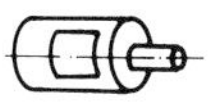

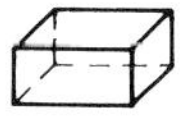

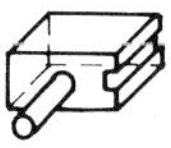

Figure 2.19 Examples of calculations for arrangement difficulty (IPA Stuttgart)

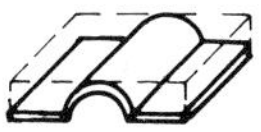

18   Product design as a requirement for economic assembly

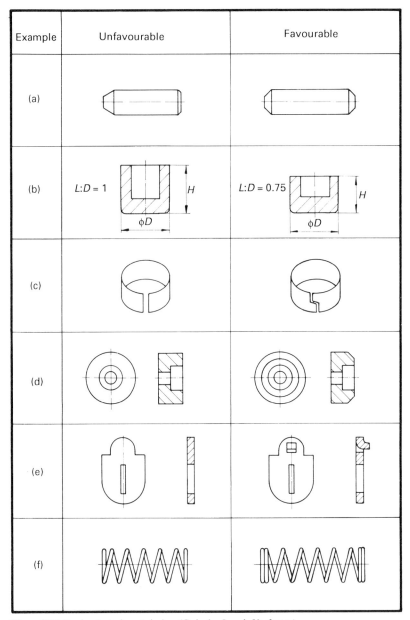

Figure 2.20 Feed-oriented part design (Gairola, Lund, Verfasser)

2.2.3.2 Material transfer

If a part is arranged correctly, it must be transferred to the work point. The work point can be a separating station for grasping an assembly or machining station. With assembly-oriented part design it must promote an arranged and non-self-locking material transfer. The effective criteria for material transfer are shown schematically in

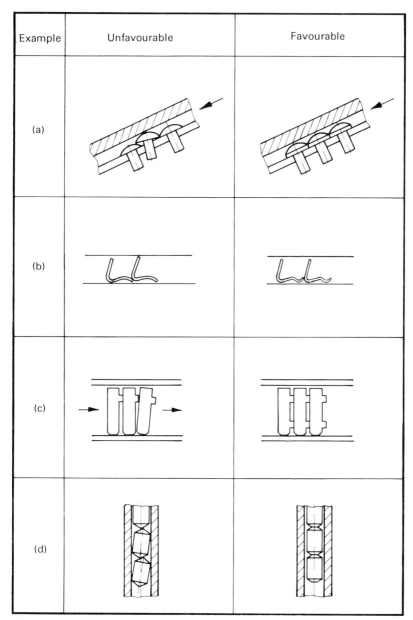

Figure 2.21 Good and bad component formations for transfer (Gairola, Verfasser)

Figure 2.21. Example (a), a mushroom-head part with a sharp head edge, leads to overriding and consequently obstructs transfer.

Overriding and blocking of adjacent parts is avoided by cylindrically forming the head edge. Example (b) shows a sheet-metal formed part. Jamming with column form sliding is avoided by redesigning to form a part with contact faces. Example (c) shows

20 Product design as a requirement for economic assembly

that an asymmetrical sheet-metal part tilts when sliding in column form and therefore, above a certain slide feeder length, results in jamming. A symmetrical design avoids tilting and therefore also jamming. Example (c) illustrates the wedging of cylindrical parts with a double end face 90° apex in a tube magazine used for transfer. Wedging is prevented by truncating the apex on both end faces.

2.2.4 Fundamental question 4: Assembly direction and ease of assembly

2.2.4.1 Assembly direction

The assembly direction of single parts should preferably be in a vertical straight line, and particularly so if the assembly procedure includes press-on or -in operations and/or re-forming operations. Horizontal assembly operations can only be undertaken without difficulty if the assembly forces which occur are low. Non-linear assembly operations are necessary if the assembled part is obstructed in a linear assembly direction. The obstruction can be by the design arrangement of the base part. Figure 2.22 shows unobstructed, linear assembly operations in a vertical and horizontal form. Two assembly directions are necessary if an enclosed assembly operation must be undertaken on account of obstruction by the base part. Figure 2.22 shows one such example. Assembly operations of this type are difficult to automate.

Differing assembly directions increase the cost of assembly and, particularly so if the assembly operations include pressing on and in or re-forming. Figure 2.23 shows an example with two assembly directions. Part 1 is assembled vertically and connected to the base part by re-forming. Part 2 has a horizontal assembly direction. The assembly forces which occur have equal magnitude. To support the forces without danger of deformation of the base part, rotation of the pre-assembled part for the assembly of part 2 is necessary.

The rotation operation of the base part increases the assembly costs.

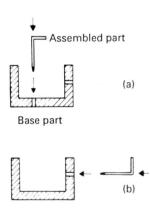

Figure 2.22 Assembly directions: (a) vertical; (b) horizontal and (c) in two directions

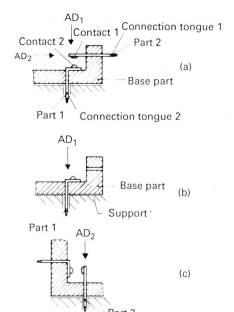

Figure 2.23 Assembly of two parts with different assembly directions (AD = assembly direction)
(a) Finished part
(b) Assembly position part 1
(c) Assembly position part 2

Under the assumption that the function of the subassembly is determined by the distance between the contact faces of parts 1 and 2 and not by the direction of the connection tongues, the base part can be redesigned so that the assembly direction for parts 1 and 2 is the same.

Figure 2.24 shows a possible solution for uniform assembly directions which avoids rotation of the base part. In comparison to the design shown in Figure 2.23, part 2 has been modified from a flat sheet-metal part into an angled part similar to part 1.

2.2.4.2 Ease of assembly

To assist assembly operations, the parts to be assembled should have assembly aids (AA) on their assembly faces in the form of lead-in chamfers. Figure 2.25 illustrates what is understood by assembly aids or assembly chamfers. It is important that at least one of the parts to be assembled has one such assembly aid. The ideal situation is if both of the parts to be assembled can be formed in this manner. It is advantageous to provide assembly aids on assembly combinations in both axes. However, this is often associated with design difficulties. It is then simpler to divide the assembly aids between the assembly partner parts. Figure 2.26 shows one such example. For the positive assembly of the plastic part (2) in the sheet metal part (1), the two tongues on

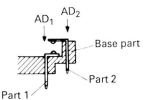

Figure 2.24 Assembly of two parts with one assembly direction

22 Product design as a requirement for economic assembly

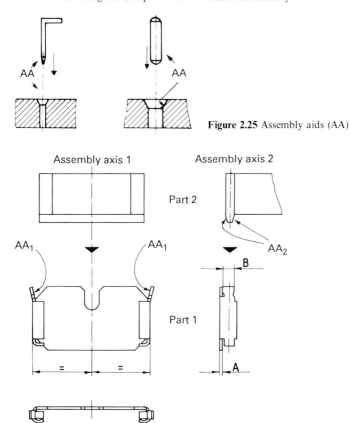

Figure 2.25 Assembly aids (AA)

Figure 2.26 Distribution of assembly aids on two parts

the sheet-metal part are bent outwards so that they form a single axis assembly aid (AA_1).

In comparison with the sheet-metal part, the plastic part (2) has two assembly chamfers (AA_2) in the second axis direction so that both parts can be positively assembled. Following assembly, the two tongues on the sheet-metal part are bent over and provide form locking fixing of the sheet-metal part on the plastic part.

2.2.4.3 Assembly working spaces

Adequate working space must be provided for the economic assembly of a part. Free space is necessary for the final assembly of a part into a base part. At the same time, the assembly procedure must be completed without any further operation, for example, secondary pressing or inserting. Adequate space must therefore be available in the base part so that the aids required for assembly, from fingers to tools, can assemble a part without obstruction. This requirement is illustrated by a simple example in Figure 2.27.

On a further example, Figure 2.28(b) shows what design modification is necessary on the base part (a) to generate the free space to make possible a screw-in operation using an automatic screwdriver. The unsatisfactory screw-in situation is shown in (a). In this situation, the screw makes contact in the corner of the base part on two faces.

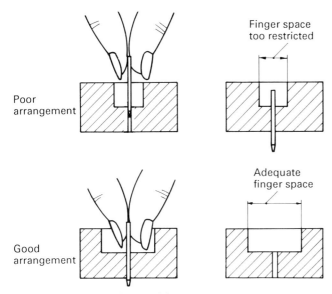

Figure 2.27 Free space for assembly

The space required for insertion of the screw with an automatic screwdriver does not exist. Diagram (b) shows the modified base part so that the tip of the automatic screwdriver with a screw guide can directly assemble the screw by a vertical, linear movement.

Figure 2.29 illustrates the assembly operation of three plug pins into a base part. As (a) shows, adequate free space is available in the base part for plug part (2). On the other hand, the assembly of plug parts (1) and (3) is difficult because the assembly slots in the base part are directly adjacent to the base part inner wall faces.

Plug parts (1) and (3) can neither be assembled directly manually nor by an assembly tool in the form of a gripper. The parts must therefore be pre-assembled so that they can be assembled into the final position in a second operation. Diagram (b) shows an assembly tool consisting of a gripper and a centrally located secondary plunger. The gripping area on the plug part is long enough to hold the part reliably. During the assembly operation at the lowest point of the assembly tool, the plug part is pressed fully home into the base part by a final stroke of the secondary plunger. The free spaces required for assembly are shown in (c). They must be dimensioned so that the assembly tool with its gripper has sufficient space for final completion of the assembly operation.

For assembly operations in which assembly is by pressing in or deforming, it is important that contact faces be provided on the base part for the support of the applied forces.

The contact faces should be at right-angles to the force flow. A differential is drawn between exposed and enclosed contact faces. In the product design it is important to ensure that the contact faces are arranged so that tilting or turning over of the parts is avoided. Figure 2.30 shows exposed contact faces for an assembly combination of two parts which are connected by pressed-in ribbed pins.

If the contact faces are not on the outer contour of the base part (so-called enclosed contact faces) a free space must be provided for the support of the assembly forces

24 Product design as a requirement for economic assembly

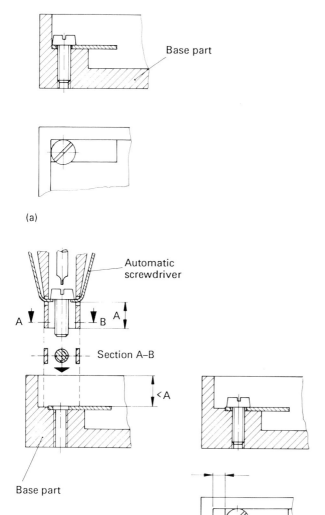

Figure 2.28 Free space for assembly of a screw: (a) base part with manual screw insertion; (b) base part with screw insertion by an automatic screwdriver

which can accept a suitably designed tool (anvil). A suitable example is illustrated by Figure 2.31.

2.2.4.4 Stability
For assembly, parts must be sufficiently stable so that residual deformations are not caused by the assembly operation. Thin-walled flat parts should be form-stabilized by raised sections or edges.

The family of so-called bending rigidity parts is particularly problematical because they are unstable in a manner such that they can deform under their own weight and

Assembly-extended ABC analysis 25

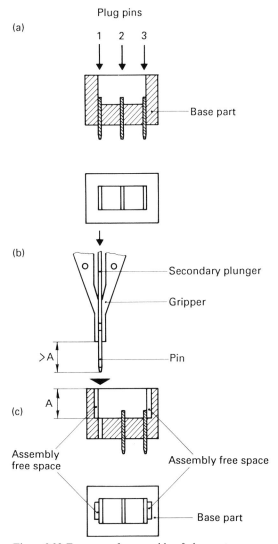

Figure 2.29 Free space for assembly of plug parts

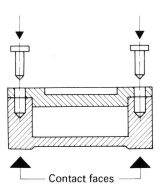

Figure 2.30 Exposed contact faces for assembly operations with applied forces

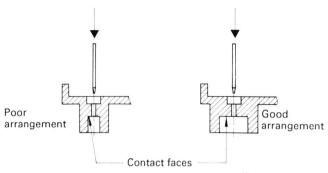

Figure 2.31 Enclosed support faces for assembly operations with applied forces

therefore do not have a stable geometrical form. For example, these include cables, wire strands, soft wires or rubber parts. Low bending rigidity parts can only be handled and assembled automatically with difficulty or even not at all and should therefore be avoided as far as possible in product design.

2.2.5 Fundamental question 5: Assembly methods

The selection of the assembly methods to be used is highly important for the cost optimization of a product and the resulting production processes. Assembly is defined as follows by DIN 8593:

> Assembly, often also defined as connecting, is the joining together of two or more workpieces of a specified regular geometrical form or similar workpieces using a formless material. In doing so, a bond is formed locally which generally enlarges. Accordingly, the act of placing together and filling is defined as assembly. The assembly of various parts of one and the same object, e.g. a ring, falls within the definition of assembly. On the other hand, the application of layers of a formless material on to workpieces falls within the main group of coating.

The subdivision of assembly processes in accordance with DIN 8593 is illustrated in Figure 2.32. With a selection of the method of assembly, a differential is generally drawn between two criteria:

1. Connections which can be dismantled and those which cannot.
2. Assembly methods with and without auxiliary parts.

It is generally concluded that permanent connections and assembly methods without the use of auxiliary parts are, as a general rule, more economic to form than breakable connections and those which employ auxiliary assembly parts.

2.2.5.1 Screw connections
The most widely used method of assembly is the screw connection. Figure 2.33 shows ten different screw connections with the object of illustrating just how widely varied the possibilities are for this method alone. By application of the extended-assembly ABC analysis and with cooperation between Product Development and Production Planning, the object is now to select the assembly technique which best corresponds to the quality requirements of a product and also by which economic assembly is technically realizable.

Assembly-extended ABC analysis   27

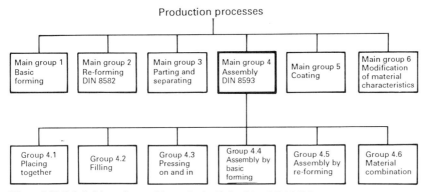

Figure 2.32 Subdivision of assembly methods (defined by DIN 8593)

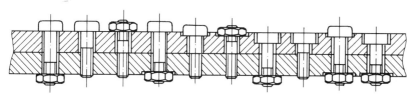

Figure 2.33 Possible design forms of screw connections (IPA Stuttgart)

2.2.5.2 Selection of assembly methods

Five possible assembly methods are shown in Figure 2.34 showing assembly combinations of two sheet-metal parts.

Diagram (a) in Figure 2.34 shows the connection of two sheet-metal parts by the auxiliary assembly parts (screw and nut). The advantage of this assembly technique is the ease of dismantling. The disadvantage is that the number of parts to be handled is increased from two to four by the two auxiliary assembly parts. Diagram (b) shows a further breakable connection of both sheet-metal parts by application of one auxiliary assembly part (screw), but with the omission of the other auxiliary assembly part (nut). This solution means that the sheet-metal part (2) must be design modified so that the mating thread can be cut in part 2 by press-forming a so-called raised boss. The

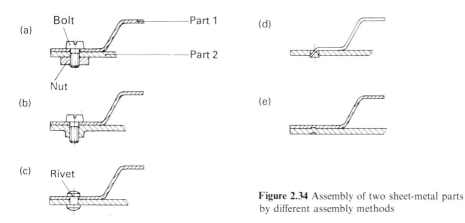

Figure 2.34 Assembly of two sheet-metal parts by different assembly methods

advantage of this solution is also to be seen in the ease of dismantling the connection and a reduction from four to three in the number of parts to be handled; however, the cost of part 2 is considerably increased. Diagram (c) shows the connection of both parts by an auxiliary assembly part (rivet) which is located by re-forming. The advantage of this solution is in the simple part design (through holes) similar to solution (a). It is also evident that, as an auxiliary assembly part, the rivet is of lower cost than a screw. The disadvantages are to be seen in that this connection is not breakable or can only be broken with difficulty. The auxiliary assembly part (rivet) means the handling of three single parts.

Diagram (d) shows the connection of both parts by an integrated re-forming technique. In this case, part 1 is design modified so that it has a free cut tongue bent through 90°. The tongue is fitted into a right-angled aperture in part 2 and is then moggled over. The advantage of this solution is the deletion of an auxiliary assembly part. The material handling is therefore reduced to the two single parts to be assembled together. The joggling operation is an easily realizable procedure. The disadvantage of this solution lies, however, in cutting free and bending of the joggling tongue on part 1 and also in the limited extent to which the connection can be dismantled. Diagram (e) shows finally the assembly or both parts by a material combination, e.g. resistance welding. The advantage of this solution is in the simple design form of the parts and in the reduction to the two parts to be assembled. The disadvantages are to be seen in the limitation of the material selection because weldability of both parts is essential, and, in addition, the connection is permanent. A further disadvantage is that the welding operations cannot always be thincluded in the assembly sequence.

The five designs of assembly methods shown are only a selection from the available possibilities illustrating the basic concepts which are important in the selection of an assembly method. Further methods such as folding, bonding, soldering, etc. can be directly related to the solutions shown, as appropriate.

2.2.5.3 Examples
The effect which the selection of the assembly methods has on the number of necessary operations in the assembly technique will be discussed on the basis of the following two examples.

A switch spring is to be force-locking-connected on a base part with a double-AMP plug pin (AMP is a standardized plug connection for the transmission of electrical power). Two different possibilities are shown in Figure 2.35. Diagram (a) shows an assembly combination consisting of four single parts – the base part (3), the switch spring (2), the double-AMP plug pin (4) and a screw (1) as the auxiliary assembly part. With this solution, four parts are to be bunkered, arranged, fed and assembled. A thread is to be cut in the double-AMP plug pin. A second possibility for this assembly combination from the aspect of a low-cost assembly method is shown in Figure 2.35(b). In this case, the screw connection is avoided and the assembly combination reduced from four to three parts. The shank length of the double-AMP plug pin (1) was increased so that this part can only be inserted through the base part (3). In place of a hole for the screw, the base part has two slots for holding the double-AMP plug pin (1). The switch spring (2) is held in the correct position on the base part by the centring pin, the double-AMP plug pin (1) is assembled, held in position and a form-locking connection formed by re-forming (notches). The screw and thread cutting in the double-AMP plug pin are eliminated by this solution. A further criterion for the evaluation of the two possible solutions shown is the higher fault susceptibility of the screwing

Assembly-extended ABC analysis 29

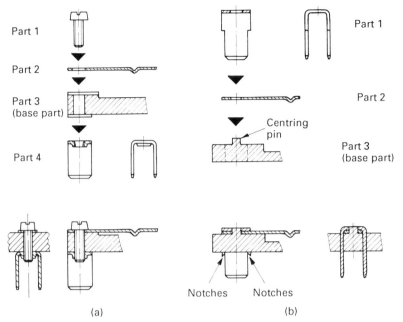

Figure 2.35 Different assembly methods for a combination of parts

operation in comparison to the notching because the screwing operation is dependent upon the quality of the screw and also the mating thread in the double-AMP plug pin. On the other hand, experience in the production operation has shown that re-forming by notches has a lower fault susceptibility than the screwing operation.

Figure 2.36 shows a second example for a comparison of assembly methods. Diagram (a) shows an assembly combination consisting of four parts. A sheet-metal part (2) is to be force-locked against rotation on the base part (3). The force-locking connection is by auxiliary assembly parts screw (1) and nut (4). Antirotation locking is provided in that the sheet-metal part (2) has a punched, angled lip which engages in a recess in the base part (3). Diagram (b) in Figure 2.36 shows a second possibility consisting of only two parts. The sheet-metal part (1) is designed so that it can be assembled into a slot in the base part (2). The form-locking connection of the parts is achieved by re-forming (in the case by twist-setting). Rotational locking of the sheet-metal part (1) in relation to the base part is achieved at the same time by the slot form. The number of parts to be handled is reduced by this solution from four to two.

2.2.6 Fundamental question 6: Quality

The quality of the parts to be processed is highly important for the economic arrangement of assembly and the product quality to be achieved. With manual assembly, substandard single parts can generally be sorted during the assembly operation without difficulty or high output loss. The same does not apply to automated assembly. In this case, substandard parts cause considerable disruption, even the stopping of assembly systems. The principal causes of disruption are exceeding tolerances, the inclusion of rogue parts and contamination.

30 Product design as a requirement for economic assembly

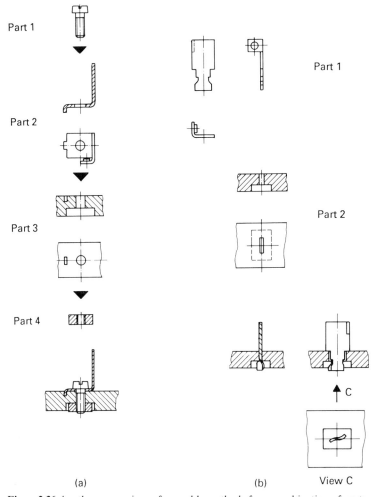

Figure 2.36 Another comparison of assembly methods for a combination of parts

2.2.6.1 Parts quality

Product development specifies the quality level of the individual parts by the technical requirements for the product. Functional dimensions can become secondary dimensions for automatic handling and functional secondary dimensions principal dimensions for automatic handling.

It is necessary that all the quality parameters of a part are specified in relation to the requirements of a product and assembly technology by close cooperation between product development and methods of planning. Tolerance limitations as shown in Figure 2.37 are necessary with an increasing level of automation of assembly methods.

A constant quality of single parts is a basic requirement for rational assembly. For economic reasons, in modern mass production, a single part can only be 100% inspected in exceptional cases. As a general rule, statistical quality control is used. Assembly is either supplied by external suppliers or in-house parts production. Just who supplies is immaterial for assembly. On the other hand, the quality level at which

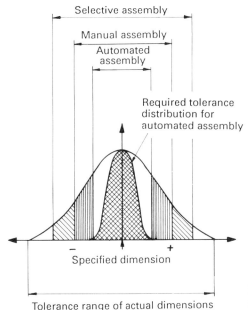

Figure 2.37 Distribution of actual dimensions of single parts about a mean value (Tipping)

the supplier provides the parts for assembly is important. As a general rule, the quality level is determined by the AQL (Acceptable Quality Level) [10]. The specified AQL value only represents a limit; exceeding the limit can mean the return of the supplied batch to the supplier. Acceptance probability by AQL varies considerably depending upon the supply batch size. Details relating to the actual defective content in a batch are not given by the AQL value. A lower AQL value means a higher probability for a smaller defective content than a higher AQL value. For manual assembly, a quality level of the parts defined by AQL 1 is adequate. The AQL value must be reduced to 0.65 for part automated assembly and at least to 0.4 for a fully automated assembly. In addition to the increasing demands for part accuracy, irrespective of the level of automation of assembly methods, care must be taken to ensure that the inclusion of foreign bodies such as waste material or other parts is avoided. With automatic feed, foreign bodies result in disruptions which, as a general rule, lead to complete stoppages of assembly installations. Contaminated or oily parts also result in disruptions [2].

2.2.6.2 Cost penalties of poor parts quality

The following example illustrates the cost penalties which can be incurred with automatic assembly by various defective contents of single parts. The quality of a screw to be fitted largely determines the availability of an automatic screw-insertion unit. By normal standards, the screws may have a defective content of 1%. With an assumed performance of an automatic screw-insertion unit of 1000 assembly operations per hour, this means a possible ten disruptions per hour caused by defective screws. The availability of the automatic screw-insertion unit is also affected by the quality of the mating thread and the exact positioning of the parts to be assembled. The effect which the screw quality has on the availability of an automatic screw-insertion unit is demonstrated by the following work-point cost calculation for a single-shift-operated automatic screw inserter (for the method of work-point cost calculation see Sections 11.13 to 11.13.3).

Work-point cost calculation for 20 000 currency units procurement costs:

1. Depreciation: 20 000 units over 5 years   4 000 units per annum
2. Interest charges: $\dfrac{20\,000 \text{ units}}{2} \times 0.08$ per annum   800 units per annum
3. Floor area cost: 7 m² × 120 units/m² per annum   840 units per annum
4. Energy costs: 0.5 kW × 1840 hours per annum × 0.12 units/kWh   110 units per annum
5. Repair, maintenance, spare parts: 2 000 units per annum
   10% of procurement cost @ 20 000 units per annum
6. Wages costs: 1 operative:
   Wage rate 11.50 units per hour × 1840 hours per annum   21 160 units per annum
7. General salary costs: 120% of 21 160 units   25 392 units per annum

Total annual operating costs: 54 302 units per annum

With an assumed working time of 230 days per annum and 8 hours per day (= 1840 hours), the cost of a working hour at this work point is:

54 302 units per annum ÷ 1840 hours per annum = 29.51 units per hour.

At a possible hourly rate of 1000 assembly operations and a screw defective content of zero percent, this gives a cycle time of 3.6 s. At an hourly rate of 29.51 units, the cost of an assembly operation is therefore 0.02951 units.

With a 1% content of defective screws and a target capacity of 1000 parts per hour, 10 disruptions per hour are to be expected.

Under the assumption that, on average, a defective screw causes one disruption and that 1 min is required for its elimination, ten defective screws mean an effective time of 50 min of 3000 seconds per hour. With this level of disruption, the hourly output rate is calculated as follows:

$$\frac{3000 \text{ available effective seconds}}{3.6 \text{ s unit cycle time}} = 833 \text{ assembly operations per hour}$$

At an hourly cost of 29.51 units and an hourly output rate of 833 parts, the assembly cost per screw is 0.0354 units.

The resultant loss of operation costs per screw attributable to inadequate screw quality per hour are:

$$0.03541 \text{ units}$$
$$-0.02951 \text{ units}$$
$$\overline{0.00590 \text{ units}}$$
$$= \text{approx. } 0.006 \text{ units per screw}$$

With a theoretical capacity of 833 screws per hour net and additional costs for down-time action of 0.006 units per screw, the annual down-time costs are:

833 screws/hour × 0.006 units/screw × 8 hours/day × 230 days
= 9196 units per annum.

With an assumed target output of 1000 screws per hour and a working time of 1840 hours per annum, the gross screw consumption is 1 840 000 per annum. In accordance with DIN 84-4.8, the procurement costs for a screw of size M3 × 9 are around 4.0 units per 1000. With an annual consumption of 1 840 000 screws, the annual procurement costs are 7360 units.

Compared with this, the annual costs for disruptions attributable to inadequate quality are 9196 units.

If the basic question of the assembly-extended ABC analysis as discussed in Section 2.2 is applied, 'What does a part or subassembly cost up to the point where its required function is achieved following final assembly?', the following is then applicable in this case:

Procurement costs for one screw	0.00400 units ($\triangleq$ 10%)
Assembly costs	0.02951 units ($\triangleq$ 75%)
Down-time costs (attributable to the defective content)	0.00590 units ($\triangleq$ 15%)
Costs per screw after achieving its functional purpose	0.03941 units (= 100%)

The example shows that the procurement costs of the screw are only 10% of the total functional costs and that the remedial costs are around 50% higher than those for the screw itself [2].

2.2.7 Fundamental question 7: Assembly costs

The answers resulting from fundamental questions 1 to 6 principally determine the assembly technique to be used and therefore the type of equipment necessary for assembly and also the resultant labour involvement.

The assembly costs are determined by the capital investment for the equipment and the wages costs for the operatives. The proportion of capital costs in relation to the assembly costs are higher with an increasing level of mechanization and automation of the assembly processes. For a calculation of the assembly costs, reference should be made to the work-point cost calculation as described in Sections 11.13 to 11.13.3.

2.2.8 Organizational implementation of the assembly-extended ABC analysis

It is difficult to subsequently modify existing, fully developed products in accordance with the requirements of assembly technology and, particularly so, if the products have already received application approval (by the VDE, TÜV or equivalent). The basic principle of the assembly-extended ABC analysis is mainly to promote a dialogue between Product Development and Production Planning by the seven fundamental questions and has the highest level of success if this occurs during the development stage of the product. The assembly-extended ABC analysis should be applied during the development of the product in a multi-stage adaptation and the range of specialist areas involved extended. Figure 2.38 shows schematically the stage-by-stage application of the assembly-extended ABC analysis and the extension of the participating areas.

The number of stages required with application depends upon the complexity of a product. A checklist as shown in Figure 2.39 is advantageous for the successful application of the assembly-extended ABC analysis. In addition to realization of the planning procedure, a checklist can also be used for the evaluation of the results relating to the individual measures.

With regard to assembly-oriented design, this evaluation should be undertaken in three stages:

Stage 1 = Highly suitable/suitable for automation
Stage 2 = Conditionally suitable/high capital cost for automation
Stage 3 = Unsuitable for automatic handling – manual handling necessary

The checklist and evaluation can be applied in detail to individual parts or for complete subassemblies and products.

Development stage Specialist area	Product preliminary design	Design	Development approval	Product release and production methods
Production management				X
Production planning	X	X	X	X
Product development	X	X	X	X
Quality assurance			X	X
Test approval institution, e.g. VDE/TÜV		X	X	X

Figure 3.28 Stage-by-stage application of the assembly-extended ABC analysis (X = team member)

Assembly-extended ABC analysis 35

Reference:	Drawing no.:	Processed by:			Date:		
		Evaluation*					
Fundamental questions	Effective factors	Fundamental questions			Effective factors		
		1	2	3	1	2	3
1. Price of single part Manufacturing costs	Material selection Manufacturing costs Batch size						
2. Supply condition	Bulk material Formatted packaging Magazined Long length format – Flow material – Belted						
3. Ease of handling	Ease of arranging Transfer						
4. Assembly direction Ease of assembly	Assembly direction Ease of assembly Assembly space Stability						
5. Assembly methods	Positioning Pressing on and in (screws) Basic forming Re-forming Material combination						
6. Quality	Part quality (AQL) Foreign body content Cleanliness Down-time costs						
Other details							

* 1 = Highly suitable/suitable for automation
  2 = Conditionally suitable/high capital cost for automation
  3 = Unsuitable for automatic handling – manual handling necessary

Figure 2.39 Checklist for supplementing the assembly-extended ABC analysis

Chapter 3

Manual assembly

3.1 Introduction

As the word 'manual' implies (Lat.: manus = hand), with manual assembly, the assembly operations are performed by people. In this case, and unlike virtually any other production process, the person is the centre point. By simply using his hands, dexterity, sense organs and intelligence, he performs assembly operations by using instruments such as tools, fixtures and gauges, etc. The output capacity of a person depends upon a number of factors such as work point and space arrangement, climate, noise and, by no means to be forgotten, the often quoted company morale.

A person is not able to work at an absolutely constant rate over an eight hour shift. The work output curve differs in relation to the type of person. As an example, Figure 3.1 shows the pattern of the individual work output curve for a certain type of person during a normal shift. The work output capacity must be carefully regulated by arranging work breaks at appropriate times. To maintain the human work capacity and to lighten the burden of work, it is not only necessary to arrange the working conditions in accordance with scientific principles but also as specified in the Health and Safety at Work Regulations. In addition, tiring of the operatives is not only reduced by a scientific ergonomic arrangement of manual assembly points but also better results can be achieved, so that in comparison to non-optimally arranged work points, the scrap quotas are reduced [2].

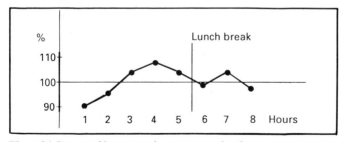

Figure 3.1 Pattern of human work output capacity shown as a percentage of the normal work output of a normal shift (Kaminsky)

3.2 Principles of work-point arrangement

Manual assembly work points must be physiologically adapted to the human body dimensions and must therefore also be variable in certain ranges such as work height, seat height and grasp range, etc. It is also important that, as far as possible, an operative can arbitrarily change her mode of working from a seated to a standing position.

Figure 3.2 shows the standard dimensions of a work point designed for seated and standing work.

The methods of work and provision of the single parts must be adapted to the working and area of vision of the human body.

The reach and field of vision of a work point [11] is shown in horizontal section at the working height in Figure 3.3. Parts which cannot be grasped without visual contact and are also difficult to bring under control must be provided within a favourable field of vision.

As far as possible, parts should be provided by manual dispensing containers fitted with a touch-release flap. Figure 3.4 illustrates touching of the touch-release flap on a parts container.

Grasping a part from a container is considerably more difficult. Figure 3.5 shows a combination of different sizes of manual dispensing containers. Manual dispensing containers must be arranged within the working area so that the parts are accessible without changing the body position. Arm movements above shoulder height should be

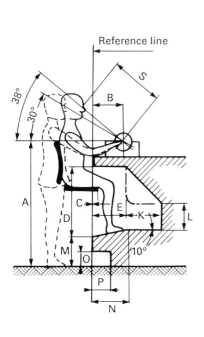

Detail	Dimensions (mm)
Working height Precision work Machine work Hand work	A (guideline) 1275 1100 – 1200 1000
Position of working part Precision work Machine work Hand work	B (guideline) 200 300 max. 325
Seat adjustment	C min. 50
Knee adjustment area	D min. 700 E min. 400
Foot movement area	K min. 350 L min. 300
Foot support	M 280 – 380 variable N min. 400
Foot adjustment area	O min. 200 P min. 200
Nose bridge – workpiece spacing Precision work Machine work Hand work	S 280 Guidelines 270 on final 450 work point

Figure 3.2 Standard dimensions of a combined manual work point (Bosch)

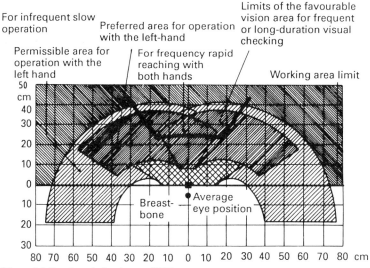

Figure 3.3 Reach and visual area (VDI)

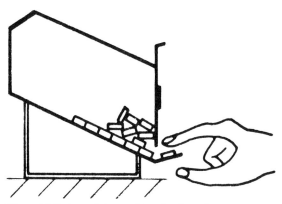

Figure 3.4 Parts container and touch release flap

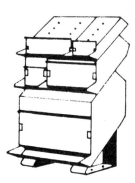

Figure 3.5 A combination of different sized manual dispensers (Bosch)

avoided on account of high fatigue. Work benches for manual assembly must be robust and have the following features:

- Adequate leg room
- Foot rest adjustable in height and inclination
- Facilities must be provided for attaching adjustable arm rests
- The facility for storing the personal effects of the operatives is desirable
- The back rest and seat height of work chairs must be adjustable.

If single parts cannot be provided in manual dispensers by reason of their size, the use of crate lifters is necessary. The containers are raised to a favourable working range level by these units at the work point. Figure 3.6 shows the manual assembly point consisting of a bench, chair, part supply by manual dispensers and crates on a crate lifting unit, and also the arrangement of an assembly tool (power screwdriver) in the working area.

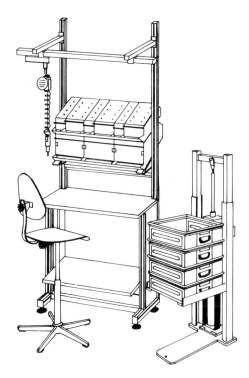

Figure 3.6 Manual assembly work point (Bosch)

3.3 Organizational forms of manual assembly

The most frequently applied organizational forms of manual assembly in the field of precision engineering and electrical equipment production are single-point and line assembly. On-site assembly is also used in some instances, e.g. for switch panel construction.

The selection of the organizational form is principally determined by the following criteria:

- Size of the product
- Complexity of the product
- Degree of difficulty of assembly
- Production volume.

3.3.1 Single-point assembly

Single-point assembly is mainly used for smaller products or assemblies with a low level of complexity and small batch size. The advantages of single-point assembly are to be seen in the high flexibility with regard to number alterations and type variety and also in the fact that disruptions do not have direct effects on other work points. The individual work output scope of the operative is large and therefore so is the handling and arrangement flexibility. The choice of individual short pauses and flexible working time is also possible. However, disadvantages arise if, by reason of the production volume or product complexity, several points must be arranged and equipped. A high capital investment is incurred by equipping several work points if expensive assembly fixtures are required.

3.3.1.1 Arrangement examples for single-point assembly
With reference to the work area and field of vision (see Figures 3.2 and 3.3), and depending upon the size of the manual dispensers for the provision of the single parts, the work involvement for single-point assembly in bench form is limited. The fact that with a very large work involvement, e.g. with the assembly of 25 different parts, the selection of the operatives for such a work point is an important criterion. In addition to the physical effort, the arrangement of such a large number of parts and assembly in the correct order calls for a high level of mental awareness. Figure 3.7 shows a methodically arranged assembly work point with the provision of ten different parts in manual dispensers arranged in 'only two tiers'. In this example, the physiological grasp area of the operatives is fully utilized in the working plane. Movements which necessitate raising of the arm from the arm supports are extensively avoided since the height difference between the manual dispensers arranged in only two tiers is so small

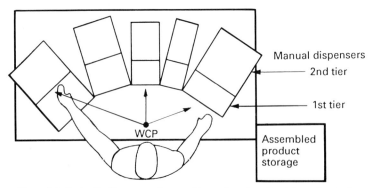

Figure 3.7 Methodically arranged assembly work point. Provision of parts by manual dispensers in a two-tier arrangement (WCP = work centre point)

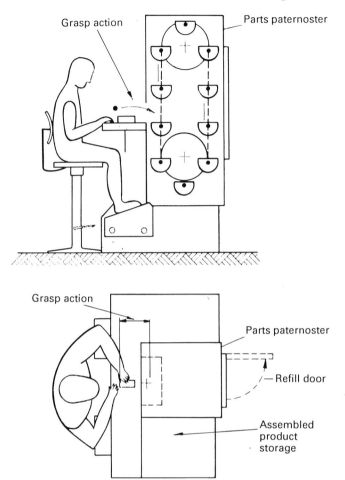

Figure 3.8 Single assembly point. Parts provision by a parts paternoster

that raising of the arm is either not or hardly necessary. An improvement in the work point arrangement to achieve constant grasp movements with the provision of a large number of single parts at the same time is realized by the application of a parts paternoster. A parts paternoster is built into the work bench as shown in Figure 3.8. The tower-type rotary magazine has an aperture directly opposite the assembly position which gives access to a container carrying single parts. The grasp action from the assembly point to the manual dispenser is the same for all parts. When a part (or even several) has been taken from the container, the drive is engaged by a foot or knee switch. The paternoster advances by one parts container and the following part is available at the same grasp point. This procedure is repeated depending upon how many different parts are required for the assembly of a product. The paternoster arrangement is particularly suitable for cases in which the provision of a large number of parts is necessary. With very small sized parts, it is possible to halve the container sections so that two parts are available at one paternoster drive stop point. For refilling the parts containers, the paternoster has a door on its rear face so that the filling operation can take place without interruption of assembly. The advantages of this type of arrangement are:

- The grip actions can be arranged optimally short in one work plane.
- The grip actions are equal for all parts.
- The single parts are automatically offered in the assembly order sequence.
- Incorrect assembly by mistaking parts is largely avoided.

Instead of a vertically arranged paternoster, the same effect of parts provision in the assembly sequence order can be achieved with horizontal action container chains.

A further possibility for a single work point arrangement is the use of rotary tables. Figure 3.9 shows one such work point.

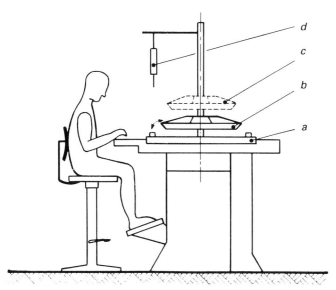

Figure 3.9 Single assembly point with rotary table operation (a = rotary table with workpiece fixtures, b = rotary table (provision of single parts), c = rotary table 2 (provision of single parts), d = power screwdriver)

The assembly operation is not undertaken in a single fixture but in a number of fixtures equal to the number of indexing positions of the rotary container.

The work point is equipped with a rotary container a with 36 workpiece fixtures (number of stations) and is indexed by the operative via a foot or knee switch. The rotary container moves by one station per switch impulse. If the so-called two-hand assembly technique is used, the switching can be arranged so that the rotary container indexes by two stations per impulse. A freely rotating rotary container b with containers for the provision of the single parts is arranged above the rotary table with the workpiece fixtures. For fairly complex products with a large number of single parts, it is also possible to arrange a second rotary container c for the provision of parts. A vertical column with a swivel arm for holding assembly tools (in this case a pneumatically operated screwdriver d) is arranged in the centre of the rotary container. container.

Organizational forms of manual assembly 43

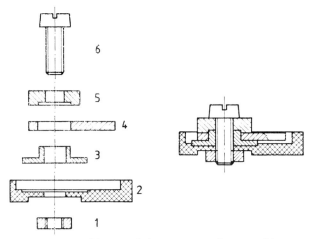

Figure 3.10 Assembly example for rotary container assembly

The work sequence at this single assembly point is explained on the basis of the assembly example as shown in Figure 3.10.

1. Six different single parts are required for the subassembly as shown in Figure 3.10 so that the operation can only be completed by a rotary table for the provision of parts. The six single parts are provided in the required quantity in the compartments of the free-rotating rotary container.
2. By reason of the size of the parts and product design, it is possible to index the 36 workpiece fixture rotary container arrangement by two stations per operation since, with two-hand operation, two parts can be inserted into two workpiece fixtures.
3. During the first operation, the rotary container together with the positioned parts is rotated by the operative so that part 1 falls within an optimum grasp area. With the two-hand method, two nuts (part 1) are grasped 18 times and placed in the rotary container workpiece fixtures.
4. After positioning of these 36 parts (part 1) in the workpiece fixtures, the rotary container is indexed by one station and part 2 optimally provided. Part 2 is now grasped 18 times in pairs and placed in the rotary container workpiece fixtures.
5. After the addition of part 2, the rotary container is indexed so that part 3 can be grasped. The operation is as part 2.
6. The procedure is repeated until part 5 of the subassembly has been positioned. The rotary container drive is then switched to single-cycle indexing and the rotary container indexed so that screw 6 is provided. The screw is then grasped 36 times, fitted and tightened using a pneumatic screwdriver.

The 36 subassemblies are removed from the rotary container workpiece fixture by the last operation and the total operational sequence is then recommenced.

A further advantage of this arrangement lies in the fact that every single operation can be performed many times, i.e. in accordance with the number of stations on the rotary container for workpiece fixtures. A higher work rate is achieved by this repeat effect.

3.3.2 Line assembly

If, by reason of its complexity or the required production volume per unit of time, it is no longer economic to assemble a product on a single assembly work point, the assembly processes must be distributed over several work points.

The linking of several single assembly points to form a line is termed line assembly. The same conditions as detailed in Section 3.2 also relate to the arrangement of the single assembly work points combined to form line assembly. The most frequently occurring forms of line assembly encountered in the fields of precision engineering and electrical equipment production can be segregated into the following three groups:

1. Line assembly by manual transfer of the assembled part
2. Line assembly by mechanical transfer of the assembled part in an unarranged form
3. Line assembly by mechanical transfer of the assembled part in an arranged form.

3.3.2.1 Line assembly by manual transfer of the assembled part
As a general rule, with line assembly by manual transfer the assembled part or workpiece is transferred from one single assembly work point to another with the formation of intermediate buffers. This manual transfer lends itself to a large number of organizational forms of line assembly and can also be very simply adapted to the prevailing space conditions. Figure 3.11 shows a line organizational form and Figure 3.12 a rectangular arrangement.

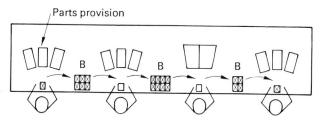

◧ = Assembled part
B = Buffer

Figure 3.11 Line assembly with manual transfer of the workpieces in a line arrangement

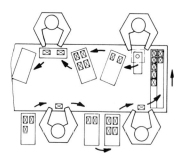

◧ = Assembled part

Figure 3.12 Manual line assembly in rectangular arrangement (IPA Stuttgart)

With line assembly, the individual assembly work points must be arranged in accordance with the assembly operation sequence. The progression of the assembly work is determined in terms of time but not, however, linked to a cycle. The work involvement of the individual assembly points must be equal. The following advantages can therefore be identified:

- Limited individually determined work rhythm of the operatives by the possibility for the formation of buffers between the individual work points.
- Work output fluctuations during a shift between individual operatives do not affect the overall performance by buffer formation.
- The training and inclusion of new operatives is relatively simple.
- In a line assembly arrangement, short duration technical difficulties with equipment do not affect the overall performance of an assembly group by buffer formation.

The disadvantages of this type of line assembly are:

- High costs for manual transfer of the assembled part (see Section 4).
- The necessity for buffer formation incurs additional costs by stock in circulation together with a high space requirement.
- Buffer formation restricts the facility for communication of the operatives working on the line.
- Very small scope for automation of the assembly operations along the assembly line [12].

3.3.2.2 Line assembly by mechanical transfer of the assembled part in an unarranged form

With line assembly and mechanical transfer of the assembled part from work point to work point, the assembled part is placed in a random form on a conveyor belt by the operatives and must be removed from the conveyor belt in order to position it in the assembly position. In comparison to line assembly with manual transfer of the assembled part, the scope for arrangement of this type of line assembly is considerably restricted by the application of conveyor belts. Figure 3.13 shows two possibilities for arrangement with the application of two types of belt. Diagram (a) shows the so-called skeleton belt arrangement and (b) the so-called lap belt arrangement.

With the skeleton belt arrangement, the single assembly work points are arranged staggered at 90° to the left and right on the conveyor belt. With the lap belt arrangement, the conveyor belt does not run along the side of the work points but along the front face, i.e. above the laps of the operatives.

The most important requirement of a connection of single assembly work points by conveyor belts to form line assembly is that the assembly process of an assembled part can be subdivided so that, in spite of random positioning on the conveyor belt, the assembly remains transferable to the next assembly work point. This means that the operations undertaken at a single assembly work point must be arranged with regard to their assembly process so that the previously achieved assembly condition is not affected by placing on the conveyor belt in a random manner (without workpiece carrier) during transport to the next single assembly work point and by removal from the conveyor belt. Every single assembly work point must therefore reach a stage in the assembly process so that the achieved condition is positively not affected by material transfer.

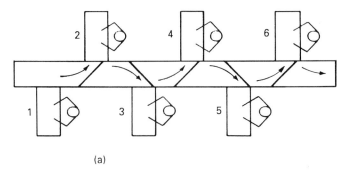

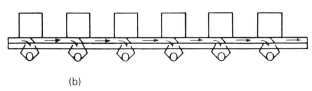

Figure 3.13 Line assembly with the application of transfer belts: (a) skeleton belt arrangement; (b) lap belt arrangement

The most important features of line assembly with mechanical transfer in a random form of the assembled part are:

- The arrangement of the assembly points must correspond to the assembly operation sequence.
- A distribution of work as uniform as possible over the individual assembly points is necessary because the buffering by a transfer system with conveyor belts is smaller than with manual transfer. The result is an almost cycle-controlled completion of assembly operations.

An important advantage lies in the short passage time of the assembled part and in the lower stock in circulation.

The disadvantages of this arrangement are:

- High additional costs are incurred by manually placing the assembled part on the conveyor belt after every completed part assembly process and removal from the conveyor belt in the assembly position at the next assembly work point.
- Short-duration disruptions at the individual work points affect the performance of the overall system by the reduced decoupling of the work points.
- The reduced possibility to form buffers means cycled working, and the overall system is therefore affected by work output fluctuations of individual operatives.

3.3.2.3 Line assembly by mechanical transfer of the assembled part in an arranged form

The transfer of the assembled part in an arranged form necessitates the use of workpiece carriers. The object of these carriers is to maintain the arranged condition of the assembled part during transfer from one assembly work point to another. They should be designed so that they also fulfil the function of an assembly fixture so that removal of the assembly part at the work point and positioning in an assembly fixture is avoided.

The lap belt arrangement is better than the skeleton belt arrangement for the transfer of workpiece carriers. With the latter type, the workpiece carriers must be removed from the belt with the assembled part, positioned on the work point and, after completion of the assembly operation, replaced on the belt.

Since, under certain circumstances, with regard to its volume and weight, the workpiece carrier can be several times that of the part to be assembled, this practice would result in considerable physical burdening of the operatives. On the other hand, the lap belt arrangement has the advantage that with a favourable work point arrangement, the workpiece carrier is located directly in the centre of the single assembly work point.

The workpiece carriers are preferably transferred on a double conveyor-belt system. The function is illustrated by the schematic diagram as shown in Figure 3.14.

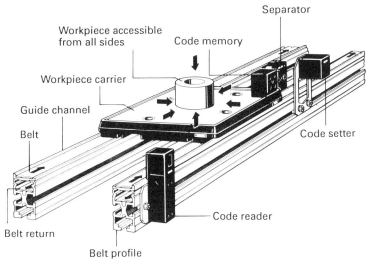

Figure 3.14 Double conveyor-belt arrangement (Bosch)

The workpiece carriers are transferred between single assembly work points on two parallel conveyor belts. The workpiece carrier is stopped at the work point by a stop device (separator); the assembly operation can then proceed. Depending upon their spacing and the work content at the single work points, workpiece carriers can accumulate and form a larger or smaller buffer capacity (see Figure 3.15).

Double conveyor-belt systems are preferably arranged in a rectangular form as shown in Figure 3.16. The forwards and return section of the double conveyor-belt system can then be fully utilized for the installation of single assembly work points and the workpiece carriers transferred to the single work points in a circular form of motion.

Figure 3.17 shows a section of a double conveyor-belt system with a single assembly work point.

On account of the limited possibility for the formation of buffer capacities between the single work points with the application of double conveyor-belt systems,

48 Manual assembly

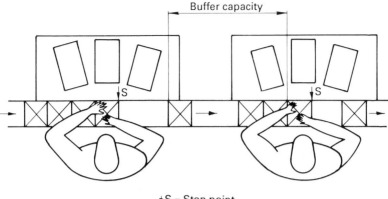

Figure 3.15 Buffer formation by the accumulation of workpiece carriers between the single work points with lap belt assembly

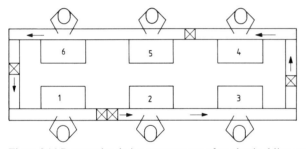

Figure 3.16 Rectangular design arrangement of mechanized line assembly

considerable importance is attached to the cycling, i.e. the formation of a uniform work cycle at all line assembly work points. With highly complex assembly operations in which the times for single operations can be a multiple of the single cycle time of the other assembly points, linking of the work points and equalization of the individual assembly times is only possible if the extremely high work contents are distributed over several work points. In this respect it is advisable to branch the workpiece carriers from the main conveyor belt on to several similar single assembly work points arranged in parallel.

Figure 3.18 shows one such solution schematically. Three parallel single assembly work points are arranged on a main conveyor belt; the oncoming workpiece carriers on the main conveyor belt are branched on to the parallel conveyor belts by transverse secondary conveyor belts. Every single assembly work point is equipped with a secondary conveyor-belt system. In order to control the distribution of workpiece carriers from the main conveyor belt so that every parallel assembly work point receives a pre-determined number, the number of workpiece carriers to be directed to every assembly point is recorded by controls. Following completion of the assembly process at the single work points, the workpiece carriers are returned from the secondary conveyor belt to the main conveyor belt and transferred to the following

Organizational forms of manual assembly 49

Figure 3.17 Single assembly work point on a double conveyor-belt system (Bosch)

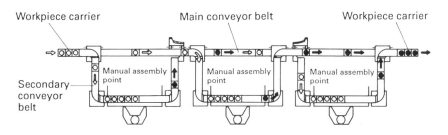

◉ = Workpiece carrier before assembly
● = Workpiece carrier after assembly

Figure 3.18 Branching and distribution of workpiece carriers from the main conveyor belt to three parallel single assembly work points (Bosch)

single assembly work points. In the example shown, the cycle time of a single assembly work point on the secondary conveyor belt can be three times the cycle time of the single assembly work points of the main conveyor belt [2].

Line assembly with mechanical transfer in an arranged form of the assembled part is economically applicable with products of higher complexity and production volume if a minimum of four work points can be linked together. The highest work rate is achieved by repeatability but does, however, lead to monotony, which should be avoided.

The problem with small work contents per work point is the cycling of a line and the autonomy from work point to work point from the conveyor-belt cycle. A sensible increase in work activity per single work point does result in a higher work content and, at the same time, increased decoupling from work point to work point by the accumulating workpiece carriers. However, it does prejudice the repeat activity required for optimum work activity increase.

An increase in work content and repeat activity can be achieved with the system as described below. It is a modular construction and consists of two basic modules, the circular assembly table and the workpiece carrier counter buffer. Figure 3.19 shows the schematic arrangement and mode of operation. The workpiece carriers are transferred between the circular assembly tables on a conveyor belt with the interaction of a counter buffer. The oncoming workpiece carriers are transferred on to the circular assembly tables by an infeed discharge unit. Depending upon the circular table diameter and the size of the workpiece carriers, a circular table can accommodate 18 to 24 workpiece carriers.

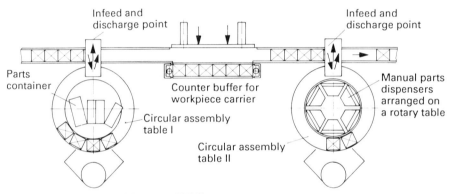

Figure 3.19 Modular assembly system (EGO)

The work content allocated to the circular assembly tables is equal. Four different workpieces are directed to circular assembly table I as shown in the diagram in parts containers and six different workpieces to circular assembly table II in parts containers arranged on a rotary container. This arrangement of work gives an optimum egonomic work point arrangement. Depending upon the number of workpiece carriers, equal numbers of parts are processed in sequence. This means that with 24 workpiece carriers on the circular assembly table, the same assembly operation is undertaken 24 times in sequence. After completion of this operation, the next part is assembled 24 times in sequence and then the next, etc. Irrespective of the work content, the workpiece carriers remain on the circular assembly table during several revolutions and are only ejected from the table under the conveyor belt after completion of all the allotted operations. Upon ejection of a workpiece carrier, another is fed in so that idle running of the circular assembly table does not occur with infeed and discharge. The work contents at the single stations should be the same. If, by coincidence, work is undertaken synchronously on two circular assembly tables, this means that with ejection on circular assembly table I, a simultaneous infeed of workpiece carriers is required on circular assembly table II so that the workpiece carriers can be fed direct to circular assembly table II via the conveyor belt. If the linked circular assembly tables do not function synchronously, the accumulated workpiece carriers are moved transversely into the counter buffer. If, after completion of assembly operations, circular assembly table II requires new workpiece carriers from circular assembly table I, for as long as circular assembly table I is not operated synchronously, they are taken and fed in from the counter buffer. Figure 3.20 shows schematically different possibilities for the arrangement of this assembly system: (a) line assembly with circular assembly tables on one side of the workpiece carrier transfer; (b) a staggered arrangement of circular assembly tables; (c) the arrangement of a rectangular system. If, for example, six operations each of two seconds' duration are allocated to a circular assembly table with 24 workpiece carriers, this results in a work content of a little under 3 min. If the counter buffer has a storage capacity of 72 workpiece carriers, a buffer capacity of approximately 12 min is achieved and therefore also decoupling of the work points relative to each other of the same magnitude.

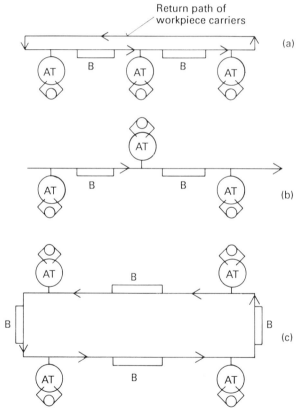

Figure 3.20 Example of possible arrangements for a modular assembly system as shown in Figure 3.19 (EGO)

Positioning of the parts containers from which the parts must be taken at the most favourable point for the assembly position is made possible by an arrangement of the parts containers on a rotary container in the centre of the circular assembly table. Figure 3.21 shows a circular assembly table with an arrangement of 18 double

52   Manual assembly

Figure 3.21 Circular assembly table in a modular assembly system as shown in Figure 3.19 (EGO)

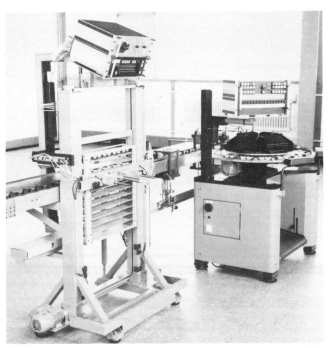

Figure 3.22 Modular assembly system consisting of a circular assembly table and workpiece carrier counter magazine (EGO)

workpiece carriers, the provision of six different parts in containers arranged on a rotary container. Figure 3.22 shows the total modular system, counter buffer and circular assembly table linked to a conveyor-belt system.

3.3.2.3.1 Workpiece carriers

With line assembly and mechanical transfer of the assembled part in an arranged form by the application of workpiece carriers, considerable importance is attached to their design. As far as possible, all necessary assembly processes in an assembly operation should be able to be performed in a workpiece carrier.

The following criteria are important in the design of a workpiece carrier and must be carefully examined:

1. How can the base part or single parts be held in a workpiece carrier: internally or externally?
2. Can the assembly operations be undertaken from one side so that the subassembly need not be rotated?
3. If a subassembly must be rotated, e.g. through 180° (top to bottom), it is necessary to check if, in both cases, i.e. during the assembly operations before and after rotation, a subassembly can be held on its outer contours. If this is not possible, what must the workpiece carrier look like after turning?
4. If pre-assembly operations are necessary which require a separate assembly fixture, these should be included on the workpiece carrier as far as possible.
5. The simpler the design of workpiece carriers, the simpler the assembly procedure. The number of workpiece carriers required is a minimum of five times the number of the single assembly work points included in the line assembly. The resultant high number of workpiece carriers necessitates a rational method of production.

Figure 3.23 shows a simple workpiece carrier on which the base part is only supported on its outer contours by the limit pins.

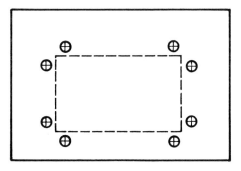

Figure 3.23 Workpiece carrier with limit pins for positioning on the outer contour

54 Manual assembly

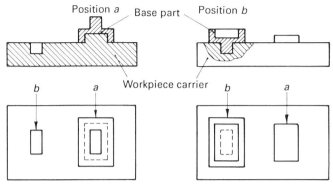

Figure 3.24 Workpiece carrier with two location points

Figure 3.24 shows a workpiece carrier which is designed for the rotation of the base part or subassembly through 180° at some point in time.

After rotation, the geometry of the base part calls for a second location point. The base part is held on its internal contour at location point *a* and after rotation through 180° on its outer contour in location point *b*. Assembly procedures are conceivable which have up to six assembly directions so that, under certain circumstances, a workpiece carrier must include six differently oriented work piece location points.

3.3.2.3.2 Cycling

The output of line assembly with mechanical transfer of the assembled part by workpiece carriers is controlled by the longest single work point time.

Figure 3.25 shows an assumed cycle time distribution for ten single work points. In comparison to the remaining single work points, the cycle time distribution shows that the single assembly work points 3 and 5 have the shortest cycle times. Single work point 5 has the shortest time at 7.2 s and work point 8 the longest at 12 s. There is therefore a difference of 4.8 s between the shortest and longest cycle times. As shown by the diagram, with the exception of single work point 8, all the other work points have an unused excess time of 2 s to 4.8 s. It must therefore be investigated if pre-assembly

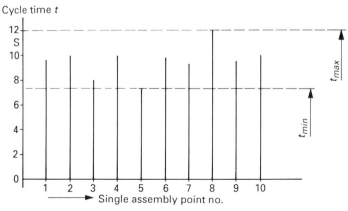

Figure 3.25 Cycling of line assembly with ten work points

work can be allocated to the work points with the shortest cycle times by the arrangement of a secondary location point on the workpiece carriers.

Single work point 8 should be relieved in terms of time by this measure.

In this respect, Figure 3.26 shows a workpiece carrier with a workpiece location point a and a secondary location point b. Assembly takes place in workpiece location point a; workpiece location point b is provided to facilitate pre-assembly operation.

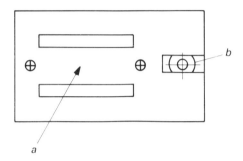

Figure 3.26 Workpiece carrier with workpiece location point a and a secondary location point b for pre-assembly operation

Figure 3.27 illustrates what is meant by a pre-assembly operation. As shown by Figure 3.25, in comparison to the other work points, single assembly work point 3 has available time, so that the annular ring part 1 as shown in Figure 3.27 can additionally be placed at this work point in the secondary location point b of the workpiece carrier as shown in Figure 3.26. Compared with the average assembly time of the other single assembly work points, work point 5 also has available time, so that in addition to the actual assembly operation the screw part 2 as shown in Figure 3.27 can be placed in the annular ring part 1 as placed in secondary location point b at single work point 3.

Single assembly point 8 is relieved by the pre-assembly operations at single assembly points 3 and 5 so that the single cycle times at the ten work points are largely equalized. The result of cycle equalization by pre-assembly operations is shown by Figure 3.28.

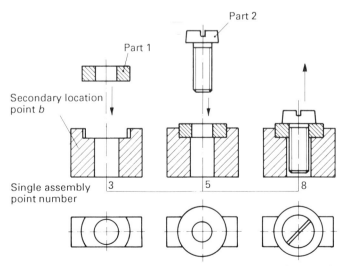

Figure 3.27 Order of pre-assembly work with secondary location point b on the workpiece carrier as shown in Figure 3.26

56   Manual assembly

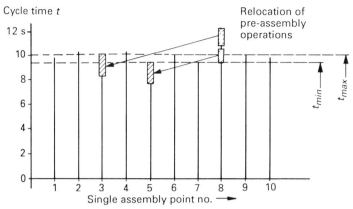

Figure 3.28 Cycling using the secondary location point in a workpiece carrier (see Figures 3.26 and 3.27)

The cycle time of the ten single work points is now between 9.3 s and 10 s. With the cycling of line assembly with transfer of the assembled part in workpiece carriers on double conveyor belts, the time difference between the shortest and the longest cycle time should not exceed 10% of the longest cycle time [2].

3.3.2.3.3 Rating of line assembly by mechanical transfer of the assembled part in an arranged form
Line assembly by mechanical transfer of the assembled part with workpiece carriers is principally used in large-scale standard production with a high level of work distribution. The arrangement of the assembly points is determined by the assembly operation sequence during which the assembly operations must be undertaken on a cycled basis since the achievable buffer capacity does not make decoupling from the cycle time possible.

The advantages of line assembly are:

- Short passage time of the assembled part
- Good possibilities to convert individual part sections from manual to automatic assembly operations
- No additional costs by manual transfer of the assembled part.

The disadvantages of line assembly by mechanical transfer of the assembled part in an arranged form are:

- Application of a large number of workpiece carriers (depending upon type)
- Low flexibility with volume fluctuations and type or design variations
- Disruptions at single assembly work points have a rapid effect on the overall performance of the system
- Cycle equalization is, to some degree, difficult. The use of a floating point operative is necessary
- No autonomy of the operatives from the system fixed cycle, only limited operative flexibility
- Short cycle activities lead to unbalanced work loading and monotony [12].

Chapter 4

Primary–secondary analysis – An aid for the determination of the economic efficiency of assembly concepts

4.1 Introduction

As was shown by the breakdown of wages costs into single production processes, 25% to 75% of the costs are attributable to the field of assembly. Based on this cost situation, the methods used up to the present for the planning of manual or semi-automatic assembly operations by REFA, MTM (Method Time Measurement), WFS (Work Factor System), etc. are no longer adequate. Piece-rate time allowances and bonus payment only give data relating to the assembly costs but not the efficiency of the overall assembly concept. In assembly operations, additional planning methods are therefore necessary in order to structure the operations economically and analyse by measurement.

The analysis of productive and idle times has been applied for some considerable time for the evaluation of the efficiency in the classical production technologies such as metal-cutting or re-forming and the determination of the time-related efficiency of the production equipment used. This time-related efficiency is a part of the calculation of the machine hourly rate.

Up to the present, such evaluation criteria have not, however, been employed in assembly technology. They should be indispensible for the optimization and rationalization of assembly operations.

4.2 Definition of the efficiency of assembly operations

The efficiency (E_A) of an assembly process is measurable with regard to economy in its entirety by an analysis in terms of primary assembly processes (PAP) and secondary assembly processes (SAP):

1. Primary assembly processes (PAP) include all operations used in the realization of a product during its assembly, i.e. all energy costs, items of information and parts for the completion of a product.
2. Secondary assembly operations (SAP) include all operations which, by reason of the selected assembly principle, necessitate secondary requirements for time, information and energy without affecting the realization of a product. Suitable examples are material transfer, handling, regrasping, etc.

The efficiency of assembly is calculated by the following formula:

$$E_A = \frac{\text{Total PAP}}{\text{Total PAP} + \text{Total SAP}} \times 100 \ (\%) \ [2, 13-15]$$

4.3 Field of application

Based on the above-mentioned definition, an evaluation of the actual assembly process is possible by an analysis in terms of primary and secondary assembly operations. With a narrower scope of the definition, it is also suitable for an evaluation of a single assembly work point. Furthermore, it can also be used for the analysis of the complete range of assembly with regard to economy, i.e. material supply, transport, storage, assembly process, inspection and packaging. Figure 4.1 shows these three fields of application.

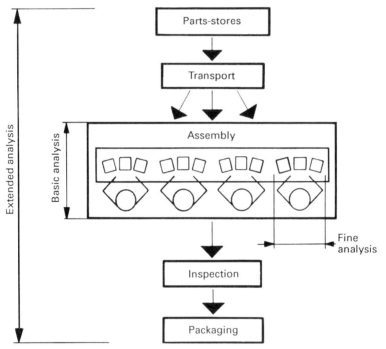

Figure 4.1 Fields of application of the analysis in terms of primary and secondary assembly processes

The arrangement of the processes required into primary and secondary requirements differs for the three fields of application as mentioned and will therefore be defined below.

4.3.1 Basic analysis

The basic analysis is limited to the assembly procedure. It commences following provision of the material to be assembled and ends before inspection and packaging.

Inspection operations which are integrated into the assembly process should, however, be included.

The following two examples are used to illustrate the definition of the terms primary and secondary assembly procedures in this field of application.

Example 1

A hypothetical product consisting of eight single parts is assembled at three manual assembly work points arranged as shown in Figure 4.2.

The base part of the subassembly to be assembled is grasped by the operative at assembly work point 1 and placed in the assembly position. Two further parts are assembled. These operations fall within the scope of the term primary assembly.

The placing of the subassembly assembled at work point 1 on the intermediate buffer (IB_1) is, on the other hand, a secondary assembly operation because it is a secondary activity which does not form part of the realization of a product but is necessary on account of the selected organizational form. Removal of the pre-assembled unit from intermediate buffer (IB_1) at assembly point 2 and arranging it in the assembly position is also a secondary assembly operation.

On the other hand, the assembly of two further parts at assembly point 2 is a primary assembly operation. The placing of the subassembly after completed assembly from work point 2 in the intermediate buffer (IB_2) is a secondary assembly operation, and likewise removal from the intermediate buffer (IB_2) at assembly point 3 to arrange the subassembly in the assembly position.

The table in Figure 4.2 shows that with this assembly process 8 out of 13 handling operations fall within the scope of the term primary assembly operations. The remaining five are secondary assembly procedures.

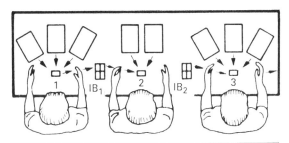

Operations	Work point 1	Work point 2	Work point 3	Total
PA number	3	2	3	8
SA number	1	2	2	5
Total number	4	4	5	13
PA time SA time	10 s 2 s	8 s 4 s	8 s 4 s	26 10
Total time	12 s	12 s	12 s	36

Figure 4.2 Examples of basic analysis by PAP–SAP

Under the assumption of the times also given in Figure 4.2 for primary and secondary operations, the work involvement to complete this assembly process is 36 s. Of this time, 26 s are allocated to primary and 10 s to secondary assembly operations. Accordingly, the time-related efficiency of this assembly process is calculated by:

$$E_A = \frac{\text{PAP}}{\text{PAP} + \text{SAP}} \times 100 \ (\%) = \frac{26}{26 + 10} \times 100 \ (\%) = 72.2\%$$

This theoretical example shows that with this hypothetical product 72·2% of the working time consumed is for product realization and the remaining 27.8% is wasted.

Example 2
On account of its complexity and the required production quantity, a hypothetical product is assembled at six single assembly work points. The organizational form is line assembly by mechanical transfer of the assembly material in an unarranged form and by a skeleton belt arrangement as shown in Figure 4.3. Without analysis of the work procedure and costs of a single assembly work point, the following PAP and SAP analysis shows that higher costs are incurred for secondary assembly procedures by the skeleton belt arrangement alone.

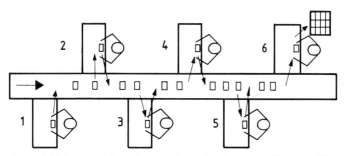

Figure 4.3 Line assembly with manual transfer of the assembly material in an unarranged form

The primary and secondary assembly procedures for this skeleton belt arrangement are as shown in Figure 4.4. Under the assumption that a primary assembly operation is undertaken at each of the six single work points, the result shows that, with this line assembly organizational form, six operations are classified as primary assembly and ten operations as secondary assembly.

With Example 1 with three assembly work points, 26 s were accounted as being primary costs and 10 s as secondary costs. With the same time-related conditions in Example 2, with six work points, 52 s are accounted as primary costs and 20 s as secondary costs. The time-related efficiency is then calculated by:

$$E_A = \frac{52 \text{ s}}{52 \text{ s} + 20 \text{ s}} \times 100 \ (\%) = 72.2\%$$

With a highly unfavourable arrangement and division of the work content over the single assembly work points, in the worst case, a ratio of PAP to SAP of 1:1 can occur, and hence, an E_A of 50%.

Field of application   61

Work point	Operation	Primary procedures (PAP)	Secondary procedures (SAP)
1	Assemble	x	
	Place on belt		x
2	Remove from belt		x
	Assemble	x	
	Place on belt		x
3	Remove from belt		x
	Assemble	x	
	Place on belt		x
4	Remove from belt		x
	Assemble	x	
	Place on belt		x
5	Remove from belt		x
	Assemble	x	
	Place on belt		x
6	Remove from belt		x
	Assemble and stack	x	
	Total	6	10

Figure 4.4 Arrangement of assembly procedures in terms of primary and secondary procedures with flow line assembly as shown in Figure 4.3

4.3.2 Fine analysis of single assembly work points in terms of primary and secondary processes

The basic movements expressed by MTM (Method Time Measurement) form the basis for the fine analysis in terms of primary and secondary processes for the planning and analysis of single assembly work points. In the field of assembly, it can be shown that up to 85% of work operations which have the potential for complete change consist of the five basic movements shown in Figure 4.5, the sequence of movement also being typical.

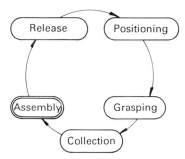

Figure 4.5 Basic movements expressed by MTM

With a narrow interpretation of the definition of primary and secondary assembly processes, from the above five, only the basic movement 'joining' can be considered as being a primary assembly operation. A calculation of the efficiency of an assembly work point in terms of this definition is fundamentally possible but is of little practical relevance because no part arrives at the assembly point of itself in an arranged form and in the correct position.

The basic movements of positioning, grasping, collection and release are necessary for the completion of a manual assembly operation, its economic arrangement being dependent upon the optimum arrangement of the basic movements. The same five basic movements are also applicable in automatic assembly.

A handling machine or an assembly robot moves to the collect position (reaching), grips a part (gripping), moves to the assembly position (collection), completes the assembly operation or transfers assembly to a second unit, for example riveting (assembly) and after completion of the assembly operation releases the part (release). For the optimum arrangement of this type of automatic operation, the parts are presented in the correct position in order to make 'blind' gripping possible. At the same time, the distances for reaching and collection are restricted to be as short as is technically possible.

The fine analysis in terms of primary and secondary processes can be by a method in which the necessary minimum amount of work is defined as a primary process and the additional amount of work as a secondary process. The limit between primary and secondary work is variable. For the fields of precision and electrical engineering, it can, for example, be determined by the following definitions which have proved to be satisfactory in actual practice.

4.3.2.1 Reaching
With the basic movement 'reaching', MTM differentiates in terms of movements A, B and C/D. A is a movement with a low, B with a medium, and C/D with a high level of control. Movement B is a widely used movement in the fields of precision and electrical engineering. The involvement for 'reaching' is dependent upon the distance between the assembly position and the gripping point of the respective part. The distance can vary considerably depending upon the complexity of an assembly operation and its work point arrangement. From the diagram as shown in Figure 4.6, the time requirement for reaching (case B) in TMU (Time Measurement Unit: 1 TMU = 0.036 s; 27.8 TMU = 1 s; 1666.7 TMU = 1 min; 100 000 TMU = 1 h) is shown in relation to the distance in centimetres. The basis for the limit to be fixed between PAP and SAP is dependent upon the size of the parts to be assembled and the distance between the assembly position and gripping point. Two different limits for the PAP are shown in Figure 4.6 which are based on the human grasping range and practical experience:

Limit for PAP *a*: Distance 35 cm with relatively small parts, corresponding to a time consumption of 14.2 TMU, equivalent to approx. 0.5 s.
Limit for PAP *b*: Distance 45 cm with relatively large parts, corresponding to a time consumption of 17 TMU, equivalent to approx. 0.6 s.

All involvements within this distance limit are defined as PAP and all involvements outside of the limit as SAP.

The application of the diagram will be explained by an example. The reaching distance (movement B) measured from the assembly point is to be 70 cm, which corresponds to a time consumption in MTM of 24.1 TMU. With limit point *a*, 14.2

Field of application   63

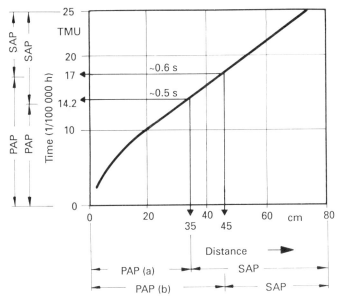

Figure 4.6 PAP = primary assembly process, SAP = secondary assembly process

TMU (58.9%) are counted as PAP and 9.9 TMU (41.1%) as SAP, and with limit b, 17 TMU (70.5%) are counted as PAP and 7.1 TMU (29.5%) as SAP.

If parts of widely varying sizes are to be handled at a single assembly work point, the PAP–SAP analysis can only be applied in a limit-type form. If, with the work point arrangement and the provision of a part, the limit a as shown in Figure 4.6 is reached, it is then taken as a basis for the whole analysis.

Figure 4.7 shows a manual single assembly work point as an example. The schematic arrangement shows that the distances for reaching above the third tier of the manual parts dispensers are classified as being in the SAP area.

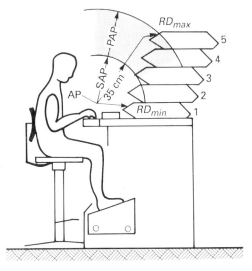

Figure 4.7 Single assembly work point – classification of the basic movement 'reaching' into PAP–SAP (AP = assembly position, RD_{min} = minimum reaching distance, RD_{max} = maximum reaching distance, 1 to 5 = manual parts dispensers)

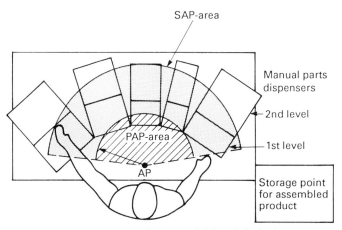

Figure 4.8 Single assembly work point – division of the basic movement 'reaching' into PAP and SAP areas

As a further example, Figure 4.8 shows the arrangement of ten manual parts dispensers into two levels of a single assembly work point at which only the manual parts dispensers at the first level are placed in the PAP area. All others require an additional SAP action for reaching and collection.

4.3.2.2 Gripping
Every action by which the operatives can grasp and bring a part under control without any additional action is classified under PAP. All other actions, for example, regripping, transferring, selecting, etc. are classified under SAP.

In accordance with MTM, the gripping of a part using a tool such as pincers or tongs is not a function of gripping but of collection. Actions for gripping tools are basically classified under the term SAP.

4.3.2.3 Collection
The distance, e.g. from the gripping point to the assembly position, is a principal factor in collection. The two alternative limits as shown in Figure 4.6 for classification in terms of PAP and SAP are also applicable for an analysis of the basic movement of collection. Accordingly, the action for distances exceeding the specified limits are classified under the term SAP. In comparison to the basic movement of reaching, the mode of working for collection is the reverse. Under the assumption of a PAP limit of 35 cm as shown in Figure 4.6, the PAP and SAP areas are shown schematically in Figure 4.9. If, with the basic movement 'collection', rearrangement of the parts is necessary using the second hand, this operation is then classified under the term SAP.

4.3.2.4 Assembly
With an analysis in terms of PAP–SAP, an assembly operation is basically a primary function. If tools, for example screwdrivers, press-in tools or similar are required, the limit values relating to the basic movement of reaching are applicable to the reach and collect movements. This means that the movements required for the collect and return movements of tools which exceed the limits as shown by Figure 4.6 are classified as secondary activities. The same rule also applies if the unit to be assembled must be

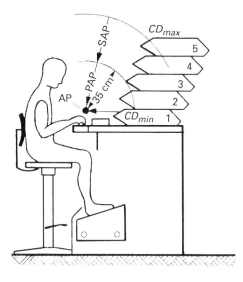

Figure 4.9 Single assembly work point – division of basic movement 'collection' in terms of PAP–SAP (AP = assembly position, CD = delivery distance)

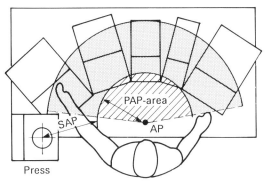

Figure 4.10 Single assembly work point with press PAP–SAP activity of assembly position AP

transferred to a tool. For purposes of illustration, Figure 4.10 shows a single assembly work point with a press positioned on one side. This diagram shows the arrangement of the PAP–SAP activity which is required to transfer the subassembly in a fixture from the assembly position to the press and vice versa.

4.3.2.5 Release
Of all the basic movements, release is the one which requires the minimum effort. For this reason and in terms of a PAP–SAP analysis, pure release is classified under the term PAP. However, if after release, a pause occurs, this is then classified under secondary activities.

4.4 Application example of assembly analysis by primary and secondary activity

Figure 4.11 shows ten parts which are to be manually assembled to form a subassembly. Part 1 is the base part in which all other parts are assembled. The

66   Primary–secondary analysis

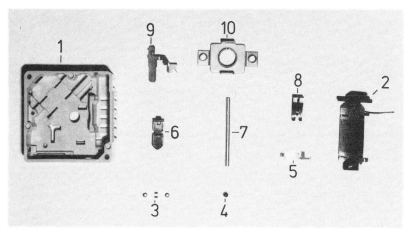

Figure 4.11 Assembly example with ten parts

handling and assembly procedures are of different levels of difficulty. Part 2, a coil assembly, is highly sensitive on account of the winding with thin resistance wire and must therefore be handled carefully.

On the other hand, part 4, a ball, and part 7, a spindle, are very simple to handle.

The following five assembly organizational forms were analysed in terms of PAP–SAP in order to determine the most economic assembly concept:

1. Single assembly work point: provision of parts by a semicircular, multi-tier arrangement of manual parts dispensers (see Figure 4.12).
2. Single assembly work point: provision of parts by a parts paternoster (see Figure 4.14).
3. Single assembly work point: provision of three parts by manual parts dispensers and seven parts by suitably arranged vibratory spiral conveyors (see Figure 4.15).
4. Line assembly: consisting of three single assembly work points with manual transfer of the assembled part from point to point and provision of parts by manual dispensers (see Figure 4.16).
5. Line assembly: with mechanical transfer of the assembled part in an arranged form and the use of workpiece carriers on a double-belt system and also the provision of parts by manual parts dispensers (see Figure 4.18).

4.4.1 Single assembly work point with provision of parts in manual parts dispensers

The arrangement of the single assembly work point is shown in Figure 4.12. The manual parts dispensers for the ten different single parts are arranged in semicircular form around the assembly position in one to three levels. The numbering of the manual parts dispensers corresponds to the single part numbers as shown in Figure 4.11. The mean distances for reaching and collecting from the assembly position to the individual gripping positions of the parts are also shown in Figure 4.12. The work procedure and activity are calculated by MTM.

The classification in terms of PAP and SAP is undertaken in accordance with the definition as given in Sections 4.3.2.1 to 4.3.2.5. As shown by Figure 4.6, the distance for a PAP activity for the principal movements of reaching and collection is assumed to

Application example of assembly analysis by primary and secondary activity 67

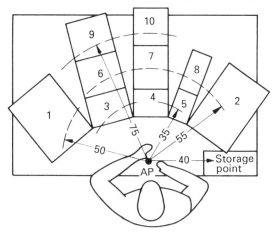

Figure 4.12 Single assembly work point for the assembly of parts as shown in Figure 4.11 (dimensions in cm)

be 35 cm. Figure 4.13 shows in excerpts the calculation of the individual assembly operations and also the classification in terms of PAP and SAP. For purposes of clarity, the process is shown in simplified form, i.e. excluding regripping, inspection, pressing and separating, etc. Parallel activities by the two-hand method are enclosed in brackets and are not included in the addition of the total activities. After completion of assembly, all operations necessary to remove the subassembly from the assembly fixture and to position it are SAP.

The following result is obtained by this method of calculation for the assembly operation with the work point arrangement as shown in Figure 4.12:

	PAP (TMU)	SAP (TMU)	Total (TMU)
Reach	90.0	29.4	119.4
Grip	12.0	51.6	63.6
Collect	99.1	52.4	151.5
Assemble	210.9	0.0	210.9
Release	16.0	2.0	18.0
Total	428.0	135.4	563.4

The overall efficiency is calculated as follows:

$$E_A = \frac{PAP}{PAP + SAP} \times 100\ (\%) = \frac{428\ \text{TMU}}{428\ \text{TMU} + 135.4\ \text{TMU}} \times 100\ (\%) = 75.97\%$$

This calculation shows that three-quarters of the total requirement of 563.4 TMU are used to create value and that one-quarter has no value-creating effect.

In detail, the analysis shows:

- Considerable secondary involvements are incurred with the basic movements of reach and collect by the arrangement of ten manual parts dispensers.
- Except for parts 4 and 7, the single parts are not easy to grasp and therefore incur a high level of secondary activities.

		Requirement in TMU (1 TMU = 0.036 s)											
		Reach		Grip		Collect		Assemble		Release		Total	
Description	MTM code	PAP	SAP	PAP	SAP	PAP	SAP	PAP	SAP	PAP	SAP	PAP	SAP
To housing (1)	R 50 C	14.2	5.4									14.2	5.4
Housing	G 4 A			2.0	5.3							2.0	5.3
To adaptor	M 50 C					16.8	5.0					16.8	5.0
In adaptor	P2SSE							19.7				19.7	
Release	R L 1									2.0		2.0	
To coil (2)	R 45C	14.2	4.0									14.2	4.0
Coil	G1 C1			2.0	3.3							2.0	3.3
To housing	M 45 C					16.8	3.3					16.8	3.3
In housing (2×)	P2SSE							50.6				50.6	
Release	R L 1									2.0		2.0	
Total (TMU)		90.0	29.4	12.0	51.6	99.1	52.4	210.9		16.0	2.0	428	135.4
Efficiency (%)		75.4		18.9		65.4		100		88.8		75.97	

Figure 4.13 Scheme for the determination of the requirement in TMU for assembly as shown in Figure 4.12 and classification in terms of PAP–SAP

- The secondary requirement of 2 TMU with the basic movement 'release' is incurred with positioning of the finished assembled subassemblies.

4.4.2 Single assembly work point with parts provision by a parts paternoster

Figure 4.14 shows the work-point arrangement for the assembly of the subassembly as shown in Figure 4.11 with the application of a parts paternoster for provision of the

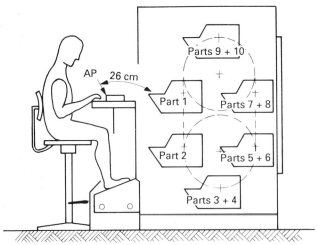

Figure 4.14 Single assembly work point with provision of parts by paternoster for assembly of the ten single parts as shown in Figure 4.11 (AP = assembly position)

single parts. The order of work at the work point as shown in Figure 4.12 is planned so that parts 3, 4, 5, 6, 7, 8, 9 and 10 can be assembled by a two-hand method. Parts 1 and 2 must be gripped and assembled individually. To maintain the two-hand method at the same level with the paternoster arrangement, the manual parts dispensers for parts 3 and 4, 5 and 6, 7 and 8 and also 9 and 10 are in the form of double manual parts dispensers. The manual parts dispensers in the paternoster arrangement also have grip flaps.

The paternoster arrangement has the effect that the distance from the assembly position to the gripping position is maintained as short as possible so that the requirement for the basic movements of reach and collect is considerably less, and equal, for all parts to be assembled.

The following result is obtained by a calculation of the work requirement and classification in terms of PAP and SAP in accordance with the scheme as shown in Figure 4.13:

	PAP (TMU)	SAP (TMU)	Total (TMU)
Reach	78.0	4.4	82.4
Grip	12.0	51.6	63.6
Collect	82.2	18.5	100.7
Assemble	210.9	0.0	210.9
Release	16.0	2.0	18.0
Total	399.1	76.5	475.6

The efficiency of the assembly work point as shown in Figure 4.13 is calculated as follows:

$$E_A = \frac{PAP}{PAP + SAP} \times 100 \, (\%) = \frac{399.1 \text{ TMU}}{399.1 \text{ TMU} + 76.5 \text{ TMU}} \times 100 \, (\%) = 83.9\%$$

In comparison with the work-point arrangement as shown in Figure 4.12, with the arrangement as shown in Figure 4.14, the overall requirement of 563.4 TMU was reduced to 475.6 TMU by the application of a parts paternoster and, at the same time, the efficiency increased from 76% to 84%. The requirements for the main movements gripping, assembly and release have not changed. On the other hand, the main movements reaching and collection were reduced by the shorter gripping and collection distances. The greatest proportion of SAP requirements which still remains results from the main movement gripping.

4.4.3 Single assembly work point, parts provision partly by manual parts dispensers and partly by vibratory spiral conveyors

With a work-point arrangement as shown in Figure 4.14, with provision of the single parts by a parts paternoster, a PAP–SAP analysis shows that, of the total remaining secondary requirement of 76.5 TMU, 51.6 TMU are incurred with the basic movement gripping. To reduce this requirement, an attempt must be made to provide as many of the ten single parts as possible pre-sorted so that they can be grasped at a common point in an arranged form. By reason of the large number of vibratory spiral conveyors required, such a work point is not feasible in actual practice, since the availability is not high enough and the time required to rectify faults would be added to the SAP requirement (the application of vibratory spiral conveyors with manual single activities of short cycle time is practical). To demonstrate the effect of an arranged presentation of single parts, it is, however, practical to investigate one such theoretical single assembly work point by a PAP–SAP analysis.

Figure 4.15 shows schematically the arrangement of one such single assembly work point. On account of their geometry or sensitivity and in accordance with the current state of the art of material feed, parts 1 and 2 as shown in Figure 4.11 are not suitable for automatic arrangement and feed. Part 4, a ball, needs no arrangement. For this reason, in the work-point arrangement as shown by Figure 4.15, these three parts are provided by manual parts dispensers. Parts 3, 5, 6, 7, 8, 9 and 10 are bunkered in vibratory spiral conveyors as loose material, separated by these items of equipment, arranged and fed to the gripping points in the correct position via discharge chutes. On account of the space requirement for the vibratory spiral conveyors and the degree of freedom required for the operatives, it is not possible to arrange the manual parts dispensers for parts 1 and 2 so that the main movements reaching and collection fall within the PAP area as defined by Sections 4.3.2.1 and 4.3.2.3, so, in comparison with the work-point arrangement with the parts paternoster as shown in Figure 4.14, an increased PAP and SAP requirement is necessary for the handling of these parts.

The following values are obtained if the requirement in TMU is calculated in accordance with the scheme as shown in Figure 4.13:

	PAP (TMU)	*SAP* (TMU)	*Total* (TMU)
Reach	72.0	8.4	80.4
Grip	12.0	10.6	22.6
Collect	94.0	27.3	121.3
Assemble	210.9	0.0	210.9
Release	16.0	2.0	18.0
Total	404.9	48.3	453.2

The efficiency is therefore calculated by:

$$E_A = \frac{PAP}{PAP + SAP} \times 100 \ (\%) = \frac{404.9 \ \text{TMU}}{404.9 \ \text{TMU} + 48.3 \ \text{TMU}} \times 100 \ (\%) = 89.34\%$$

By arranging and presenting the seven parts in the correct position, in comparison with the previously described work-point arrangements, the SAP requirement for gripping is reduced from 51.6 to 10.6 TMU and therefore the total requirement for the completion of assembly is reduced from 563.4 TMU to 453.2 TMU. This increases the efficiency to around 89%, and by 5% in comparison with the paternoster solution.

4.4.4 Linking of three single assembly work points to form a line assembly with manual transfer of the assembled part

With this organizational form, the work requirement for assembly of the ten single parts as shown in Figure 4.11 is distributed in the form of line assembly over three single assembly work points (see Figure 4.16). The work content per point is consequently smaller (three to four parts) and makes an optimum single point arrangement possible with the positioning of manual parts dispensers for the single parts. The assembled part is manually transferred from work point to work point with

Application example of assembly analysis by primary and secondary activity 71

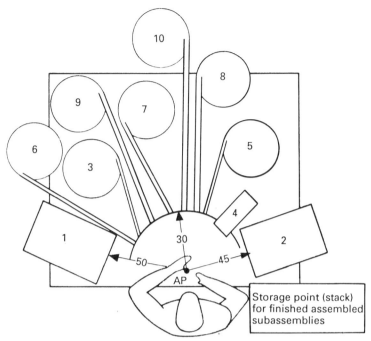

Figure 4.15 Single assembly work point with partial provision of parts by vibratory spiral conveyors for assembly of the ten single parts as shown in Figure 4.11 (AP = assembly position)

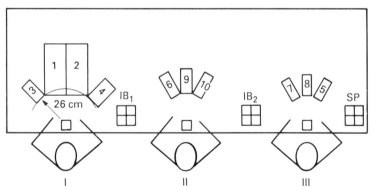

Figure 4.16 Line assembly with manual transfer of the assembled part for the assembly of the ten parts as shown in Figure 4.11 (IB = intermediate buffer, SP = storage point)

the formation of intermediate buffers between the individual work points. In order to distribute the work involvement as uniformly as possible over the three single assembly work points, in comparison with the one work point (Figures 4.12, 4.14 and 4.15), the assembly order has been somewhat rearranged. An additional SAP requirement for manual transfer between the work points is satisfied by distribution over three single work points.

72 Primary–secondary analysis

The following is obtained by a calculation of the total requirement for this workpoint arrangement in accordance with the scheme as shown in Figure 4.13:

	PAP (TMU)	SAP (TMU)	Total (TMU)
Reach	78.0	41.6	119.6
Grip	12.0	57.6	69.6
Collect	82.2	92.4	174.6
Assemble	210.9	78.8	289.7
Release	14.0	8.0	22.0
Total	397.1	278.4	675.5

Since an exactly equal distribution of the work content over the three single assembly work points is not possible, cycle equalization allowances as given in Figure 4.17 of 34.9 TMU are necessary for single assembly work points I and II.

Work point	Parts to be assembled (no.)	Requirement (TMU)	Allowance for cycle equalization CE (TMU)	Total (TMU)
I	1–2–3–4	223.7	13.1	236.8
II	6–9–10	215.0	21.8	236.8
III	5–7–8	236.8	—	236.8
Total	10	675.5	34.9	710.4

Figure 4.17 Determination of the SAP allowance for cycle equalization (CE) for line assembly as shown in Figure 4.16

These allowances represent a secondary requirement and must be taken into consideration in the calculation of the efficiency:

$$E_A = \frac{PAP}{PAP + SAP + SAP_{CE}} \times 100 \; (\%)$$

$$= \frac{397.1 \text{ TMU}}{397.1 + 278.4 + 34.9 \text{ TMU}} \times 100 \; (\%) = 55.9\%$$

The total requirement with this assembly concept is 710.4 TMU with a secondary requirement of 278.4 TMU and 34.9 TMU for cycle equalization. The magnitude of this secondary requirement principally results from the manual activity for placing down and regripping of the assembled part. This is also the reason for the low efficiency of around 56%.

4.4.5 Linking of three single assembly work points to form a line assembly with mechanical transfer of the assembled part in workpiece carriers

The high total requirement and low efficiency of the work point arrangement shown in Figure 4.16 can be improved if the transfer of the assembled part from work point to work point is mechanical with the application of workpiece carriers.

Application example of assembly analysis by primary and secondary activity   73

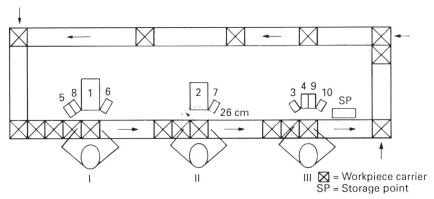

Figure 4.18 Line assembly with mechanical transfer of the assembled part on workpiece carriers for the assembly of the ten single parts as shown in Figure 4.11

This arrangement is shown by Figure 4.18. In this case, a double belt system is used in a rectangular arrangement with circulating workpiece carriers (e.g. Bosch FMS). Restructuring of the work which permits a higher degree of two-hand activity results from the elimination of the secondary requirement for manual transfer between the individual work points. Parts 1, 5, 6 and 8 are allocated to work point 1, parts 2 and 7 are allocated to work point II and parts 3, 4, 9 and 10 to work point III.

A requirement calculation gives the following result:

	PAP (TMU)	SAP (TMU)	Total (TMU)
Reach	65.0	13.2	78.2
Grip	10.0	49.2	59.2
Collect	70.5	18.5	89.0
Assemble	210.9	0.0	210.9
Release	14.0	2.0	16.0
Total	370.4	82.9	453.3

Since uniform cycling for the three single work points is not possible, an allowance for cycle equalization as shown in Figure 4.19 must be added to the work points with a lower work content as a secondary procedure. With this solution, the total requirement is 502.5 TMU with a content of 82.9 TMU for secondary procedures and 49.2 TMU for cycle equalization.

Work point	Parts to be assembled (no.)	Requirement (TMU)	Allowance for cycle equalization CE (TMU)	Total (TMU)
I	1–5–6–8	139.2	28.3	167.5
II	2–7	146.6	20.9	167.5
III	3–4–9–10	167.5	—	167.5
Total	10	453.3	49.2	502.5

Figure 4.19 Determination of the SAP allowance for cycle equalization (CE) for line assembly as shown in Figure 4.18

74 Primary–secondary analysis

The efficiency is therefore calculated as follows:

$$E_A = \frac{PAP}{PAP + SAP + SAP_{CE}} \times 100 \ (\%)$$

$$= \frac{370.4 \text{ TMU}}{370.4 + 82.9 + 49.2 \text{ TMU}} \times 100 \ (\%) = 73.7\%$$

In comparison with the line assembly with manual transfer of the assembled part, in the arrangement with mechanical transfer of workpiece carriers, the efficiency has increased from 56% to 74%. At the same time, the total requirement of 710.4 TMU has decreased to 502.5 TMU.

4.4.6 Summary and efficiency consideration

A summary of the results of the three analysed differently arranged single assembly work points as shown in Figures 4.12, 4.14 and 4.15 is given in Figure 4.20. The results for the two analysed line assembly work-point arrangements are given in Figure 4.21.

Two values given in these tables are important:

1. The total in TMU shows the necessary requirement for assembly of the ten parts as shown in Figure 4.11.
2. The efficiency shows the rationalization potential of the assembly concept.

Ref. no.	Work-point arrangement: single work points	Diagram no.	Requirement (TMU)		Total requirement (TMU)	Efficiency (%)
			PAP	SAP		
1	Manual work point, manual parts dispensers	4.12	428.0	135.4	563.4	75.97
2	Manual work point, parts provision by paternoster	4.14	399.1	76.5	475.6	83.9
3	Manual work point, parts provision partly by feed equipment	4.15	404.9	48.3	453.2	89.34

Figure 4.20 Comparison of single work point arrangements for the assembly of ten single parts as shown in Figure 4.11

Ref. no.	Work-point arrangement: line assembly	Diagram no.	Requirement (TMU)			Total requirement (TMU)	Efficiency (%)
			PAP	SAP	SAP_{CE}		
1	Manual work point, manual parts dispensers, manual transfer	4.16	397.1	278.4	34.9	710.4	55.9
2	Manual work point, manual parts dispensers, automatic transfer	4.18	370.4	82.9	49.2	502.5	73.7

Figure 4.21 Comparison of line assembly arrangements for the assembly of ten single parts as shown in Figure 4.11

Application example of assembly analysis by primary and secondary activity 75

An efficiency below 80% must be considered as being uneconomic.

The arrangement as shown in Figure 4.18 clearly demonstrates that, in spite of the relatively favourable overall requirements of 502.5 TMU, a considerable rationalization potential exists at an efficiency of only 74%. The proportion of secondary requirements of 82.9 TMU (excluding the necessary cycle equalization of 49.2 TMU) is principally attributable to the main movement of gripping. If, with this line assembly with mechanized transfer of the assembled part, vibratory spiral conveyors were to be used for a positionally correct arrangement of single parts 3, 5, 6, 7, 8, 9 and 10, this proportion of secondary assembly processes could be reduced to approximately 10 TMU to 12 TMU. This would also result in a different distribution of the work contents for the three single assembly work points so that the secondary requirement for cycle equalization could also be reduced with certainty.

To decide which assembly organizational form is the most economic, the following factors must be considered in relation to the results obtained:

- Product production volume
- Product service life
- Possibility of reuse of the equipment in excess of the service life of a product.

Under consideration of the relevant costs and with the same planned quantity and method of calculation, economic investigations should be undertaken by application of the cost calculation involving machine hours and floor area rates in accordance with VDI-guideline 3258 and as also described in Sections 11.13 to 11.13.3. For the five analysed organizational forms, the following common assumptions are made in the investigation of the economy:

1. A supplement of 20% is added to the determined TMU-requirement as a piece-work time allowance.
2. Planned quantity: approx. 500 to 600 subassemblies per hour.
3. Estimated depreciation $E_D = \dfrac{\text{Reprocurement value}}{\text{Service life } S_L}$

    Service life in this case = 8 years.

4. Estimated interest $E_I = \dfrac{\text{Reprocurement value}}{2} \times \text{Annual interest}$

    Interest in this case = 10%.
5. Space costs $C_S = m^2 \times$ currency units/$m^2 \times$ years
    Floor space costs in this case = 120 units per annum.
6. Energy costs $C_E = kW \times$ currency units/kWh $\times S_L$
    Current costs in this case = 0.20 units/kWh
    Compressed air $m^3/h \times$ units/$m^3 \times S_L$
    Compressed air costs in this case = 0.08 units/m^3.
7. Maintenance costs C_M: Variable according to work point arrangement.
8. Utilization time $S_L = 200$ days $\times 8$ hours = 1760 hours per annum.
9. Machine hourly rate $C_{MH} = \dfrac{E_D + E_I + C_S + C_E + C_M}{S_L}$.
10. Personnel costs per hour: Productive personnel costs: Wage + 110% (wage overhead costs
    unproductive personnel (supervisors, setters):
    wage + 110% wage overhead costs).

The reprocurement value with a service life of 8 years and an assumed price rise of 5% per annum is calculated as follows:

Reprocurement value $= 1.05^8 \times$ procurement value $= 1.48 \times$ procurement value.

To achieve the specified production performance and by reason of the differing total costs for the work-point arrangements, a different number of work points is also necessary. The reprocurement value is calculated as follows for the floor space cost estimation of the various work-point arrangements and the resultant number of work points:

4 work points as shown in Figure 4.12
10 400 × 1.48 = 15 392 currency units

3 work points as shown in Figure 4.14
16 650 × 1.48 = 24 642 currency units

3 work points as shown in Figure 4.15
63 000 × 1.48 = 93 240 currency units

1 assembly line as shown in Figure 4.16
10 400 × 1.48 = 15 392 currency units

1 assembly line as shown in Figure 4.18
30 000 × 1.48 = 44 400 currency units

The data relating to the economy calculation are summarized in Figure 4.22. The result shows that, with a total time requirement of 475.6 TMU and a realizable efficiency of 83.9% with 0.161 currency units (CU) per part assembly costs, the work-point arrangement as shown in Figure 4.14, namely a manual work point with parts provision by a parts paternoster, is the most economic organizational form for the assembly of the ten single parts.

With the application of these organizational forms, for the required production rate, three single assembly work points with a parts paternoster are required. The relatively good result for the work-point arrangement as shown in Figure 4.15 with the application of vibratory spiral conveyors for the positionally correct provision of the single parts must be considered purely theoretically since such a concentrated application of these items of feed equipment would result in poor parts availability on account of the susceptibility to breakdown. As shown by experience, a reduction of around 20% would have to be made from the achievable production rate as indicated by the analysis of 550 units for the rectification of breakdowns on the parts feed equipment. Accordingly, the assembly costs would increase from 0.167 CU to 0.208 CU.

Reference must be made to the fact that the assembly examples for these ten single parts do not necessitate the use of expensive assembly fixtures and tools. If these were necessary, in comparison to line assembly, where a single procurement only would be necessary, multiple procurement with parallel single assembly work points would be disadvantageous. If the production rate per hour were to be doubled, line assembly as shown in Figure 4.18 with mechanical transfer of the assembled part by workpiece carriers, with the possibility of utilization of the return belt track for equipping a parallel work-point arrangement, would be the most economic solution [2, 13–15].

			Work point arrangement as per Figure:				
			4.12	4.14	4.15	4.16	4.18
	No. of single assembly work points		4	3	3	—	—
	No. of assembly lines		—	—	—	1	1
	Parts assembly rate/h		592	525	550	352	498
Work-point costs (CU = currency units)	E_D	(CU/a)	1924	3080	11655	1924	5550
	E_I	(CU/a)	770	1232	4662	770	2220
	C_S [30 m]	(CU/a)	3600	3600	3600	3600	3600
	C_E	(CU/a)		110	150		200
	C_M	(CU/a)	1000	2000	2000	1000	2000
	S_L	(h/a)	7294 / 1760	10022 / 1760	22067 / 1760	7294 / 1760	13570 / 1760
	C_{MH}	(CU/h)	4.14	5.69	12.54	4.14	7.71
Personnel costs	4 operatives 12 CU/h + 110%		100.80				
	3 operatives 12 CU/h + 110%			75.60	75.60	75.60	75.60
	0.1 supervisor (proportion) 16 CU/h + 110%		3.36	3.36	3.36	3.36	3.36
	Personnel costs (CU/h)		104.16	78.96	78.96	78.96	78.96
Hourly rate	1. Work point costs		4.14	5.69	12.54	4.14	7.71
	2. Personnel costs		104.16	78.96	78.96	78.96	78.96
	Hourly rate (CU/h)		108.30	84.65	91.50	83.10	86.67
Assembly costs/unit	Work point as per Figure 4.12, 108.30 CU : 592 units = (CU/unit)		0.183				
	Work point as per Figure 4.14, 84.65 CU : 525 units = (CU/unit)			0.161			
	Work point as per Figure 4.15, 91.50 CU : 550 units = (CU/unit)				0.166		
	Assembly line as per Figure 4.16, 83.10 CU : 352 units = (CU/unit)					0.236	
	Assembly line as per Figure 4.18, 86.67 CU : 498 units = (CU/unit)						0.174

Figure 4.22 Calculation of assembly costs for the assembly of the ten parts as shown in Figure 4.11 with six different assembly concepts

4.4.7 Primary–secondary fine analysis for the handling and assembly of a single part

If manual handling and assembly of a single part is necessary within an automated assembly plant at a high production rate, the most economic arrangement of the single assembly work point to be integrated in the system can be determined by the PAP–SAP fine analysis. Part 9 as shown in Figure 4.11 serves as an example.

Three variants of the work-point arrangement of this single assembly work point which is to be integrated in an automated assembly system are shown schematically in Figure 4.23.

In variant I, part 9 is provided by a manual parts dispenser. The distance between the assembly and gripping position is 70 cm (system-dependent).

In variant II, part 9 is also provided by a manual parts dispenser. The distance between the assembly and gripping position is 26 cm.

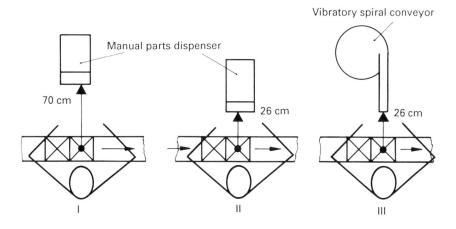

Main movement \ Requirement	PAP	SAP	PAP+SAP	E_A%	PAP	SAP	PAP+SAP	E_A%	PAP	SAP	PAP+SAP	E_A%
	I				II				III			
Reach	15.5	9.5	25.0		13.0		13.0		8.8		8.8	
Grasp	2.0	7.9	9.9		2.0	7.9	9.9		2.0		2.0	
Collect	16.8	11.8	28.6		13.7		13.7		13.7		13.7	
Assemble	25.3		25.3		25.3		25.3		25.3		25.3	
Release	2.0		2.0		2.0		2.0		2.0		2.0	
Total	61.6	29.2	90.8	67.8	56.0	7.9	63.9	91.6	51.8	-	51.8	100

Figure 4.23 Requirement for the assembly of part 9 as shown in Figure 4.11 with different methods of parts provision (PAP and SAP in TMU, E_A in %)

In variant III, part 9 is finally arranged and fed via a vibratory spiral conveyor and presented in the correct position via a discharge chute with a distance between the assembly and gripping position of 26 cm.

The actual work content of these three different variants was calculated by MTM and is shown in detail in Figure 4.23. In accordance with Figure 4.6, a limit value of 35 cm was fixed for the principal movements of reach and collect. An efficiency of 100% is achieved by the application of a vibratory spiral conveyor as shown by variant III. The total requirement of 90.8 TMU for variant I is reduced by around 33% to 51.8 TMU with variant III.

Depending upon the agreed tariffs, an allowance of approximately 20% should be added to the determined requirement in TMU as a piece-work time allowance. For an automated assembly line, this would permit the following single cycle times with the integration of one such single assembly work point:

Variant I: 90.8 TMU × 1.2 × 0.036 s/TMU = 3.92 s
Variant II: 63.9 TMU × 1.2 × 0.036 s/TMU = 2.76 s
Variant III: 51.8 TMU × 1.2 × 0.036 s/TMU = 2.24 s

With the addition of personal and also actual loss and rest times, with an automated assembly time, the arrangement of one such manual single assembly work point as shown in variant III would have to be with a 3 s cycle time [14].

4.5 Extended analysis in terms of primary and secondary requirements for the total sequence of an assembly operation

The compilation of a basic analysis in the assembly area and fine analyses for the single assembly work points is a basic requirement for the drafting of an extended analysis in terms of PAP–SAP for the evaluation of the overall efficiency of an assembly operation.

The limits for the extended analysis are variable. For example, the requirement for the goods inwards inspection can be included or the storage requirement excluded completely from consideration for finished products in cases in which dispatch is direct.

From case to case, packaging is either a primary or secondary requirement. If packaging is only used for intermediate transport (e.g. supply), the requirement is a secondary process. The packaging requirement for a finished product (e.g. pocket calculators) is, on the other hand, a primary process, since it was included in the manufacturing costs and represents a desirable increase in value. If, with the basic and fine analyses, calculation is by time factors, it is recommended that fiscal factors be used for the calculation of the extended analysis since, in comparison to the primary requirements of assembly with the secondary requirements, in particular, considerable wage group differences can occur.

Figure 4.24 shows schematically the total sequence of an assembly operation including the required personnel deployment. Under the assumption that basic and fine analysis has already been undertaken, the overall efficiency is calculated as follows:

$$E_A = \frac{PAP_B + PAP_F}{PAP_B + PAP_F + SAP_B + SAP_F + SAP_E}$$

in which:

PAP_B = PAP requirement based on the basic analysis of assembly
PAP_F = total PAP requirement based on the fine analysis of single assembly work points
SAP_B = SAP requirement based on the basic analysis of assembly
SAP_F = total SAP requirement based on the fine analysis of single assembly work points
SAP_E = total SAP requirement based on the extended analysis of the total assembly sequence.

The assumption of a direct wage is adequate for the conversion of time factors into fiscal factors, since the direct wage costs are approximately equal in percentage terms and therefore are neutral in terms of cost in the calculation of the overall efficiency.

The individual values for the schematic example as shown in Figure 4.24 are given in summarized form in Figure 4.25, from which the overall efficiency is calculated as follows:

$$E_A = \frac{\text{Total PAP}}{\text{Total PAP} + \text{Total SAP}} \times 100\% = \frac{209.25 \text{ CU/h}}{(209.25 + 159.25) \text{ CU/h}} \times 100\% = 56.8\%$$

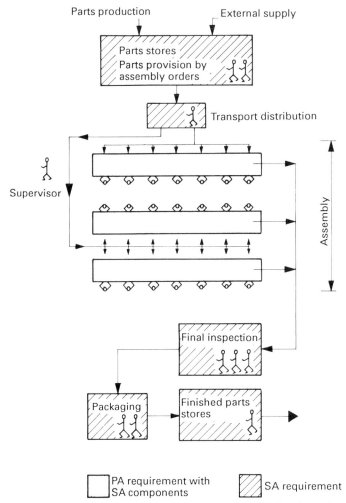

Figure 4.24 Area of extended analysis in terms of PAP–SAP for the total sequence of an assembly operation

This example demonstrates that, from the total personnel cost requirement of 368.5 currency units (CU)/hour for assembly of this product, only 57% is used for an increase in value of the product, the remaining 43% considered as being a secondary requirement.

Of this secondary requirement at 159·25 CU, 15.75 CU (equivalent to approximately 10%) are allocated to the primary area of assembly without the addition of the supervisor required for the assembly operation.

90% of the secondary requirement falls outside the actual assembly area, including the supervisor in the assembly area.

If one considers the overall efficiency of 57%, the following basic questions are useful with regard to its improvement.

1. Is the operational organization of assembly correct?

Area	Requirement	Personnel		Wage costs (CU/h)	PAP		SAP	
		No.	Wage (CU/h)		(%)	(CU/h)	(%)	(CU/h)
Parts stores		2	14.00	28.00	—	—	100	28.00
Transport distribution		1	13.50	13.50	—	—	100	13.50
Assembly		18	12.50	225.00	93	209.25	7	15.75
Supervisor		1	16.00	16.00	—	—	100	16.00
Final inspection		3	15.00	45.00	—	—	100	45.00
Packaging (intermediate transport)		2	13.50	27.00	—	—	100	27.00
Stores/finished products		1	14.00	14.00	—	—	100	14.00
Total:		28	—	368.50	—	209.25	—	159.25

Figure 4.25 Cost determination for the total sequence of an assembly as shown in Figure 4.24

2. Are the parts stores, material provision and distribution technically and organizationally state of the art?
3. Is the final inspection rationally structured?
4. Does the assembly technology match the quality requirements?
5. Can inspection operations be integrated into the assembly process at low cost?
6. Is the packaging technology optimally structured or can packaging be undertaken at lower personnel costs by improved structuring?

It is virtually impossible to set guidelines for the economic limits of efficiency in the overall cycle of an assembly because this depends upon the complexity of a product, batch sizes, production volume, type and variant diversity. When applied to the overall cycle of an assembly, the extended analysis in terms of PAP–SAP does, however, highlight the secondary requirement and consequently identifies weak points in the cycle [14].

4.6 Practical examples

In order to illustrate the application of the primary and secondary assembly analysis, the economic organizational forms of a manual assembly operation are demonstrated below based on three practical examples.

4.6.1 Example 1: Switch assembly

4.6.1.1 Initial basis
A switch shown in Figure 4.26 consists of two subassemblies, a 'drive' with four single parts and a 'switch unit' of 11 single parts, and is finally assembled by two screws. The production rate is 2300 units per day.
   Assembly was initially arranged in the form of a line assembly as shown in Figure 4.27 with manual transfer of the assembled parts with the formation of intermediate buffers and included five single assembly work points. The time allowance in this case was 183 min for 100 units. At an output rate of, on average, 134% and a working time of 52 min per hour (wage rate agreement II, Baden-Württemberg), to achieve the

82 Primary–secondary analysis

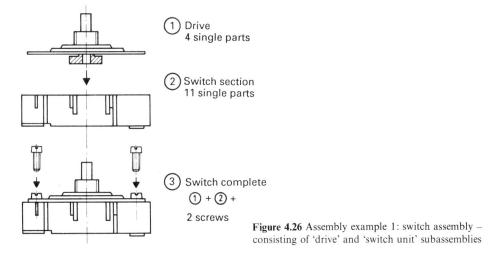

Figure 4.26 Assembly example 1: switch assembly – consisting of 'drive' and 'switch unit' subassemblies

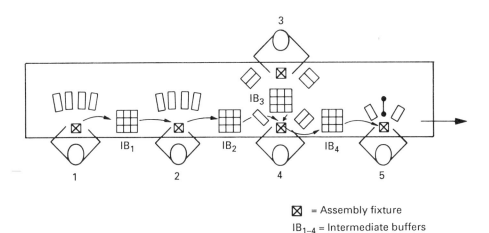

⊠ = Assembly fixture
IB_{1-4} = Intermediate buffers

Figure 4.27 Line assembly with manual transfer of the assembled switch as shown in Figure 4.26

required production volume it was necessary to implement a day shift of 8 h and a part-time shift of 4 h. The motivation for a primary–secondary analysis was that a production increase from 2300 to 3500 units per day was required and, at the same time, a cost reduction of the assembly operation.

The basic analysis showed that by manual transfer of the assembled part and intermediate buffer formation (IB_1 to IB_4 as shown in Figure 4.27), more than 20% of the time used is classified as an SAP requirement. Furthermore, a fine analysis of the individual work points demonstrated that, in particular, the arrangement and positive assembly of a clamping spring, shown in Figure 4.28, which was needed six times, was the cause of a high secondary requirement.

To reduce the high secondary requirement by manual transfer from work point to work point, a complete assembly at one single assembly work point would appear to be appropriate.

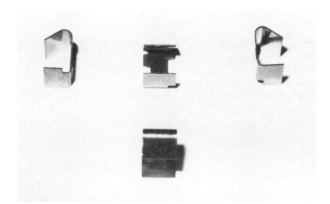

Figure 4.28 Clamping springs for the switch as shown in Figure 4.26

The large number of parts and the assembly tools required all concentrated at one single work point would, however, incur a high secondary requirement for the handling of the parts and assembly tools. Furthermore, the arrangement of a vibratory spiral conveyor for the correct arrangement and presentation of the clamping springs is not optimally possible in terms of space. A high secondary requirement for handling of these parts cannot be avoided without an arranged presentation of the clamping springs. For these reasons, a cost reduction cannot be realized by a single-point assembly organizational form.

4.6.1.2 Solution proposal
The size of the switch $60 \times 40 \times 25$ mm and its weight of approximately 100 g render the following solution possible. The assembly cycle remains distributed over five work points. The single assembly work points are equipped with indexing units for holding plate-form multi-workpiece carriers. The loose interlinking of the five work points is achieved by manual transfer of the plate-form workpiece carriers. The arrangement of the single assembly work points is optimized by a fine analysis. The clamping springs are presented by previously arranged vibratory sprial conveyors.

The assembly holders or workpiece carriers are a plate-form design as shown in Figure 4.29 and have a 24-workpiece capacity.

Every effort must be made in the design of these workpiece carriers to achieve the lowest possible weight because they must be transferred from work point to work point together with the loaded workpieces. The single workpiece holders are preferably made from plastic as segments. The single elements are mounted on alloy carrier rings. The single assembly work points are equipped with indexing units for holding the plate-form workpiece carriers. The workpiece carriers are centred on the indexing units by their centring hole and indexed by carrier pins. Figure 4.30 shows the layout and Figure 4.31 a photograph of the assembly line.

The 11 single parts of the switch assembly are distributed thus: four to each of the single assembly points 1 and 2, and three parts to point 4. Assembly of the subassembly 'drive' is at work point 3. Final assembly is at work point 5.

4.6.1.3 Assembly cycle
The operative at single assembly work point 1 removes a plate-form workpiece carrier from the buffer for empty workpiece carriers (EB_1), centres the carrier on the indexing

84 Primary–secondary analysis

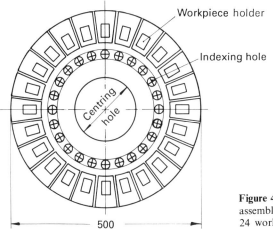

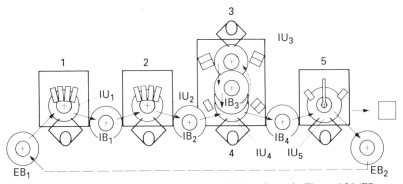

Figure 4.29 Plate-form workpiece carrier for the assembly of a switch as shown in Figure 4.26 with 24 workpiece holders

Figure 4.30 Layout of the assembly line for the switch as shown in Figure 4.26 (EB = empty buffer, IB = intermediate buffer, IU = indexing unit)

Figure 4.31 View on the assembly line for the switch as shown in Figure 4.26

unit (IU_1) at the work point and assembles in the order of 4 times 24 single parts. Upon completion of these grasping and assembly operations, the plate-form workpiece carrier is removed from the indexing unit (IU_1) by the operative and placed on the intermediate buffer IB_1 in front of work point 2. Efforts are made to ensure that the intermediate buffers between the single work points hold on average three or four workpiece carriers. With an average cycle time per workpiece carrier per work point of 5 min, three workpiece carriers in the intermediate buffer represent a buffer capacity of 15 min. This is adequate for the decoupling of the work points. The personal time requirements of the operatives and differences in the output rate during a shift can be covered by this arrangement.

Figure 4.32 shows a section of the arrangement of work point 1. The manual parts dispensers are positioned directly above the assembly point.

The requirement of the main movements reaching and collecting therefore falls within the limits between a primary and secondary requirement. The clamping springs are arranged and presented by a vibratory spiral conveyor so that there is no secondary requirement for reaching, gripping and collection. So as not to impair the removal of the plate-form workpiece carriers by an optimum arrangement of the manual parts dispensers, they are placed on a height-adjustable table. This arrangement is shown schematically in Figure 4.33.

A workpiece carrier is taken from intermediate buffer IB_1 by the operative at single assembly work point 2, centred on the indexing unit IU_2 at work point 2 so that four further parts can be assembled in sequence 24 times.

After completion of the assembly operations, the workpiece carrier is transferred to intermediate buffer IB_2.

As shown by the sketch (Figure 4.34), work points 3 and 4 are combined. Work point 3 is positioned opposite work point 4 because the drive unit of the switch is pre-assembled at work point 3.

At work point 3, the operative withdraws an empty workpiece carrier from intermediate buffer IB_3 between work points 3 and 4 in order to locate it on indexing unit IU_3.

Figure 4.32 Single assembly work point 1 of the assembly line as shown in Figure 4.30

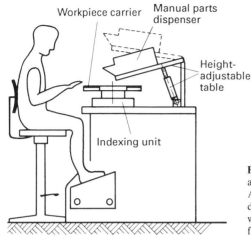

Figure 4.33 Single assembly work point 1 of the assembly line as shown in Figure 4.30. Arrangement of the manual parts dispenser with a device for elevation so that the plate-form workpiece carriers can be placed on or removed from the indexing units without difficulty

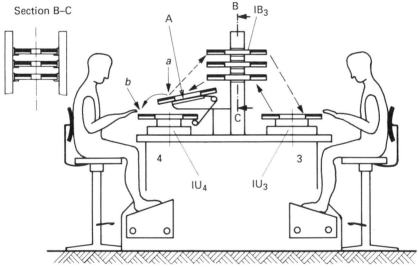

Figure 4.34 Work point 3 for assembly of drive unit and work point 4 of the assembly line as shown in Figure 4.30 (A = location for workpiece carrier for drive unit, a = gripping position for drive unit, b = assembly position for drive unit on switch unit, IB_3 = intermediate buffer 3)

The four single parts of the drive unit of the switch are now assembled 24 times at this work point. Upon completion of the assembly process, the loaded workpiece carriers are returned to intermediate buffer IB_3.

At work point 4, the operative takes a workpiece carrier from intermediate buffer IB_2 between work points 2 and 4 which is loaded with the parts pre-assembled at point 1 and point 2 and locates the workpiece carrier on indexing unit IU_4 at work point 4. At this point, three further single parts are fitted 24 times in sequence. The operative then withdraws and positions the workpiece carrier from the intermediate buffer between point 3 and point 4 loaded with the pre-assembled drive unit on the carrier flange A positioned above indexing unit IU_4 in order to grasp the pre-assembled unit from this point and assemble it into the switch unit. The workpiece carrier above with the drive unit is automatically and synchronously cycled by the indexing of the workpiece carrier so that, with every indexing operation, a pre-assembled drive unit is positioned within easy gripping range for assembly into the switch unit. As shown by Figure 4.34, the pre-assembled drive unit is only moved from point a to point b for assembly. After completion of the last operation at assembly point 4, the empty workpiece carrier for the pre-assembled drive unit is returned to intermediate buffer IB_3 between points 3 and 4 for reuse at work point 3. The finished switch assemblies are removed from point 4 together with the workpiece carrier and placed in intermediate buffer IB_4 between work points 4 and 5. Figure 4.35 shows work point 3 with intermediate buffer IB_3.

Figure 4.35 Single assembly work point 3 of the assembly line as shown in Figures 4.30 to 4.34

The operative removes a loaded carrier from intermediate buffer IB_4 at work point 5 in order to position it on the indexing unit IU_5 at this point.

In this case and by the two-hand method, two screws are inserted through the drive cover plate and the socket of the switch unit. Subsequently, the 48 screws are tightened to a specified torque by an electric screwdriver and the finish assembled switches removed from the workpiece carrier and placed in cartons. The empty workpiece carrier is placed by the operative on empty buffer EB_2 for workpiece carriers. From time to time, the stack is removed from this point to the start position EB_1 on the assembly line.

4.6.1.4 PAP–SAP analysis
The following listing of the operations at the individual work points is used for illustration of the components of primary and secondary assembly operations.

88 Primary–secondary analysis

Point no.	Operation	Number of operations	
		PAP	SAP
1	Remove workpiece carrier from EB_1 and locate, grip and assemble 4×24 parts (parts 1, 2, 3 and 4); place workpiece carrier on intermediate buffer (IB_1).	96	1 1
2	Remove workpiece carrier from IB_1 and locate, grip and assemble 4×24 parts (parts 5, 6, 7 and 8); place workpiece carrier on intermediate buffer (IB_2).	96	1 1
3	Remove workpiece carrier from IB_3 and locate, grip and assemble 4×24 parts (parts 9, 10, 11 and 12); place workpiece carrier in IB_3.	96	1 1
4	Remove workpiece carrier from intermediate buffer IB_2. Grip and assemble 3×24 parts (parts 13, 14 and 15). Remove workpiece carrier from IB_3 and grip and assemble $24 \times$ pre-assemblies from work point 3. Place workpiece carrier with subassembly on IB_3. Place workpiece carrier on intermediate buffer IB_4.	72 24	1 1 1 1
5	Remove workpiece carrier from IB_4, grip and fit 24×2 screws. $48 \times$ screws. Remove $24 \times$ finish assembled product. Place workpiece carrier on EB_2.	48 48 24	1 1
	Total	504	12

With this organizational form, there are 504 primary and 12 secondary assembly processes.

A piece-work time allowance of 1512 s is necessary for the 504 primary processes and 84 s for the 12 secondary processes. The total time allowance for the work content of a workpiece carrier is therefore 1596 s.

24 switches are assembled per workpiece carrier. The single time allowance per switch is therefore 1596 s divided by 24 switches, equivalent to 66.5 s per switch. In comparison to the previous time allowance of 110 s, this represents a reduction of almost 40%.

The efficiency for this assembly organizational form is therefore calculated as follows:

$$E_A = \frac{PAP}{PAP + SAP} \times 100\% = \frac{1512 \text{ s}}{1512 \text{ s} + 84 \text{ s}} \times 100\% = 94.7\%$$

The fact that a reduction in the assembly time of approximately 40% was achieved by this assembly technique is principally attributable to a reduction in the secondary requirement. It should also be noted that with the 24-part plate-form workpiece carriers, it is possible to undertake 24 identical operations in succession and to achieve a more rapid method of working by the repeat effect [2].

4.6.2 Example 2: Switch element

4.6.2.1 Initial basis

A switch element as shown in Figure 4.36 consists of six single parts and is to be assembled at a production rate of approximately 400 units per hour (effective work time = 52 min). In addition to the inserting operations, certain parts must be pressed

Practical examples 89

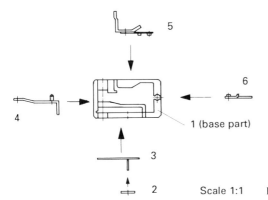

Scale 1:1   **Figure 4.36** Assembly example 2: switch element

into the base and connected positively by re-forming. The contact gap between parts 5 and 6 must be set, the switch function checked and the individual parts sealed in the base part with sealing paint.

The MTM planning of a single assembly work-point arrangement indicated a time allowance of 87.5 min for 100 switch elements. Under the assumption of an average output rate of 135%, 4 to 5 work points are necessary. An analysis in terms of PAP–SAP indicated that the calculated time allowance of 87.5 min is related to 62.5 min for primary and 25 min for secondary processes.

The efficiency is then calculated as follows:

$$E_A = \frac{62.5 \text{ min}}{62.5 \text{ min} + 25 \text{ min}} \times 100\% = 71.4\%$$

In addition to this low efficiency, further disadvantages in the single assembly work-point arrangement are to be seen in the high investment costs for the multiple provision of relatively expensive equipment for pressing in, re-forming and contact gap setting.

The distribution of the assembly processes over several single assembly work points as a line assembly with manual transfer of the assembled part would only incur an additional increase in the secondary assembly processes and also increase the total time allowance.

4.6.2.2 Solution proposal

The organizational form described below was obtained by the application of a basic and fine analysis for the single work-point arrangement.

Figure 4.37 shows the layout of the overall assembly consisting of four loosely connected single assembly work points. Three operatives P1, P2 and P3 work on this assembly line at four stations I to IV. The assembly of the six single parts is distributed between stations I and II.

The operations, press-in parts, re-form, set contact gap, check switch function and seal with paint are concentrated on the automatic functioning station III. Visual inspection and manual testing is undertaken at station IV and also the finished assembled switch element packaged. Stations III and IV are operated together by operative P3.

The assembly sequence is as follows. The plate-form workpiece carriers as already described, each with 24 workpiece holders, form the basis of this organizational form in line assembly and are used for the arranged transport of the assembled part from work point to work point. An empty workpiece carrier is taken by operative P1 at station I

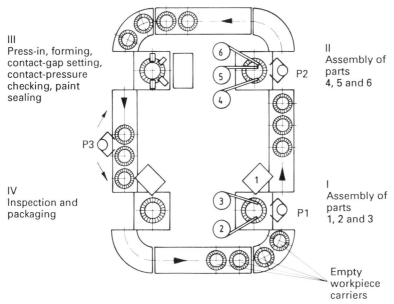

Figure 4.37 Overall arrangement of four loosely connected single assembly work points for the assembly of switch elements as shown in Figure 4.36

and placed on the indexing unit as already described. At this point, the base part 1 is placed 24 times in the workpiece carrier. To avoid a secondary activity, the part is provided packaged in an arranged form in a carton. Two further sheet-metal formed parts 2 and 3 are then inserted 24 times in succession. To avoid secondary processes on the sheet-metal formed parts, they are provided in an arranged form via vibratory spiral conveyors. After completion of these operations, the plate-form workpiece carrier is removed from the index unit and placed on a roller conveyor for transfer to work point II. At work point II, operative P2 grasps the workpiece carrier and places it on the indexing unit in order to fit parts 4, 5 and 6, 24 times in succession. To avoid a secondary activity, these sheet-metal formed parts are also presented in an arranged form by a vibratory spiral conveyor. After completion of these operations, the workpiece carrier is removed from the index unit and placed on the roller conveyor for transfer to work point III.

By the arrangement of three single parts at work points I and II, their erognomic and material handling arrangement can be optimized so that, with the exception of gripping, inserting, removal and stacking of the plate-form workpiece carriers, no other secondary activity occurs.

The third operative P3 alternates between work points II and IV. P3 places the workpiece carriers loaded with pre-assembled subassemblies at work point III in the indexing unit. A switch impulse given by the operative then starts the automatic sequence of operations, press-in, re-form, set contact gap, check switch function and seal with paint. Upon completion of these operations, the workpiece carrier is removed from indexing unit III and placed on the roller conveyor for transfer to work point IV. During the automatic operations at work point III, the same operative undertakes a visual inspection at work point IV and manually checks the packaging of the finished assembly. These operations are undertaken as at all other stations in an indexing unit.

Figure 4.38 Single assembly work point I of the assembly line as shown in Figure 4.37

After removal of the assembly from the workpiece carrier, it is placed on the roller conveyor for transfer to work point I.

Figure 4.38 shows single assembly work point I on this assembly line. The basic and fine analysis showed that at work point I, two secondary operations, namely gripping and positioning of the workpiece carrier and removal and placing on the roller track, were accompanied by 72 primary operations, namely the assembly of three parts 24 times. A circuit arrangement of the plate-form workpiece carriers and therefore automatic return to work point I was achieved by a rectangular overall arrangement.

The economic overall result of the solution as shown in Figure 4.37 is shown by Figure 4.39. The time allowances of a single assembly work-point arrangement are shown in comparison to this line assembly.

The total requirement of 87.5 min for 100 switch elements was reduced to 52.26 min. The secondary requirement content of 25 min with a single assembly work-point arrangement is reduced to 1.3 min by line assembly. In addition, the primary requirement of 62.5 min was reduced to 50.96 min by the improved single assembly work point arrangement:

$$E_A = \frac{50.96 \text{ min}}{50.96 \text{ min} + 1.3 \text{ min}} \times 100\% = 97.5\%$$

The efficiency of the solution as shown by Figure 4.37 is therefore calculated as above.

In comparison to a single assembly work-point solution, this represents an improvement of around 72% to approximately 98%.

Assembly method	Piece-work (minutes per 100 units)	PA (min)	SA (min)	Efficiency %
Single assembly work points	87.5	62.5	25.0	71.4
Line assembly as per Figure 4.37	52.3	51	1.3	97.5
Saving in minutes	35.2	11.5	23.7	
Saving in percent	40.2	18.4	94.8	

Figure 4.39 Assembly times for switch element as shown in Figure 4.36; comparison of achievable results, single assembly work points; line assembly as shown in Figure 4.37

The principal reasons for the time allowance reduction and increase in efficiency are:

1. The completion of 24 identical operations in succession results in an increased familiarization process and more rapid work.
2. The sheet-metal formed parts are presented by a vibratory spiral conveyor so that they can be grasped in the correct position and assembled. Consequently, there is no secondary activity incurred by using the other hand to manipulate the parts correctly.
3. Automation was made possible by combining the mechanical operations of press-in, re-form, set contact gap, check switch function and seal with paint so that no personnel costs are incurred for these operations.

The work content at points I and II is of the order of 4.3 min. A buffer capacity of at least five plate-form workpiece carriers is therefore provided. The resulting buffer time capacity of approximately 20 min to 21 min is adequate to cover a certain individual output performance variation of the individual operatives. A synchronization of the overall output performance must, however, be achieved over a full 8 h shift.

The result as shown in Figure 4.39 also indicates that with the solution as shown and in comparison to single assembly work points, a reduction in the time allowance of 40.27% is achieved. In conclusion, the following is achieved:

- A considerable reduction in secondary activity.
- A limited increase in work interest at the individual work points without overloading the operatives.
- In spite of line assembly, decoupling between work points by intermediate buffering of the plate-form workpiece carriers.
- Optimum ergonomic arrangement of the individual work points by restricting the number of operations undertaken.
- By arrangement of the buffers and decoupling of the work points, output synchronization is only necessary over a shift. During the course of a day and within certain limits, work can be undertaken at differing rates at the individual work points.
- At a factor of 1.5, the total investment is above the investment required for the arrangement of an equivalent number of single assembly work points. A shorter amortization period is, however, achieved by the larger work time reduction.
- A further advantage of this organizational form is the flexibility in the adaption to the production rate. For example, with a part shift, work can be undertaken on a rotational basis at the work points by reason of the buffer capacity [15].

4.6.3 Example 3: Headlight assembly

4.6.3.1 Initial basis

Car headlights are to be assembled in five basic types each with 12 to 24 variants. Depending upon the headlight type, the number of single parts varies between 40 and 100, and 80 to 200 units can be assembled per hour. Two design examples are shown in Figures 4.40 and 4.41.

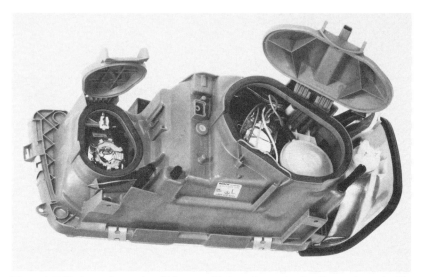

Figure 4.40 Variants of a car headlight assembled on the production line as shown in Figures 4.42 and 4.43 (Bosch)

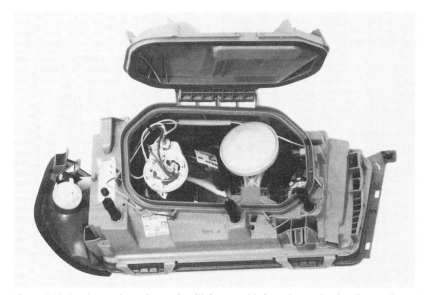

Figure 4.41 Another variant of a car headlight assembled on the production line as shown in Figures 4.42 and 4.43 (Bosch)

94 Primary–secondary analysis

4.6.3.2 Solution proposal
Manual line assembly with mechanical transfer of the assembled part in an arranged form was selected for the economic assembly of the large number of basic types and variants with widely varying levels of complexity, i.e. with extensive avoidance of secondary assembly activities. A double-belt system arranged in rectangular form with 16 work points comprised as follows is used:

    4 Pre-assembly points
  10 Assembly points
    1 Reserve point
    1 Rectification point.

Figure 4.42 shows the start of the assembly line with the infeed point for the empty workpiece carriers. The work content per assembly point is 4 min, so that the passage of two workpiece carriers between the work points forms a decoupling of 8 min. Depending upon the size of the assembled parts, the work points are arranged so that in terms of a PAP–SAP only a small requirement is incurred for secondary requirements. There is no secondary requirement for positioning, transfer or regripping of the subassembly with mechanical transfer of the assembled part. On account of the large number of variants, the workpiece carriers are coded and fed to the respective assembly points from the main conveyor belt by transverse acting units. With this arrangement, differing variants can be assembled at the same time.

Figure 4.43 shows the rear side of the assembly line with the transverse acting units for feeding the workpiece carriers to the individual assembly points and to an additional double flat belt which acts as a single-level buffer in the event of breakdowns within the assembly operation (see Section 5.4.2 and Figures 5.90 and 5.91).

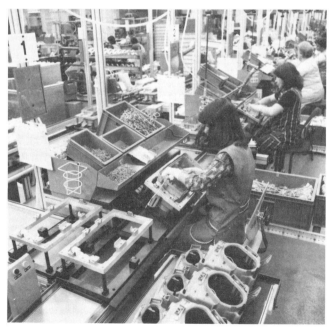

Figure 4.42 Section of the assembly line for car headlights (Bosch)

Figure 4.43 Return conveyor belt and surface buffer for workpiece carriers on the assembly line for car headlights (Bosch)

Chapter 5
Modules for the automation of assembly processes

5.1 Introduction

Items of automatic assembly equipment are technical devices by which assembly processes can either be undertaken fully automatically or semi-automatically with the aid of a person. They are offered on the market in the form of mechanical assembly equipment, assembly machines, assembly lines and automatic modules in many different designs, performance ranges and fields of application. Although every assembly operation consists of different assembly processes, experience with the application of automatic assembly equipment has shown that, with comparable products, many processes are repeated in a similar form. The outcome of this experience is that modules have been devised and continually further developed for mechanical assembly equipment. In the field of assembly technology, the unit construction principle as successfully applied in many branches of technology has made possible the construction of automatic assembly systems without the necessity for a new development from first principles. The principle of the equipment module, which is comprised of so-called basic and *ad hoc* modules, has proven to be advantageous for the construction of such equipment. Because different technical solutions are always conceivable for every assembly programme, it is necessary to possess good knowledge of the available modules. The objective selection of the most suitable modules for the solution of a particular assembly problem under specified boundary conditions is an important and directly effective measure for the economic arrangement of assembly installations.

The modules described in this Section are only a condensed summary and are only given for the purposes of understanding the solutions as described in later Sections. In addition, it is recommended that relevant specialist literature should be studied, such as:

- Loose-leaf volume MHI – Montieren – Handhaben – Industrieroboter (Assembly – Handling – Industrial robots) – (with a comprehensive catalogue section) by H. J. Warnecke *et al.* published by MI-Verlag, Landsberg – Publication 85/86.
- Handbuch der Fertigungstechnik (Handbook of production technology), Volume 5, Fügen, Handhaben und Montieren (Assembly, handling and erection), edited by G. Spur and Th. Stöfferle, Carl Hanser Verlag, Munich/Vienna 1986.

As far as possible, the function of the modules is to undertake the handling and assembly operations in one unit. A fundamental knowledge of handling is a prerequisite for correct selection, especially in the field of material conveying technology.

5.1.1 Handling

In accordance with VDI-guideline 2860, handling is a part function of material flow and is principally defined as follows [17]:

Handling is the realization, defined change or temporary maintenance of a specified three-dimensional arrangement of geometrical bodies in a reference coordinate system. Other conditions, for example time, quantity and path of movement, may be specified.

As shown in Figure 5.1, the part function 'handling' is subdivided into the following areas:

- Storage, stocking quantities
- Variation of quantities

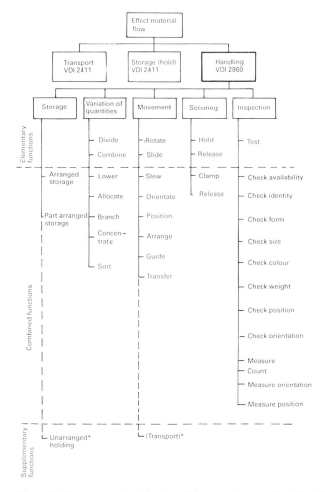

Figure 5.1 Structure and subdivision of the part function 'handling' (VDI). *Strictly speaking, these functions are not handling functions, but are included here for completeness

- Movement (realization and change of a defined three-dimensional arrangement)
- Securing (maintenance of a defined three-dimensional arrangement)
- Inspection.

A further subdivision therefore produces elementary and combined functions.

Modules for the automation of assembly processes largely include many of these functional areas. For example, a vibratory spiral conveyor fulfils the functions of holding, varying quantity and movement. A knowledge of the workpiece parameters is important for correct selection. As shown in Figure 5.2, the workpiece parameters are comprised of the workpiece properties and behaviour. The workpiece properties are subdivided into the behavioural type (form), geometrical workpiece data, characteristic form elements and physical properties. With the workpiece behaviour, a distinction is made between behaviour when stationary and behaviour when being moved. Furthermore, the reliability of automated handling is affected by the quality of the workpieces, cleanliness and the proportion of foreign bodies. The behaviour of workpieces, when moving and when stationary, both individually and in combined form, and the resulting preferred axial positions of the workpieces, have an effect on automated handling. In this respect, Figure 5.3 illustrates the behavioural groups of workpieces and their preferred axial positions. In accordance with Wiendahl, workpieces which are difficult to handle include those which are heavy, large, have coarse tolerances, interlocking parts, workpieces of high sensitivity, low rigidity, undefined geometry and extreme physical and chemical characteristics.

Handling and production operations are easily represented by the basic symbols as shown in Figure 5.4 and the part function symbols as shown in Figure 5.5. Production processes including several production, handling and, if necessary, inspection stages, are illustrated by interconnecting lines between the symbols of the individual part functions, as shown in Figure 5.6(a).

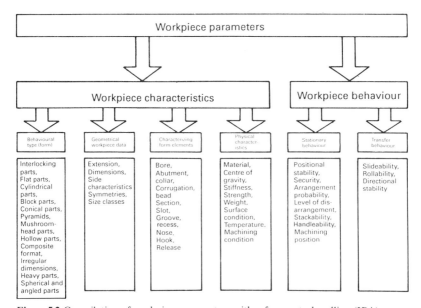

Figure 5.2 Compilation of workpiece parameters with reference to handling (IPA)

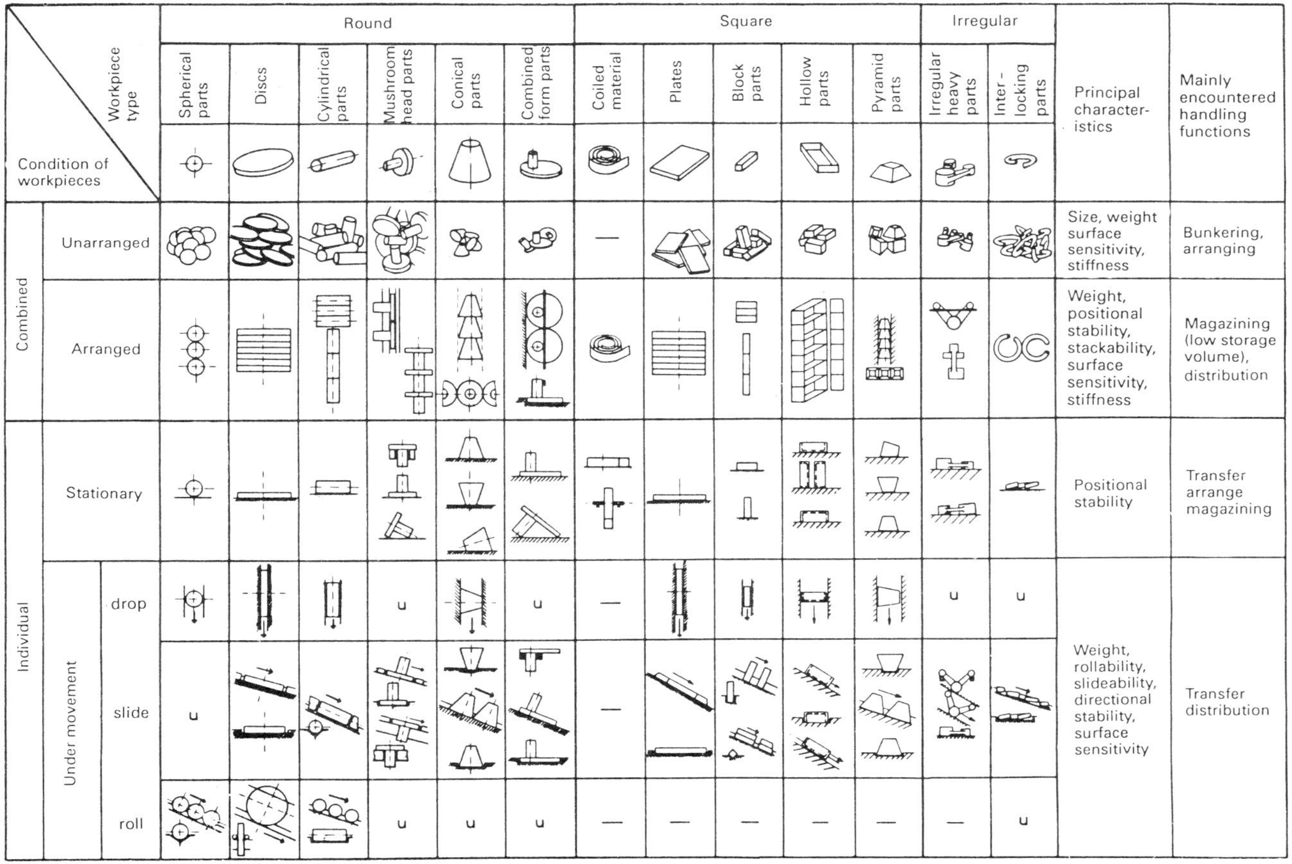

Introduction 99

Figure 5.3 Behavioural groups of workpieces, preferred axial positions (Wiendahl, IFA, Hanover University)

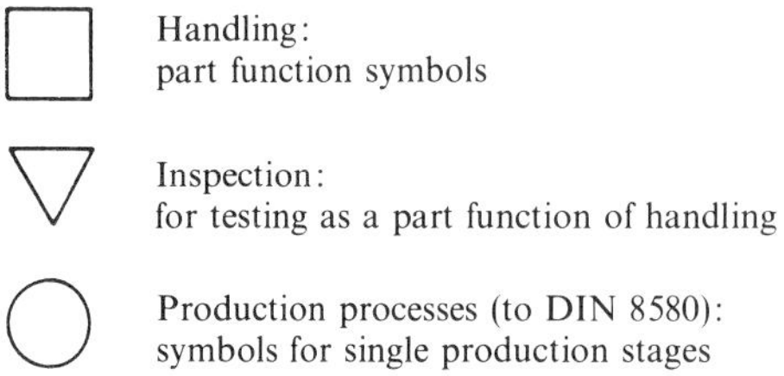

100 Modules for the automation of assembly processes

▢ Handling:
   part function symbols

▽ Inspection:
   for testing as a part function of handling

○ Production processes (to DIN 8580):
   symbols for single production stages

Figure 5.4 Basic symbols for the representation of production processes (VDI)

The process-controlled sequence of part functions is identified by arrows on the connecting lines. Simultaneously occurring functions are shown by the combination of symbols without a connection line (diagram (b)). An equipment-technical combination of functions is represented by the dotted line around the symbols, as can be seen in diagram (c). Figure 5.7 shows a symbolic verbal representation and an operating sequence of the arrangement of small parts on a vibratory spiral conveyor (compare Section 5.2.2.1) [17].

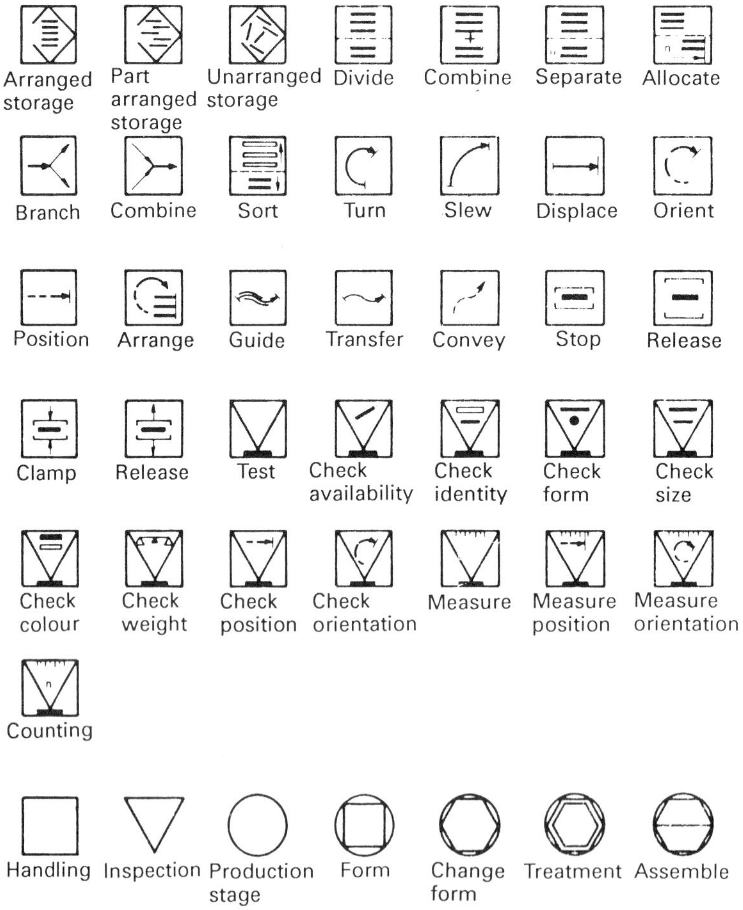

Arranged storage	Part arranged storage	Unarranged storage	Divide	Combine	Separate	Allocate
Branch	Combine	Sort	Turn	Slew	Displace	Orient
Position	Arrange	Guide	Transfer	Convey	Stop	Release
Clamp	Release	Test	Check availability	Check identity	Check form	Check size
Check colour	Check weight	Check position	Check orientation	Measure	Measure position	Measure orientation
Counting						

| Handling | Inspection | Production stage | Form | Change form | Treatment | Assemble |

Figure 5.5 Part function symbols for handling (VDI)

Introduction 101

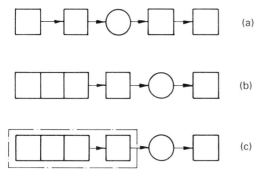

Figure 5.6 Representation of production stages with the application of basic and part function symbols as shown in Figures 5.4 and 5.5 (VDI)

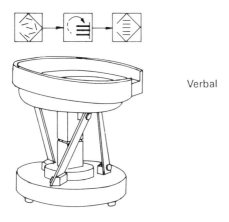

Verbal

Functional sequence

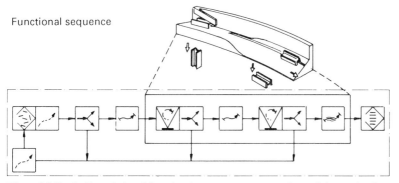

Figure 5.7 Symbolic verbal and functional sequence representations of the mode of operation of a vibratory spiral conveyor (VDI)

5.2 Feeder units

Feeder units are defined in VDI-guideline 3240, pages 1 and 2. In the field of precision engineering and electrical engineering, feeder units must principally fulfil the functions of storing, ranging, feeding or placing in a magazine of individual parts.

The availability of feeder units is principally dependent upon two factors:

1. The selection of the correct feeder process depending upon the part to be handled
2. Constant quality of the single part to be handled.

Exceeding tolerance limits, particularly in the geometrical shape of a part to be handled, results in considerable difficulties. A further important criterion in the availability of feeder units is the cleanliness of and the proportion of foreign bodies in the parts to be handled.

An explanation of cleanliness and proportion of foreign bodies is given by the examples below:

1. *Turned parts*
   Turned parts are manufactured with the use of cutting oils. It is important that such parts are cleaned and degreased. Uncleaned parts carry adherent contaminant material into the functional parts of the feeder units and cause faults, with resultant breakdowns. Furthermore, with turned parts, care must be taken to ensure that the first and last pieces of the base material bars are not found between good parts. If this occurs, such parts with completely different geometrical shapes get mixed up with the stored parts and cause faults on the feeder units and, as a general rule, breakdowns.
2. *Stamped parts*
   With stamped parts, the punchings or the gradually increasing burr is problematic. The same applies to cleaning of the parts as for turned parts.
3. *Plastic formed parts*
   In this case, care must be taken to ensure that, when separating the formed parts from the remnant material in the injection channels, no remnant material falls amongst the form parts. Deburring is a further point to be considered with formed parts. Injection-moulding burrs which occur by incomplete closing of the moulds must be removed such that the formed part falls within the specified tolerance limits. In this case also, a high level of cleanliness is necessary.
4. *Galvanised parts*
   When galvanizing parts, care must be taken to ensure that the galvanizing drums are cleared of residues from the previously treated parts in order to prevent the inclusion of foreign matter during the treatment of new parts.
5. *Easily deformable parts*
   Precautions must be taken to ensure that easily deformable parts made from very thin materials are stored in containers which are dimensioned so that parts at lower levels are not deformed by the dead weight.

The feeder units as used in the field of precision engineering and electrical engineering can be segregated into three principal groups in relation to the part to be handled and its ability to be arranged:

1. Feeder units which can only arrange parts by one arrangement feature, e.g. for the feeding of balls, cylindrical equal-sided parts, or mushroom-head parts with a clear centre of gravity formation, for example rivets and screws with an appropriately

long shank. As shown in Figure 2.19, these are the parts with an arrangement difficulty of ≤2.
2. Feeder units by which a part can be arranged by several arrangement features.
3. Feeder units with additional secondary sorting equipment.

5.2.1 Feeder units for parts with one arrangement feature

From the large number of items of equipment offered in the field of precision engineering and electrical engineering, principally only those described below are used for the arranged feed of parts which must be arranged in terms of one arrangement feature.

5.2.1.1 Hopper with scoop segment

Figure 5.8 shows the design of a hopper with a scoop segment. For storing single parts, the hopper is designed with faces inclined towards the feeder unit scoop segment so that, under the effect of their dead weight, the single parts move towards the centre. The scoop makes a swinging movement about a point of rotation in a manner such that in the lower position its top edge is parallel to the hopper base. It is then swung upwards from the initial position, the movement being dependent on the angle of rotation selected (up to a maximum of 45°). The width and design of the top edge of the scoop is dependent upon the geometry of the parts to be aligned, and is shown schematically in Figure 5.9.

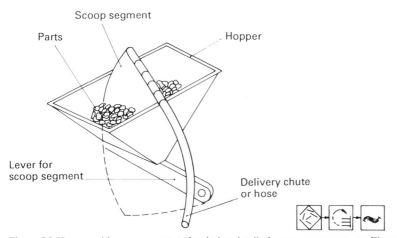

Figure 5.8 Hopper with scoop segment (for design details for scoop segment, see Figure 5.9)

If the scoop segment is swung in the start position, that is in the base plane of the hopper, the parts to be handled fall on to its top edge. Because of the swinging movement, only the correctly aligned single parts remain on the scoop and, upon reaching the highest stop point following the swing movement, can then slide down the delivery chute along a rail to the removal point. On the other hand, parts which are not correctly aligned fall from the scoop back into the hopper during the swinging motion.

The motion of the scoop must not be jerky but smooth. Motion generated either by a crank or cam drive is therefore preferred. With a kinematically correct drive of the

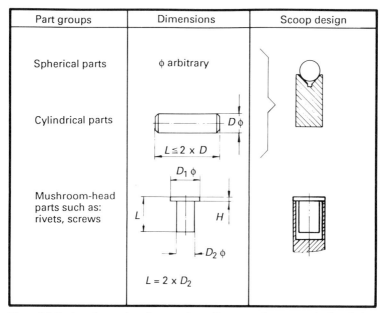

Figure 5.9 Design of top edge of scoop, depending upon the parts to be handled

scoop, this feeder unit is also suitable for the aligning and feeding of sensitive components, e.g. glass or ceramic parts. The capacities of these feed units are dependent on the size of the parts to be handled, the length of the scoop and the time taken for the movement cycle. With a part edge length of 10 mm and a scoop length of 200 mm, the theoretical capacity of approximately 8000 parts per hour can be expected.

5.2.1.2 Hopper with blade wheel

Hoppers which remove parts by means of a blade wheel as shown in Figure 5.11 are suitable for the arranging and feeding of simple, large parts with only one arrangement criterion as shown in Figure 5.10. The blade wheel rotates among the single parts which are stored in the hopper as bulk material.

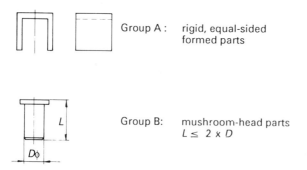

Figure 5.10 Part groups, suitable for feeding from a hopper with blade wheel removal system (Figure 5.11)

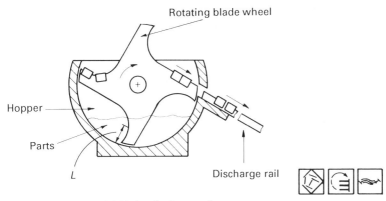

Figure 5.11 Hopper with blade wheel removal system

During rotation, the active area L of the blade wheel contacts single and correctly located parts at random and which are then discharged via a slide rail during the course of further rotation. Easily deformable parts cannot be handled with this type of feed equipment. Furthermore, only parts can be arranged which can be aligned in a suspended position. The design of the upper side of the blade wheel in the active area is dependent upon the geometry of the parts to be arranged, as shown by Figure 5.9. The capacity of this type of feed equipment is dependent upon the filling height of the parts in the hopper, the length of the active part L of the blade wheel and its rotational speed.

5.2.1.3 Hopper with magnetic plate discharging system

A hopper with a magnetic plate discharging system can be used for the arrangement and filling of simple flat parts, for example rings, round blanks, discs and symmetrical flat parts and also for mushroom-head parts for which the shank length is significant in the determination of the point of centre of gravity. The basic requirement is, however, that the parts to be handled are responsive to a magnetic field. Figure 5.12 shows schematically the design of a hopper with a magnetic plate discharge system.

A magnetic plate wheel, positioned vertically and continuously driven, engages to a depth of approximately one-half of its diameter into the parts hopper. The hopper is designed so that it tapers towards its base.

Even with a low fill height, the parts in the hopper automatically move in the direction of the magnetic plate down the inclined walls by the effect of gravity.

The magnetic plate is fitted with a number of permanent magnets around its circumference depending upon the diameter of the parts to be handled. Their magnetic field draws a part flat on to the wheel which is then removed from the hopper by the rotation of the magnetic plate. An inclined discharge rail designed according to the thickness and diameter of the removed round parts is arranged so that the infeed aperture is below the inner pitch diameter of the permanent magnets on the magnetic plate. A magnetically held part is brought into contact with the release surface of the discharge rail by the rotating plate and is detached by a pivot rotation of the retaining magnet. A part so arranged then slides or rolls off into the discharge rail. Round parts can only be arranged if they are symmetrical on both faces. With the arrangement and feed of mushroom-headed parts, the discharge rail must have a slot form so that the part's shank can pass through the slot. The capacity of feed units with magnetic plate

106 Modules for the automation of assembly processes

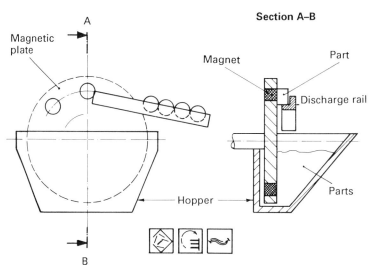

Figure 5.12 Hopper with magnetic plate discharge

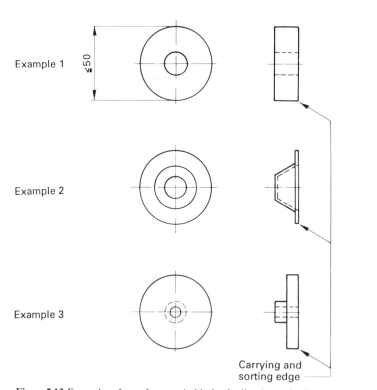

Figure 5.13 Examples of part forms suitable for feeding by an inclined conveyor

discharge is dependent upon the pitch circle diameter of the magnetic plate, the number of permanent magnets and the rotational speed of the magnetic plate.

5.2.1.4 Inclined conveyors

Inclined conveyors, which are also known as plate chain conveyors, are particularly suitable for the high-capacity arrangement and feed of simple, cylindrical flat parts and also, with certain limitations, for cylindrical large size flat formed parts. Figure 5.13 shows three different part shapes which are suitable for arranging and feeding by these items of equipment. Figure 5.14 shows the schematic arrangement and the method of operation of an inclined conveyor.

A vertical slightly inclined conveyor belt is driven by an electric motor; an overload clutch should be included between the drive and conveyor belt drive wheel. The conveyor drive wheel is located in the hopper housing for the parts. The carrier and sorting strips which are slightly inclined in the direction of conveyance are riveted or screwed to the conveyor belt.

The degree of inclination of the conveyor belt is dependent upon the centres of gravity of the parts to be handled. Depending upon their form, the carrier and sorting strips of the conveyor belt withdraw a particular number of workpieces in a specific position from the randomly arranged workpiece reserve. Incorrectly located parts fall into the arranged workpiece reserve. Incorrectly located parts fall into the hopper

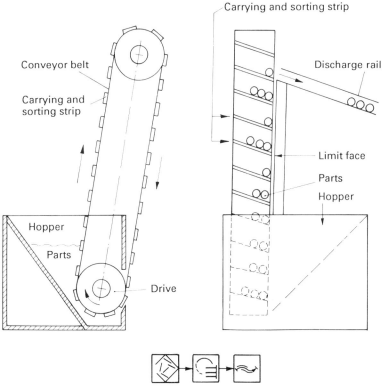

Figure 5.14 An inclined conveyor

when the carrier and sorting strips move out of the hopper. The inclined arrangement has the effect that the withdrawn parts move to the bottom of the incline against the end face of the conveyor. The moment the continuously running conveyor passes the discharge rail, the end face of the conveyor is interrupted and the parts can slide on to the discharge rail in an orderly fashion. Inclined conveyors can be equipped with a right- or left-hand discharge or with discharge over the head of the conveyor belt. Their advantage lies in the large storage and conveying capacity.

5.2.2 Feeder units for parts with several arrangement criteria

In practice, the most widely used feed unit is a vibratory spiral conveyor, which is also known as the oscillatory conveyor or vibratory hopper. This unit is suitable for the ordering and feed of parts with several ordering features. Its basic design is shown in Figure 5.15. The vibratory spiral conveyor consists of the following basic parts: a heavy base plate (1), which is used for isolation of the oscillations and is mounted on three rubber feet; three inclined leaf spring packages (2), arranged around the circumference connect the base plate to the carrier plate for the parts container; the spiral channels (5), which lead to the outlet (6) of the vibratory spiral conveyor, are located in the parts holder (4); the oscillation exciter is an AC-driven magnet (3). The arrangement elements, which are also known as baffle plates for arranging the parts, are positioned on the spiral channel or the parts container wall. With the vibratory spiral conveyor, workpieces can be both stored and arranged. The parts can be placed in a magazine and transferred in conjunction with discharge rails.

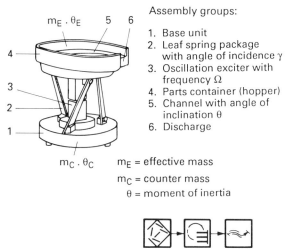

Assembly groups:
1. Base unit
2. Leaf spring package with angle of incidence γ
3. Oscillation exciter with frequency Ω
4. Parts container (hopper)
5. Channel with angle of inclination θ
6. Discharge

m_E = effective mass
m_C = counter mass
θ = moment of inertia

Figure 5.15 Design of a vibratory spiral conveyor (with a fixed phase angle) (Wiendahl, IFA, Hanover University)

The randomly arranged parts in the parts holder are moved towards the edge of the holder by the arched form of the base and the mechanical oscillations which are induced by the electromagnet. The AC-driven magnet generates periodic tensile forces so that the parts holder which is mounted on the leaf springs vibrates close to resonance. During oscillation, the inclined position of the springs generates a short helical movement so that the resulting impulse forces the parts to move from the hopper in a so-called micro-jump movement (not visible to the naked eye), along the helical channels to the hopper outlet point. The parts are arranged by the arrangement elements which are positioned in the helical channels or on the hopper wall so that the parts leave the hopper discharge point correctly arranged. When connected to a 50 Hz mains, the vibratory spiral conveyor oscillates at 100 Hz. The oscillation amplitudes are considerably less than 1 mm. The amplitude is controlled by voltage change and therefore also the drive force and consequently the feed speed of the parts which are being fed.

The operation of the micro-jump movement on the channel is shown in Figure 5.16.

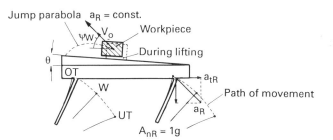

Figure 5.16 Generation of a micro-jump movement on a straight line (Wiendahl, IFA, Hanover University)

5.2.2.1 Arrangement of parts in vibratory spiral conveyors

Arrangement elements are used for the arrangement of parts in vibratory spiral conveyors. The parts to be fed are arranged in the required position by rotation, turning, suspension, aligning, fitting and aligning on edges, etc. A distinction is made between active and passive arrangements. Passive arrangement means that incorrectly positioned parts are rejected and fall back into the hopper. Active arrangement, on the other hand, means that parts are correctly arranged by a series of arrangement elements without falling back into the hopper. The ordering elements are generally sheet-metal plates. Their form and type must be matched to the arrangement requirements of the workpiece. Malfunctions on the vibratory spiral conveyor can be caused by slight variations on the workpiece attributable to form or tolerance variations or the arrangement elements.

Work has been in progress for some considerable time to develop standard type arrangement elements for certain parts families. A distinction is made between irregular, flat, cylindrical, block, taper, pyramid, mushroom-head, hollow, composite-formed, irregular large parts, spherical and long parts. An analysis of workpiece forms in mass production indicates that, with this grouping arrangement, a large proportion of the parts fall into the large irregular and composite form category.

Standard arrangement elements are not suitable for these parts; they must be specially developed and positioned for every part to be handled. All the same, it can be assumed that a range of standard elements is available which can be used individually or in combinations and adaptations for the solution of numerous feed and arrangement functions on vibratory spiral conveyors. A selection of arrangement elements is shown in Figures 5.17, 5.18 and 5.19. A prerequisite for arranging parts is that they can move freely in the proximity of the arranging elements in the vibratory spiral conveyor without causing marked obstruction during arrangement.

110 Modules for the automation of assembly processes

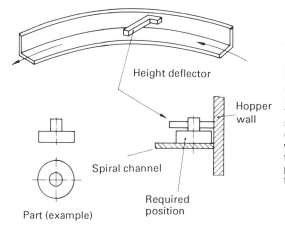

Elements for preventing parts blockages. Rejection by the position of the part's centre of gravity.

Arrangement element by part height and centre of gravity.
Correctly positioned parts are deflected by the arrangement element from the hopper wall on the spiral channel in the direction of the edge without falling back into the hopper. Incorrectly positioned parts therefore fall back into the hopper.

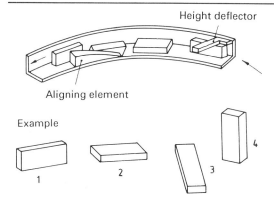

Alignment element combined with a height deflector.
Example:
Part 1 correct, passes the height deflector and aligning element.
Part 2 arrives incorrect, passes the height deflector, is moved into the correct position by the aligning element.
Part 3 arrives incorrect, is rejected because of its centre of gravity position by the aligning element and falls back into the bunker.
Part 4 arrives incorrect, is contacted by the height deflector and falls back into the hopper.

Figure 5.17 Arrangement elements for vibratory spiral conveyors

An excessively high feed rate of the parts is avoided by the installation of an element to prevent blockages (see Figure 5.18) [7].

5.2.2.2 Types of vibratory spiral conveyors
The design of vibratory spiral conveyors is divided into two groups according to the

Feeder units   111

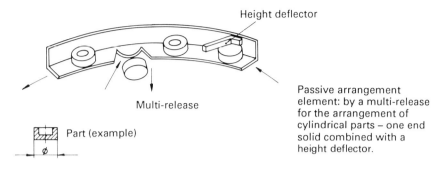

Height deflector

Multi-release

Part (example)

Passive arrangement element: by a multi-release for the arrangement of cylindrical parts – one end solid combined with a height deflector.

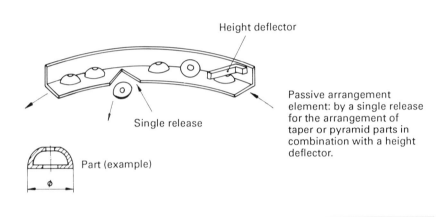

Height deflector

Single release

Part (example)

Passive arrangement element: by a single release for the arrangement of taper or pyramid parts in combination with a height deflector.

Blockage prevention element

Height deflector

Longitudinal slot

Part (example)

Passive arrangement element: by a longitudinal slot for the arrangement of mushroom-head parts (suspended) in combination with a height deflector and a component to prevent blockages.

Figure 5.18 Arrangement elements for vibratory spiral conveyors

shape of the parts container: vibratory spiral conveyors with a cylindrical hopper and those with a stepped hopper. With both designs the discharge direction can be either right (clockwise) or left (anticlockwise).

Both types are shown in Figure 5.20. There is no general rule which states which is the most suitable design. The choice depends rather upon the geometry and material of the parts. One advantage of a stepped parts container is that parts which fall back into

112 Modules for the automation of assembly processes

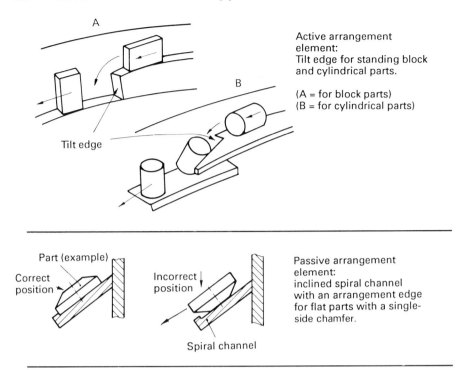

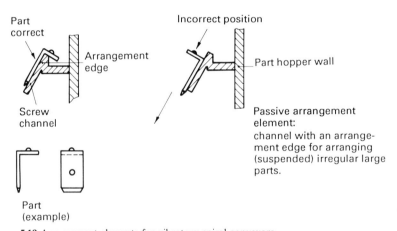

Figure 5.19 Arrangement elements for vibratory spiral conveyors

the hopper on account of an excessively high feed rate or deflection by an arrangement element are not severely stressed because they fall down only one step. With a cylindrical hopper, a deflected part falls the total height of the hopper on to its base. A further advantage of the stepped arrangement is that jamming by parts passing over each other above the screw channels is avoided because there is no upper cover provided in the form of the screw channels.

Feeder units 113

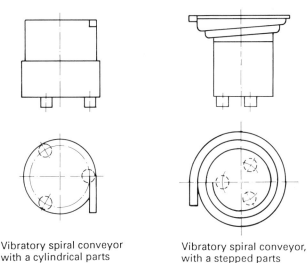

Vibratory spiral conveyor with a cylindrical parts hopper, right-hand discharge (clockwise)

Vibratory spiral conveyor, with a stepped parts hopper, left-hand discharge (anticlockwise)

Figure 5.20 Types of vibratory spiral conveyors

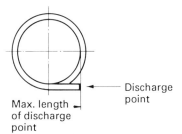

Max. length of discharge point

Discharge point

Figure 5.21 Correctly dimensioned discharge point

The length of the discharge rail tangential to the hopper on vibratory spiral conveyors is limited since the oscillation is generated by the plate springs in a circular arrangement and not along a linear path. This has the effect that as far as possible the discharge point on a vibratory spiral conveyor should not protrude over the diameter of the parts container. Figure 5.21 shows a correctly dimensioned discharge point. Figure 5.22 shows a photograph of a vibratory spiral conveyor.

5.2.2.3 Discharge rails

When connected to the discharge point of vibratory spiral conveyors, discharge rails perform the function of arranged storage and transfer of parts. They can be divided into three principal types which are shown in comparison in Figure 5.23.

Figure 5.23(a) shows the rigid discharge rail in the form of a slide with feed of the arranged parts by gravity. This design is used if, with regard to their geometry, the parts can be fed by a slide and remain in an arranged position and do not obstruct each other when sliding. The most important criterion with this design is the form of the transition radii at the entry of the discharge rail from the vibratory spiral conveyor into the incline and then from the incline into a horizontal position to the discharge point.

114   Modules for the automation of assembly processes

Figure 5.22 Vibratory spiral conveyor

Very thin parts have a tendency to ride over each other in these areas, which results in discharge disruptions.

Figure (b) shows a power-driven discharge rail which is positioned horizontally or with a slight incline and also excited periodically by an electromagnetic drive. The arranged parts which emerge from the vibratory spiral conveyor are both arranged and stored in the discharge rail and fed to the discharge point by the micro-jump principle.

Figure (c) shows an electromagnetically driven discharge rail in the form of a conveyor belt. Flat or round belts can be used, depending upon the geometry of the parts to be stored and transferred.

Discharge rails must not be permanently connected with vibratory spiral conveyors. Permanently connected, rigid discharge rails would oscillate over their entire length by the screw-form oscillations; this would have the effect that the parts could not be fed along the rail, and also positive gripping would not be possible at the discharge point.

With electromagnetically driven discharge rails, a space must be provided at vibratory spiral conveyor discharge points so that mutual contact does not occur as a result of both oscillations. With parts whose length in the feed direction is several times greater than the gap width between the vibratory spiral conveyor and discharge rail, the transition gap can be at right-angles to the linear channel; an inclined gap must be provided for shorter parts (see Figure 5.24). Comb-type transition gaps are also common.

The form of the discharge rail is dependent upon the geometry of the parts to be transferred. The design must ensure that the parts do not lose their ordered position during transfer. Figure 5.25 shows the cross-sections of different types of discharge rails. Type 1 – an open solid channel – is suitable for flat parts whose transfer in a line does not result in jamming even with the generation of a build-up pressure in the discharge rail.

Feeder units 115

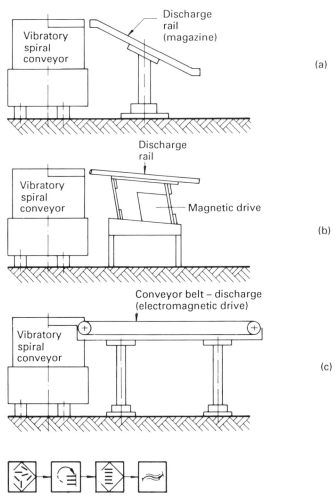

Figure 5.23 Design types of discharge rails

Thin parts or parts whose contacting edges during transfer are very narrow are not suitable for this type. Type 2 shows a solid rail which is completely covered. Its disadvantage is that free access is not possible and disruption cannot be identified. Type 3 with half-side covering of the solid rail is more suitable than type 2 for immediate access and the identification of faults. Type 4 shows a solid discharge rail with double side covering for the transfer of mushroom-head parts which cannot be transferred suspended but with the shank upwards on the mushroom head. Type 5 shows a solid discharge rail with a bent wire bar covering. In this case, riding of the parts over each other is prevented by a round wire bar positioned in the direction of motion of the discharge rail. This type enables rapid identification of faults and good accessability. Type 6 shows a wire construction discharge rail. Its advantage is in the low friction contact area and the insensitivity to foreign bodies and contamination. These types are not suitable for an electromagnetic drive because the design is too unstable. Type 7

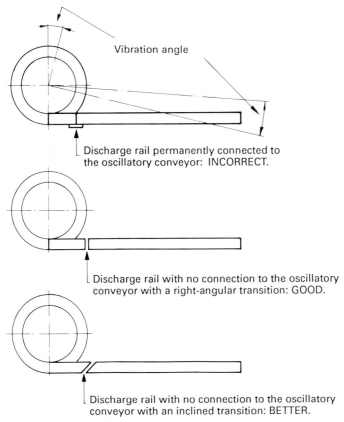

Figure 5.24 Transition between a vibratory spiral conveyor and an electromagnetically driven discharge rail

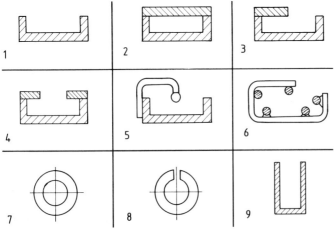

Figure 5.25 Design examples of discharge rails

shows an enclosed tubular type for smooth cylindrical parts but also with the disadvantage that, as with type 2, the identification of faults and access is not possible. Type 8, a tubular discharge rail with one side open, enables the identification of faults and gives easy access for their elimination. Type 9 shows an open solid U-form channel suitable for the transfer of suspended mushroom-head parts. A retainer or cover as shown in type 5 should be provided if very flat mushroom-head parts are transferred.

With reference to the maximum and minimum tolerances of the parts to be stored and transferred, adequate clearance should be provided between the part and cover on the discharge rail so that jamming and build-up cannot occur either with a low level of contamination or when the tolerances have been exceeded slightly. On the other hand, the space clearance must not be too large so that the parts can ride over each other. Figure 5.26 shows that the clearance a must be less than dimension b of the part to be transferred.

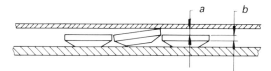

Figure 5.26 Tolerance determination in relation to the transferred part in a covered discharge rail

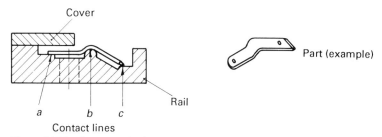

Figure 5.27 Design example of a discharge rail for a sheet-metal formed part

In the design of discharge rails, care must be taken to ensure that the contact areas between the part and rail are kept as small as possible. Figure 5.27 shows the cross-section of a design of a discharge rail for a contact part. This part is a sheet-metal formed part, virtually equal-sided and sent at an angle of 30°. The bent edge is in the form of a radius. As known from practice, bent formed parts are relatively inaccurate over their length; it is therefore advantageous to transfer this form part along the discharge rail on the radius in order to avoid a precise limitation in the discharge rail via the part length. With this design, the part is in contact at three points (a, b and c) and is half-covered so that checks and access are possible at all times.

5.2.2.4 Parts separation – removal – distribution

With the direct removal of a part from a discharge rail without separation, a static removal point must be provided so that the part can be removed from a stationary position. Figure 5.28 shows the overall arrangement of a feed unit consisting of a vibratory spiral conveyor, an electromagnetically driven discharge rail and a static removal point. It is in the form of an extension of the discharge rail profile and

118 Modules for the automation of assembly processes

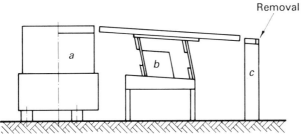

Figure 5.28 Complete arrangement of a feeder unit (a = vibratory spiral conveyor, b = electromagnetically driven discharge rail, c = a static removal point)

dimensioned so that only one part can be removed. The gap width between the oscillating discharge rail and the static removal point should be the same as that between the vibratory spiral conveyor and the discharge rail.

Dynamic pressure is generated on the part to be removed by the effect of the feed of the following parts on the discharge rail. To remove the part reliably, the dynamic pressure is removed from it by suitable separating devices. During separation, the part is moved or slewed at right-angles to the direction of feed. The direction of movement is dependent upon the part and should be parallel to the longest area of contact of the parts which are in contact in a line. Figure 5.29 shows examples of separating equipment which are dependent upon the part form.

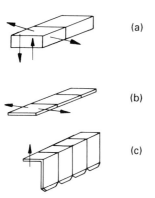

Figure 5.29 Directions for the separation of parts transferred in a line

The block part, example (a), with a relatively large height, can be slid laterally and also upwards and downwards because the face contact between the part to be moved and the following part is also large enough in terms of height. Example (b) shows that thinner-type flat parts must be moved laterally. During separation, formed parts, such as the unequal-sided angled part shown in example (c), must be moved in the direction of the long side of the part.

Figure 5.30 shows the schematic arrangement of separation by the transverse movement of disc-form parts. Figure (a) shows the position of the separating slider (1) in the start position ready for the reception of a part from the discharge rail (2). Figure (b) shows the separation slider (1) after separation and the positioning of a part for removal in the removal position. To assist gripping of the part in this position, the

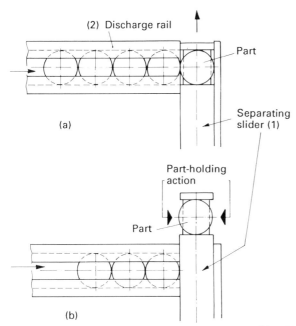

Figure 5.30 Separation of parts by a transverse action slider: (a) start position; (b) separating position

separation slider is recessed on both faces so that a single part can be easily gripped at its diameter.

In order to move a part for removal into a position away from the discharge rail, the linear motion can be replaced by an angular motion. In this case and as shown in Figure 5.31, the separating slider is replaced by a segment.

With vertical or inclined discharge rails (e.g. tubular-type) separation as shown in Figure 5.32 is preferable. The separation slider has two pins (a) and (b). The spacing between the pins is equal to the size of the part to be separated. The parts drop by movement of the separation slider in one direction and are held by pin (b). A part is released by pin (b) with movement of the separation slider in the other direction and, at the same time, the following part is held in the discharge rail by pin (a).

In the case of parts which can be fed directly into the workpiece fixture of a circular indexing table without the assistance of a handling unit, the separating operation can be performed by the indexing table itself.

Figure 5.33 shows one such solution. A circular flat part is to be fed and separated. In the indexed position, the workpiece holder in the circular indexing table forms an extension of the discharge rail profile; its length is limited so that only one part can be fed in. Feeding of further parts from the discharge rail is prevented by the indexing table timing until the next workpiece holder is aligned with the discharge rail in the new indexed position.

With a high feed unit feed rate and a requirement to feed the separated parts to several points at the same time, a distribution can also be undertaken with separation. Figure 5.34 shows a design arrangement for separation and distribution from a discharge rail into two discharge rails with alternate operation. The feed unit discharge rail is sealed by a slider which has two separate workpiece holders of the same profile

120   Modules for the automation of assembly processes

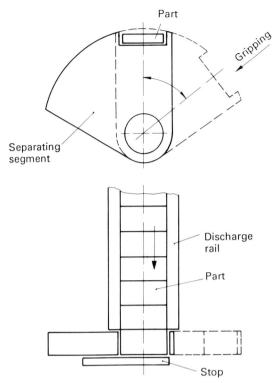

Figure 5.31 Separation by a segment

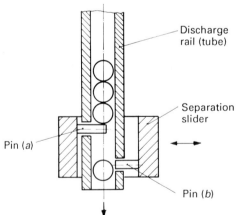

Figure 5.32 Separation by a separation slider for a vertical or horizontal separation rail

on the discharge rail. If the slider is aligned with the feed unit discharge rail, a part can drop into the respective holder. In the same position, the other slider holder releases the previously fed part into one of the two discharge rails. If the slider is now moved, the empty holder is aligned with the feed unit discharge rail for repeat reception of a part.

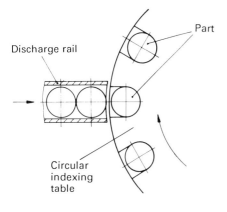

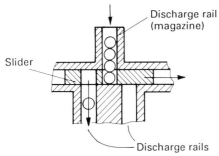

Figure 5.33 Direct feed and separation of parts into a circular indexing table workpiece holder

Figure 5.34 A device for separation and distribution of individual parts

The part in the slider of the first holder is aligned with the second discharge rail and releases the part. With this mode of operation, a workpiece is taken from the feed unit discharge rail every cycle and alternately distributed to the two discharge rails.

If, instead of alternate, a simultaneous distribution is required, this can be achieved by an additionally connected cover slider. Figure 5.35 shows schematically the simultaneous distribution of the parts fed from the feed unit discharge rail into three discharge rails. The distribution slider connected to the cover slider accepts three parts

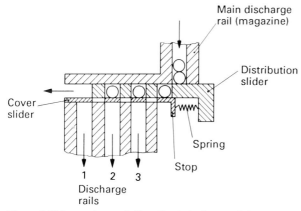

Figure 5.35 Separation and allocation into three outlets

during its movement transverse to the feed unit discharge rail or magazine. The distribution slider apertures are closed by the connected cover slider. This slider is spring-loaded relative to the distribution slider. If the stop is opened during the transverse movement of the cover slider, the distributor slider makes an additional movement to overcome the spring force between both elements. The apertures of the distributor slider, cover slider and discharge rails are then aligned so that three parts are released into discharge rails 1 to 3. In this position, the feed unit discharge rail is closed by the distributor slider.

5.2.2.5 Secondary sorting equipment

Parts which cannot be arranged in their final position by a feed unit must be brought into this position by secondary sorting equipment. At the same time, a check must be made to determine if, with an arranged part position, repositioning in the form of turning over or rotation is necessary or if additional position identification is necessary. A distinction is made here, too, between active and passive sorting.

If parts are only to be rearranged by turning over or rotation, this is undertaken by appropriate active items of equipment. Figure 5.36 shows one such example by turning over. The mode of operation is as follows. The parts are arranged in a vibratory spiral conveyor (1), stored and transferred by a linear conveyor (2). A turn-over wheel (3) cyclically driven by a suitable drive (4) is dimensioned so that before every movement a parts holding aperture is aligned with the linear conveyor (2). The holding aperture is an extension of the profile of the linear conveyor (2). The holding aperture in the turn-over wheel is aligned to discharge rail (6) after two cycles (displaced by 180° to the feed point). The single part is pushed into this rail by an ejector (5) and then transferred to

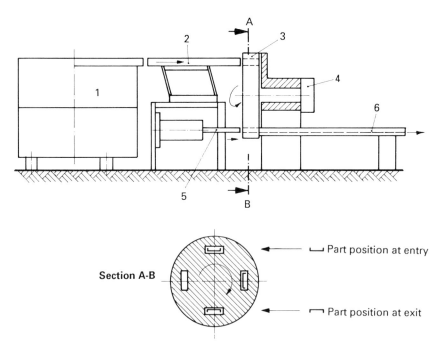

Figure 5.36 Secondary sorting unit (1 = vibratory spiral conveyor, 2 = linear conveyor, 3 = turn-over wheel, 4 = turn-over drive, 5 = ejector, 6 = discharge rail)

the removal point together with the previously loaded parts. Every part is turned through 180° about the horizontal axis between the linear conveyor and removal point by the turn-over wheel.

If position identification is necessary, repositioning can be either active or passive. With passive repositioning, parts identified as being incorrectly positioned are separated and returned to the parts pile. With active repositioning, the parts are brought into the required position by a turn-over device.

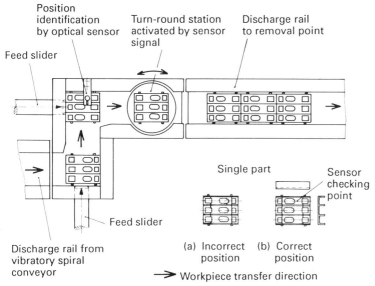

Figure 5.37 Secondary sorting with position identification and active arrangement

Figure 5.37 shows schematically the design of a secondary sorting unit with position identification and repositioning by a turn-round unit. The single part shown is, for example, placed in a vibratory spiral conveyor so that it can be held and transferred with its base on the discharge rail. It cannot, however, be arranged at this point by apertures in the base. The discharge rail was therefore split into two in order to accommodate the secondary sorting with position identification and the turn-around station and the unit installed in the centre. During transfer from the vibratory spiral conveyor, the parts are moved transversely to the position identification station. The actual position is identified by an optical sensor. If the part is correctly positioned, the sensor light beam does not contact the receiver; the signal contacts the receiver, however, if the position is incorrect. This information is stored. The part is transferred from the position identification station to the turn-round station. If the part was correctly positioned, it is transferred on to the discharge rail to the removal point, by-passing the turn-round station. If an incorrect position is identified by the position identification, the turn-round station is activated and the part turned through 180°. The parts are then transferred to the removal point along the second section of the discharge rail.

5.2.2.6 Capacity of vibratory spiral conveyors

The material transfer speed of vibratory spiral conveyors is, to a large degree, dependent upon the helix angle of the helical channel. Since it decreases with increasing

124 Modules for the automation of assembly processes

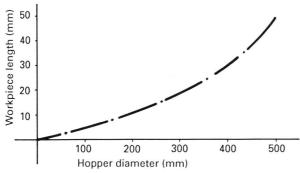

Figure 5.38 Hopper diagram of vibratory spiral conveyors in relation to the workpiece length (empirical values)

helix angle, the bunker diameter should be as large as possible. Empirical values for the transfer speed fall between 0.8 m/min and 2.5 m/min.

The output rate in workpieces per minute is dependent upon the part length and the number of arrangement features. Figure 5.38 shows empirical values for the determination of minimum diameters of the parts containers in relation to the maximum length of the parts to be transferred.

The capacity decreases with an increase in the ordering features and therefore also the arrangement elements and, particularly so, if the arrangement is passive. The output capacity is affected by the weight of parts in the hopper and should be held constant by the fill height by dispensing parts into the vibratory spiral conveyor via a separate auxiliary hopper. The maximum output capacity is achieved if the hopper is not filled above one helix pitch of the channel. Figure 5.39 shows a comparison between a correct and incorrect fill height.

Additional hoppers as shown in Figure 5.40 can be used to increase the hopper capacity at optimum fill height. The additional hopper has its own vibratory drive. The auxiliary hopper drive is switched on by a parts level monitoring device on the main feed vibratory spiral conveyor if the level of parts falls below a set minimum, and is switched off again upon reaching the maximum fill level.

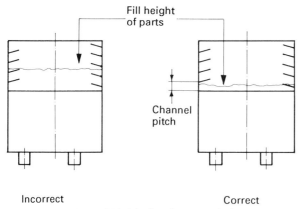

Figure 5.39 Optimum fill height for vibratory spiral conveyors

Feeder units 125

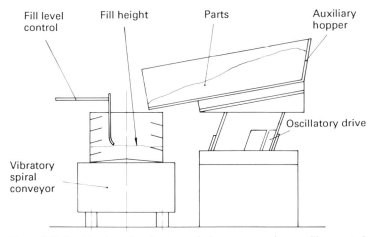

Figure 5.40 Combination of a vibratory spiral conveyor and an auxiliary parts hopper

5.2.3 Electronic position identification of parts

Attempts are in progress to supplement the limited possibilities for part identification by conventional arrangement equipment by electronic systems. Developments have been in progress at research institutes and many companies for many years with the object of identifying workpieces and their positions by television cameras and other pick-ups.

The feed unit manufactured by IBP–Pietzsch is an example of a combined solution involving mechanical and electronic arrangement elements. In this case, the mechanically arranged parts are fed up to a stop and scanned by initiators placed beneath the transfer plane. The scanning produces a signal pattern depending upon the position of the part and which can be compared with stored patterns. The momentary position of the part is given by this comparison. This result can then be used to obtain a position change by turning-over or rotation. The application of this equipment is restricted to flat parts with a few defined positions and pronounced workpiece features [3, 21].

Another method is to identify parts or their position by television cameras. In the meantime, such systems have found an industrial application [19, 21]. In spite of some successes, the immense difficulties associated with this identification system should not be underestimated.

In brief, they can be summarized as follows:

1. *Grey image processing*
   On account of differing illumination, surface finish, etc. and also differing 'grey shades', every workpiece has a mean value between black and white. The processing of grey shades is exceptionally difficult and is currently avoided by suitable illumination techniques or contrast colouring.

2. *Data reduction*
   The large number of items of information in a picture must be reduced to a processable scale. The object must therefore be to identify a workpiece by the minimum number of significant features.

3. *Artificial intelligence*
As previously, only clearly defined 'patterns' can be identified. The result is that only patterns of separated workpieces can be identified by the image processing computer. With partial overlapping which occurs in every 'box', the workpieces (patterns) can no longer be identified at an economic price level.

Progress is expected to be made in this field in future years. However, the application of image processing identification systems currently remains limited on account of both technical and economic reasons.

5.2.4 The feed of interlocking parts

Interlocking parts include wire and punched bent parts with interlocking elements which can result in hooking, locking and workpiece clustering when used as bulk material. Typical examples of interlocking parts are coil springs, locking rings and fixing hooks, etc.

With manual or automatic handling of interlocking parts, as a general rule, separating or else loosening of a part from an entangled pile requires more effort than the assembly of the part. That is, of course, if the parts can be separated at all.

There are virtually no limits for the forming of coil springs which are difficult to feed. Whether or not a particular type of coil spring is suitable for automatic disentanglement and feed can be determined by a simple test. Simply allow a handful of coil springs to fall as a group on to a flat plate from a height of 15 cm to 20 cm. If more than 90% of the helical springs separate, they can be automatically disentangled. If a higher percentage remain interlocked in this test, automatic disentangling and feed is not possible without a design change to the springs. Another possibility for checking the disentangling property is shown in Figure 5.41. The coil springs are held on their ends, and an attempt is then made to interlock the free ends. If the ends cannot be interlocked or if it is only possible with the application of force, the springs are suitable for automatic disentangling and feed.

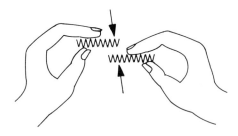

Figure 5.41 Manual checking for the disentanglement property of helical springs

In the design of coil springs, many factors are decisive as to whether or not it is possible to feed them and disentangle them. Firstly, helical springs must be designed so that the gap between the coils is less than the wire diameter.

Interlocking is therefore made impossible. In addition, Figure 5.42 shows some design features for the design of helical springs which make disentanglement and automatic feed possible.

Disentanglement and feed equipment for helical springs as shown in Figure 5.43 are mostly compressed-air impulse-driven. The helical springs are placed in the glass vessel

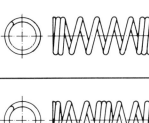

Example 1
Two contacting coils on both spring ends to prevent interlocking of the springs

Example 2
Approximately three contacting coils in the centre of the spring to increase stability

Example 3
End of the wire on the larger outer diameter of the taper spring bent inwards over the centre to prevent interlocking

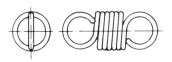

Example 4
Both hooks are tightly closed to prevent interlocking

Example 5
Double taper end springs require contacting coils on the taper section (at least 3 or 4)

Figure 5.42 Guidelines for the design of helical springs suitable for automatic feed (Eversheim; WZL, TH Aachen)

of the spring feed unit. The coil springs are circulated at a set rhythm in the glass vessel by the effect of a cyclic air flow pulse and disentangled by impacting on the vessel walls. At the same time, separated springs are forced by the air flow into the feed hoses for holding and transfer to the point of removal. Any springs jammed in the nozzle are blown back into the glass vessel by the rhythmic change in air flow direction. Items of equipment of this type have several outlets so that several removal points can be fed by one feed unit.

It is advisable to integrate the forming of springs which are not suitable for automatic disentangling and feed into the automated assembly process.

A condition for this application is that the capital investment is justified by the production quantity demand. Quite often, the manufacturing process requires thermal treatment of coiled springs. Figure 5.44 shows the schematic arrangement of one such item of equipment. It consists of a spring-coiling machine and an additional item of equipment for the heat-treatment and separation of the coiled springs. The mode of operation is as follows.

After coiling and separation of the helical springs from the wire, they do not drop into a container as bulk material but are fed to a rotary table with holding apertures positioned downstream from the coiling machine. The rotary table moves synchronously with the spring-coiling machine and transports the springs to the heat-treatment station. At this point they are held by a tong arrangement between two

128 Modules for the automation of assembly processes

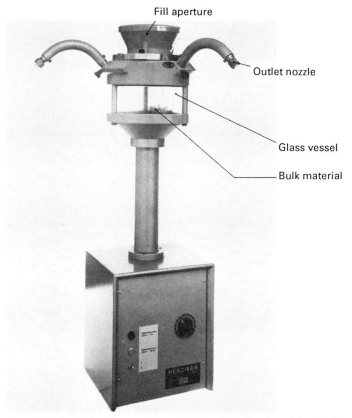

Figure 5.43 A unit for disentangling and feeding helical springs (Menziken)

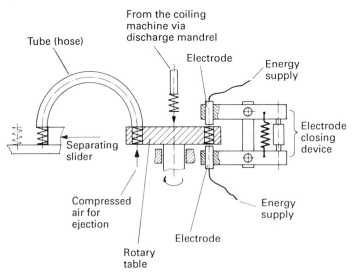

Figure 5.44 Additional equipment for the heat-treatment of helical springs following production in an automatic spring-coiling machine

electrodes and subjected to an induction current impulse. The short-duration treatment and subsequent cooling causes the required grain structure change in the spring material. During the next cycle at a further station the heat-treated coil springs are air-ejected from the rotary table by a compressed-air device into a tube (hose) and transferred to the separating station.

Figure 5.45 shows a further possibility for disentangling these types of easily interlocking parts. This is a so-called vibration-plate, disentangling unit developed by Kettner and Hütter at the Institute for Factory Installations at Hanover Univeristy. It separates the three functions required for disentangling, i.e. storage, disentangling and removal, into a single chain. The parts in a pile are transported intermittently on a conveyor to a vibrator plate. The plate vibrates approximately sinusoidally at 25 Hz and at an amplitude of around 8 mm to 10 mm. The parts are released by the effect of the relative movements of the parts upon impact with the pile, then jump over a barrier by reason of the slightly inclined plate position on to the adjoining transport system. Virtually any large quantities can be stored. A change to different parts is by a change in amplitude. Those items of equipment are produced by Bosch Industrieausrüstrung, Stuttgart.

- Energy feed oscillator plate
- Stationary stored workpieces
- Defined disentangling area

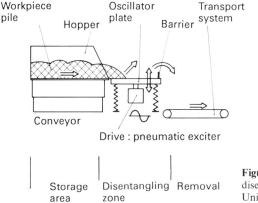

Figure 5.45 Disentangling unit; oscillatory plate disentangler (Kettner, Hutter; IFA, Hanover University)

5.3 Handling equipment

As shown in Figure 5.46, the items of handling equipment used in the field of precision and electrical engineering can be classified by their design, control and type of programming into positioning units and industrial robots.

Application equipment is items of mechanical equipment which perform pre-programmed movements by a fixed program. Only the strokes can be adjusted. These units of equipment are used in dedicated items of equipment for large-scale production.

Programmable industrial robots are used for small production batches or frequent production changes. Industrial robots are universally applicable automatic

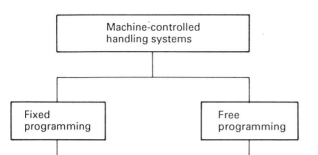

Figure 5.46 Classification of material handling systems for the field of application: precision and electrical engineering

manipulation machines with several uses whose movements in relation to sequence, distances and angles are optionally programmable (i.e. can be changed without mechanical action and, if necessary, controlled by sensors). They can be equipped with grippers, tools or other items of production equipment and can perform handling and/or production functions [18].

5.3.1 Positioning units

Positioning units consist of a drive, control, the kinematic structure and the gripper (see Figure 5.47).

5.3.1.1 Drives

The following types of drive are possible with positioning equipment:

- External drive
- Pneumatic drive
- Electrical drive
- Hydraulic drive.

The hydraulic drive is used for high-load-capacity positioning equipment but is not widely used in the fields of precision and electrical engineering.

Positioning units with an external drive are mechanical, mostly cam-controlled, units. The drive is by a drive shaft which functions synchronously with the working

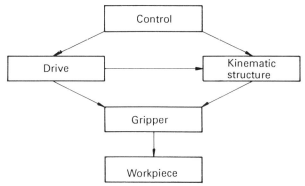

Figure 5.47 Structure of positioning equipment (Warnecke, Schraft, IPA Stuttgart)

cycle of the unit. This type of drive is only possible with rigidly constructed dedicated units. The position of the handling unit to the drive shaft must ensure a positive transmission of movement.

Most of the positioning units used in precision and electrical engineering are equipped with a pneumatic drive. The design of these units is simple and can be segregated into single modules. Translatory and rotational movements can be made simply and quickly via cylinders. Interlocking with simple electrical and electronic controls is quite simple using electropneumatic valves [20]. A disadvantage is that with a pneumatic drive the movements must be performed in a time sequence.

The compressibility of the compressed air, friction ratios and outer edge damping have an effect on the speed, so that the distance–time ratios cannot be exactly predicted, and they are also not reproducible.

The electric drive is increasingly gaining in importance with motor development. It requires virtually no maintenance and can also be easily connected to electrical or electronic controls. A distinction is made between two design types:

- Electric motor drive with a drive shaft fitted with cams for the generation of individual movements.
- Direct drive of the individual axis with variable motors (e.g. stepper motors, disc armature motors, etc.).

With cam drives a drive motor, and with a direct drive a motor for each movement axis, is required. The electric drive, particularly with the application of cams, has the advantage that the sequence of movements can be exactly pre-set. The profile of the cam makes a sinoidal sequence of movements possible and also the synchronization of all operations. The cams can be designed so that overlapping operations can be performed without obstructing any other movements. The total cycle time is then shorter. Figure 5.48 shows a comparison of the movements of a pneumatically and electric motor driven linear positioning unit with cam control by a distance–time diagram [2, 20] and also shows a possibility for the reduction of the overall cycle time by overlapping movements by cams.

5.3.1.2 Kinematics

The kinematic structure of positioning units is on the whole quite simple. Three movements are required in order to be able to move to any particular point. These are:

132   Modules for the automation of assembly processes

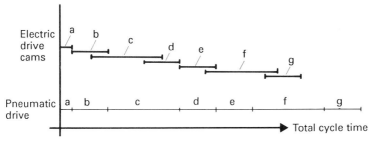

Figure 5.48 Comparison of movements of positioning units with cam and pneumatic drives

- Three translational movements, or
- Two translational movements and one rotational movement, or
- One translational movement and two rotational movements, or
- Three rotational movements.

The translational directions of movement are designated by X, Y and Z. The axis designations are shown by Figure 5.49. The X- and Y-axes are always horizontal. The Z-axis is always vertical and is vertical on the X- and Y-axes. The rotational movements about these translational axes are designated as A, B and C. A is the rotational axis about the X-axis, B the rotation about the Y-axis and C the rotation about the Z-axis [20]. As a general rule, positioning units have two axes, linear functioning units the X- and Z-axes and units with a swing rotation the Y- and Z-axes.

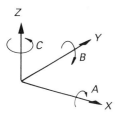

Figure 5.49 Axis designation for handling equipment

Figure 5.50(a) shows the movement sequence of a positioning unit with translational X- and Z-axes, and (b) an application with a rotational axis C and a translational axis Z. If the Z-axis in the positioning unit is supplemented by a rotational movement C, the possibility then exists to effect a position change of the workpiece between the workpiece pick-up and put-down points [2, 20].

5.3.1.3 Control

The symbol for a control is the open sequence of operation over the single transmission element or control chain. The term 'control' is not only generally used for the function of control but also for the overall system in which the control takes place (definition as given by DIN 1926).

The control of a positioning unit includes:

(a) Signal detection: interrogation of the signal carrier.
(b) Signal processing: input and storing of programs; also the reception of input signals and transmission as output signals.

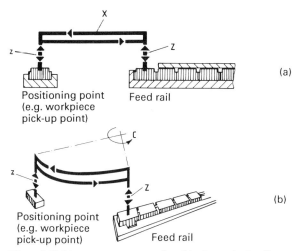

Figure 5.50 Movement sequence of positioning units (a = linear motion, b = swing motion)

(c) Signal conversion: the conversion of an input signal as clearly as possible into a related output signal, if necessary with the application of auxiliary energy.
(d) Energy control: to transfer signals from the information section into a control section and to release movement energy.
(e) Energy conversion: to convert energy into motion.

Figure 5.51 shows schematically an arrangement example of a control in accordance with the explained part functions.

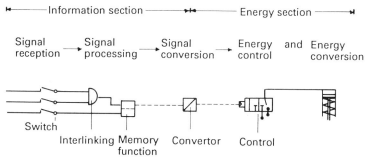

Figure 5.51 Design example of a control for a positioning unit (Warnecke, Schraft [20]; IPA Stuttgart)

5.3.1.4 Grippers

In order to manipulate workpieces with positioning units, the units must be equipped with grippers. The grippers must also form a force- or form-fit connection between the workpiece and positioning unit. A distinction is made between mechanical, vacuum and magnetic grippers.

Mechanical grippers are preferably pneumatically operated. They are in the form of parallel or tong grippers (see Figure 5.52). Parallel grippers are suitable for the external or internal gripping of workpieces. In general tong grippers are only used for external gripping. The type of gripper used is determined by the workpiece.

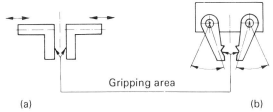

Figure 5.52 Design of a gripper with (a) parallel clamping or (b) tong clamping of the workpieces

With a drive, a distinction is made between a drive for opening and closing the gripper and the application of the holding force. With small workpieces in the fields of precision and electrical engineering, it is advisable to apply the holding force by springs and to operate the gripper by compressed air. If the holding force is operated by springs, the grippers do not lose their parts in the event of a compressed-air failure.

With relatively heavy workpieces, the gripper operation and application of the holding force must be by an external energy supply.

Vacuum grippers have the advantage that they are small and have a low weight. To operate the required vacuum, the installation of a vacuum pump is, however, necessary. The part to be handled must have an adequately large and smooth gripping area for the vacuum.

The use of magnetic grippers requires that the parts to be handled are made from a magnetic material.

5.3.1.5 Design of positioning units

The advantage of translationally constructed and pneumatically operated positioning units is to be seen in their modular design. Figure 5.53 shows a modular design schematically. The individual modules are available in different sizes so that optimum operating conditions are achieved for an application-related assembly. The Z-axis can perform a rotational movement by additional modules, e.g. a gripper rotation unit, so that a gripped part can be moved into another workpiece position by rotation between the gripping and assembly positions. Variable X-axis movements can be achieved by fitting a variable swivel-action stop.

Figure 5.54 shows another positioning unit constructed from modules. Figure 5.55 shows a portal design positioning unit constructed from the same modules. Figure 5.56 shows a positioning unit with a rotational C-axis and a translational Z-axis. The rotational range of the C-axis is set by stop cams (a) and, depending upon the Z-axis movement, can be between 5° and 225°. The Z-axis movement is variable between 0 mm and 20 mm.

A vibration-free sequence of movement is required for cycle times shorter than 2 s and also for a constant repeat accuracy. Cam-controlled positioning units are particularly suitable for this purpose. Figure 5.57 shows one such unit. The drive is by a gear reduction motor. Two cams designed for the application are fitted on one face of the gear drive on a stub shaft. The sequence of movement is determined by gearbox drive speed and the cams. The X-axis movement is controlled by one cam and the Z-axis movement by the other. The cams can be Archimedean spirals, sinusoidal or higher sinusoidals. Jerk-free starting and stopping and exact intermediate stops are possible by this mechanism.

Handling equipment 135

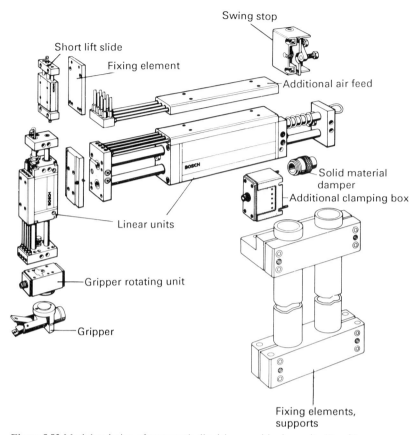

Figure 5.53 Modular design of pneumatically driven positioning units (Bosch)

The control drum is fitted on the opposite side of the gearbox on the stub shaft (see Figure 5.58). It is used for the control of subfunctions, e.g. operation of the gripper. The time sequence of the subfunctions is controlled by metal foil strips bonded to the control drum and the signal transmitter. The workpiece can also be rotated during movement along the X-axis by fitting a rotary unit to the gripper slide bar (see Figure 5.59).

Figure 5.60 shows schematically three different path examples for movements between the start position O at rest, the gripping position A and various placing positions B.

5.3.2 Industrial robots

On account of the specific requirement made on such items of equipment, namely that they must be universally applicable and programmable, the design of industrial robots is considerably more complex than that of positioning units. In order to fulfil its function, an industrial robot comprises several part systems. Figure 5.61 shows these part systems and the correlation of their functions. They are discussed in further detail below.

136 Modules for the automation of assembly processes

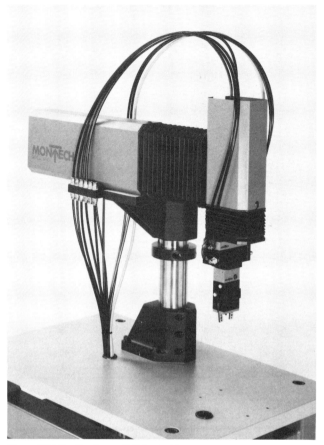

Figure 5.54 Positioning unit (Montech)

5.3.2.1 Kinematics, arm and gripper

As already mentioned with the positioning units, in three-dimensional space, the position of a rigid body is defined by its positional vector and its orientation. If a body is freely movable, it has six degrees of freedom and can be brought into another arbitrary position by three rotations and three translations.

With industrial robots, a distinction is therefore made between two basic movements:

(a) Translational movements
(b) Rotational movements.

The most well-known industrial robots comprise a combination of three movement axes (the degrees of freedom) for positioning and up to three movement axes (manual axes) for the orientation of the workpieces. Figure 5.62 shows a selection of the most frequently occurring combinations of axes of movement [18, 19].

The stiffness and accuracy of an industrial robot is affected by the design and position of the principal axes. The stiffness of the overall system decreases with increasing numbers of joints and guides and also with increasing lengths of the

Handling equipment 137

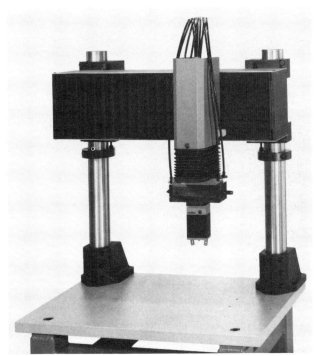

Figure 5.55 Portal-type positioning unit (Montech)

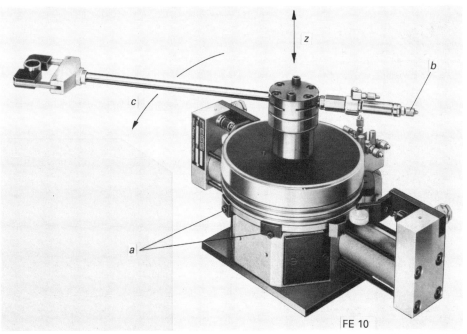

Figure 5.56 Fels-type, pneumatically operated swing positioning unit (a = stop cams for limitation of the swing movement, b = air connection for operation of the gripper)

138 Modules for the automation of assembly processes

Figure 5.57 Cam-controlled positioning unit (Siemens)

Figure 5.58 Controller drum and electronic signal generator of a positioning unit (Siemens)

Handling equipment   139

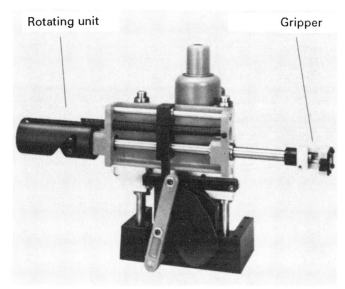

Figure 5.59 Cam-driven positioning unit with a rotating unit for rotation of the workpiece between gripping and positioning (Siemens)

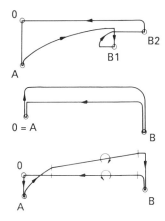

Figure 5.60 Path examples for a cam-driven positioning unit (Siemens) (O = start position at rest, A = gripping position, B = placing position)

individual arms. The positional accuracy of robots with principally translational axes is higher than for those with mainly rotational axes.

The correct selection of gripper is of decisive importance with the application of industrial robots for assembly operations. The completion of several assembly functions is seldom possible with a single gripper, so that the fitting of multiple grippers or a gripper change system must be possible.

5.3.2.2 Control

An industrial robot control must store, control and monitor the program sequence. Figure 5.63 shows the arrangement of an industrial robot control. Logic connections with production and peripheral arrangement equipment must be made.

140 Modules for the automation of assembly processes

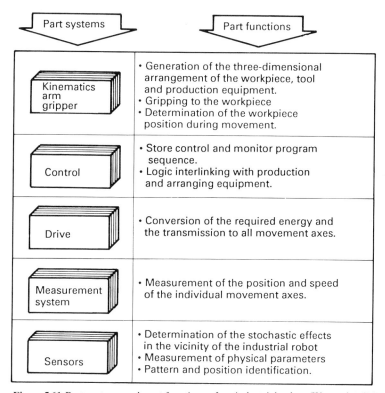

Figure 5.61 Part systems and part functions of an industrial robot (Warnecke, Schraft; IPA Stuttgart)

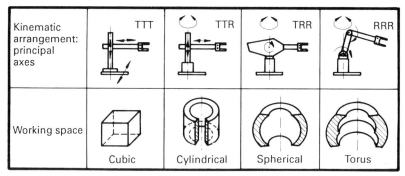

Figure 5.62 Combination examples of movement axes with industrial robots (T = translational axis, R = rotational axis) (Schraft; IPA Stuttgart)

Basically, a distinction is made between two control types:

- PTP (point-to-point) control
- CP (continuous-path) control.

With PTP control, during movement there is no functional interrelationship between the individual axes. The gripper movement path is therefore not accurately defined. Apart from the start and end points, the gripper path is also dependent upon the programmed speed and loading. Intermediate points can be programmed if certain obstructions have to be avoided along the path between the start and end points. So as

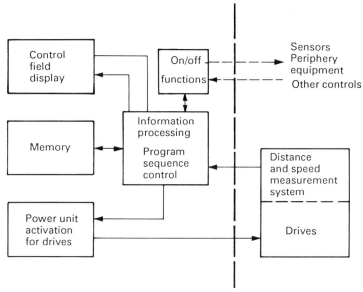

Figure 5.63 Arrangement of an industrial robot control (Warnecke, Schraft; IPA Stuttgart)

not to interrupt movements at these points, PTP controls should be structured so that these points are identified as auxiliary points in the program. Multi-point-controls (MP) are a further development of PTP controls. They have a large memory capacity and permit close approximation following of a path with closely spaced points.

CP controls permit the following of a defined path in space and whereby a functional interrelationship exists between the axes. Intermediate positions can be calculated by various interpolations.

The programming, i.e. compilation and input of the programs, is principally undertaken on site by approaching positions and their memory recording or by manually guiding the robot arm with simultaneous memory recording of the movement coordinates. This type of programming is known as 'teach-in'. 'Off-line programming' is quite rare because it requires the use of higher-level programming languages [18, 19].

5.3.2.3 Drive

The drives of controllable axes in industrial robots must have an adequately high dynamic and torsional stiffness. The part functions of a robot drive are shown in Figure 5.64. The drive types principally differ in accordance with the type of drive energy.

Three systems are available:

- Pneumatic systems
- Hydraulic systems
- Electrical systems.

Hydraulic systems are only used for manipulating heavy weights, and accordingly they are not discussed in this book, which is for applications in precision and electrical engineering.

142 Modules for the automation of assembly processes

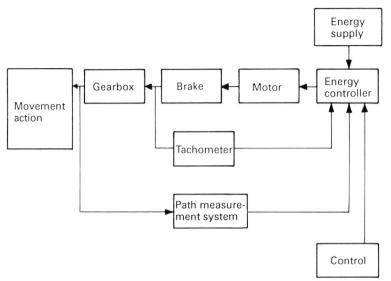

Figure 5.64 Principal arrangement of a robot drive (Wiendahl; IFA, Hanover University)

Pneumatic systems are characterised by a simple design. The energy in the form of compressed air is converted directly into movement of the axes via pneumatic cylinders. Controlled speeds are problematical on account of the compressibility of the air and friction condition in the cylinders. Pneumatic drives are suitable for point-to-point control, i.e. from stop to stop. The programmability is, however, accordingly limited.

Direct current and stepping motors are principally used for electrical drive systems. DC motors are characterized by short reaction times. This means rapid starting and braking. Their speed is infinitely variable. The requirement for small sizes, low weight, wide control range with regard to speed and torque and also short reaction times is largely fulfilled by disc armature motors. DC motors can also be used at low speeds and at high torques by interconnecting precision gears (gears with high efficiency and low backlash gearboxes) [21, 22].

5.3.2.4 Measurement system
Position and speed measurement systems are not a part of the control but are indispensable for the function of control because they determine the input parameters. The systematics of distance measurement systems is shown in Figure 5.65. A distinction is made between analogue and digital distance measurement systems. Analogue systems mainly use potentiometers. With an appropriate circuit arrangement on their output, they generate a direct current which is proportional to the distance to be measured. This is a low-cost, but also low-accuracy, solution. Digital systems measure absolutely or incrementally. The distance to be measured is split into distance elements whose length is determined by the resolution of the measurement system. A high positional accuracy of the unit is achieved with a high resolution capability.

Figure 5.66 shows the design arrangement of an angle coder. It can have one or several coding discs which are connected to the rotational axis. Coding discs are

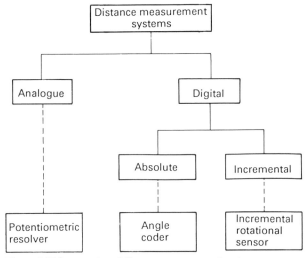

Figure 5.65 Systematics of distance measurement systems

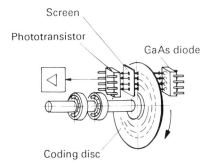

Figure 5.66 Design arrangement of an angle coder (Warnecke, Schraft; IPA Stuttgart)

generally scanned photoelectrically. Consequently, a clear relationship is given at any time between the angle of rotation of the coding disc and the distance to be measured as generated by a spindle-nut system of known pitch or a lever arm of known length.

With a high resolving power, these measurement systems are large and therefore expensive. Incremental measurement systems are quite often used in place of angle coders. For example, by photoelectric scanning of a rotating slotted plate, incremental measurement systems only count the distance increments covered during a movement. The position of an axis is given by the counter reading. A disadvantage is that their value is lost by switching off the unit or with a current failure. The unit must then be retraversed to its reference point. The movement speeds are generally measured by special tachometer generators [18].

5.3.2.5 Sensors

The field of application of sensors on industrial robots is divided into two groups:

- Sensors as part systems of internal distance measurement systems
- Recording of environmental conditions for the compensation of errors or faults and the identification of workpiece positions.

144 Modules for the automation of assembly processes

Sensors are basically part of measuring equipment for the recording of values. They can be divided into three groups in terms of their complexity:

- Simple switching binary sensors in the form of light barriers or inductive limit switches
- Sensors for single or multidimensional analogue parameters, e.g. for inductive, capacitive or optical procedures for distance measurement, availability checking and slip recognition, etc.
- Image processing sensors. These sensors calculate the position of workpieces two-dimensionally by contact or optically, with and without grey-value grading [22].

5.3.2.6 Types of industrial robots
The design form of widely varying industrial robots, dependent on the task to be performed, is made possible by the arrangement and combination of translational and rotational axes. Figure 5.67 shows the most frequently occurring types. The achievable repeat accuracy of these items of equipment is highly important for assembly operations. Horizontal articulated-arm units and units with translational X-, Y- and Z-axes, some in the portal design, are predominantly used in the fields of precision and electrical engineering.

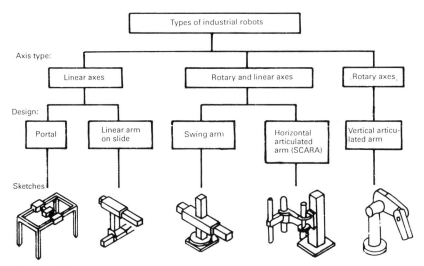

Figure 5.67 Design types of industrial robots (Spingler, Bässler)

5.3.2.6.1 SCARA horizontal articulated-arm robots
The SCARA horizontal articulated-arm robots were first introduced in 1980 in Japan and are based on a fundamental development by Professor Makino of Yamanashi University, Japan. The term SCARA is the abbreviation for Selective, Compliance Assembly Robot Arm.

The object of this development was an application for assembly. One of the fundamental problems of automated assembly is achieving the required degree of alignment accuracy of the parts to be assembled. The principle of development of horizontal articulated arm units was therefore to arrange the flexibility of the unit, which is required to compensate positional inaccuracy and to prevent tipping over during assembly, transversely to the direction of assembly. On the other hand, the unit must have an adequate level of stiffness in order to support the assembly forces in its joints. The horizontal arrangement of the articulated arm is suitable for achieving both these requirements.

Figure 5.68 shows schematically the design arrangement of a SCARA horizontal articulated-arm robot and Figure 5.69 one such unit in service.

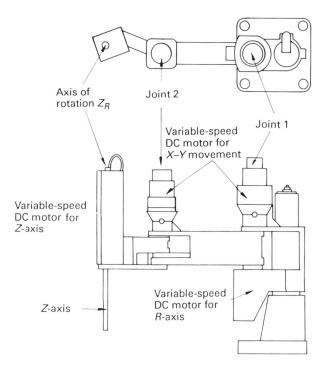

Figure 5.68 Design arrangement of a SCARA horizontal articulated-arm robot (IBM)

SCARA units are also referred to as swivel robots on account of the simulation of the human arm. Two joints, which are also referred to as shoulder and elbow joints, are arranged horizontally. The X-Y movements are generated by these two joints. The drive is generally by variable-speed DC motors and reduction gearing direct into the arms; distance and speed measuring systems are both integrated into the drive. The Z-axis is operated both pneumatically and by variable direct current motors. With a pneumatic drive, the Z-axis travel is limited by stops and has no intermediate control. The Z-axis can also perform a rotary movement about the rotational axis R which with most units is limited to $\pm 180°$. The drive for the rotary movement is in line with the drive of the first axis and is quite often transmitted to a second axis and then from this point to the Z-axis by toothed belts. The mass of the Z-axis rotary movement drive is transferred to the stationary part of the robot by this design and consequently reduces the moment of inertia.

The working range of a SCARA robot type is shown in Figure 5.70. The double-sided support bearing for the articulated arms of the SCARA robot systems, IBM-type, restricts the movement of the arms on one side in the X-Y direction, which results in an

146 Modules for the automation of assembly processes

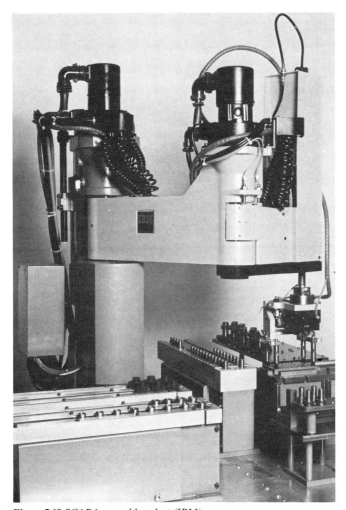

Figure 5.69 SCARA assembly robot (IBM)

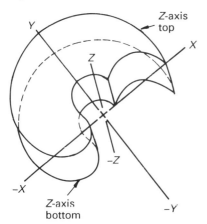

Figure 5.70 Working area of a SCARA robot as shown in Figures 5.68 and 5.69 (IBM)

Handling equipment 147

Figure 5.71 SCARA assembly robot (Bosch)

asymmetrical working area. Figure 5.71 shows a SCARA robot Bosch system. In this case the one-sided support bearing for the articulared arms gives a symmetrical working area. The two different articulated-arm mounting arrangements are shown in Figure 5.72.

5.3.2.6.2 Robots with translational X-, Y- and Z-axes
Portal robots have translational X-, Y- and Z-axes. Depending upon the movement ranges of the individual axes, the resultant working area is more or less cubic. The Z-axis is generally equipped with an additional rotary axis A for the gripper. Figure 5.73 shows an assembly portal robot Bosch system. With an 800 mm X-axis travel and a 1200 mm Y-axis travel, a fully effective right-angular working area is available for the provision of parts, the arrangement of workpiece carrier-transport systems and assembly fixtures. The individual axes are generally driven by DC motors with gear reduction drives which act on toothed-gear racks via pinions.

Figure 5.74 shows a different design of assembly robot (by PIA), equipped with translational axes. The movement elements are supported by a self-supporting light alloy structure. The X- and Y-axis drive is by stepping or DC motors which are stationary. The drive torque is transmitted to the movement axes by toothed belts.

To avoid drift from the starting positions, after every one hundred cycles the axes traverse automatically to a reference point of the path-measurement control for recalibration. In spite of the use of toothed V-belts, a high reproducible positional accuracy is achieved by this procedure.

Figure 5.75 shows a DEA assembly robot type with translational X-Y-Z-axes. The X movement is achieved by moving the robot along the machine column table.

The Cardan principle is a new type developed by ASEA. Figure 5.76 shows the design of the working arm of this assembly robot. It has six programmable axes. The

148 Modules for the automation of assembly processes

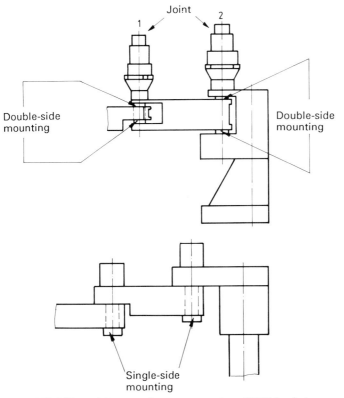

Figure 5.72 Different joint mounting arrangements on SCARA robots

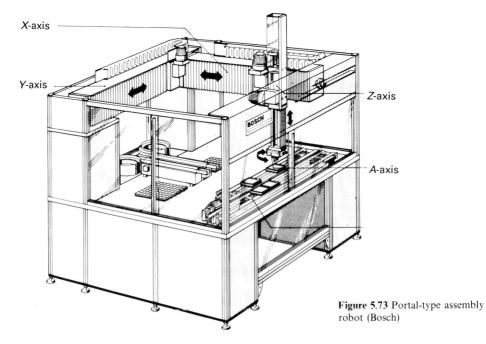

Figure 5.73 Portal-type assembly robot (Bosch)

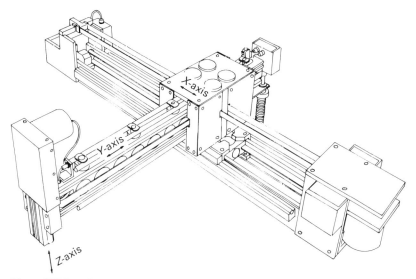

Figure 5.74 Portal-type robot system (Preh Industrieausrüstung (PIA))

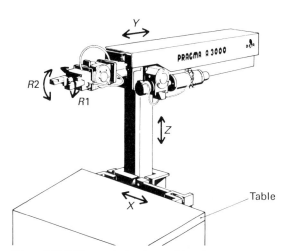

Figure 5.75 DEA translational assembly robot

robot arm has translational movement in the Z direction. The X-Y movements are performed as pendulum movements by reason of the Cardan mounting of the robot arm. High accelerations are made possible by the positioning in the centre of gravity of mass. The robot arm can make a rotary movement. The wrist joint can be inclined; the hand flange also has a rotary movement. If the working range of the robot is to be extended, it is fitted with a traversing unit so that the number of axes can be increased to seven. The kinetmatic design permits traversing in cartesian, cylindrical and wrist-joint oriented cartesian coordinates.

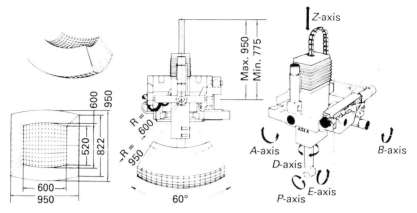

Figure 5.76 ASEA assembly robot system

5.4 Transfer equipment

Transfer units are mechanical units by which the assembly part is transferred from work point to work point in the sequence of the assembly operations. A distinction is made between cycled and uncycled systems. In cycled systems the workpiece carrier or assembly fixture is rigidly connected to the transfer unit; there is no rigid connection with uncycled systems.

5.4.1 Cycled transfer equipment

Cycled transfer units are classified by their design, e.g. circular and longitudinal transfer arrangements. A decision must be taken between the two types depending upon the application.

The principal criteria in this respect are:

- The number of assembly or machining directions required
- The number of stations required
- The possibility for interlinking several machines.

The design of circular cyclic transfer units principally limits the assembly or machining directions to two, namely vertical from above downwards and horizontal from outside in the direction of the centre. If three assembly or machining directions are required, longitudinal transfer units are generally more suitable.

The number of stations possible with circular transfer equipment is limited to 16 to 24 on account of the masses to be accelerated and for reasons of availability; with longitudinal transfer systems the number is practically unlimited. The linking of several machines is dependent on their arrangement. If they are arranged in a straight line, then longitudinal transfer equipment is more suitable. The interlinking of systems in a circular and longitudinal transfer arrangement is also possible.

With cycled transfer units, the cycle time comprises the shift time t_s to the further transport and the holding or stationary time t_h for processing, e.g. for an assembly operation.

5.4.1.1 Circular cyclic transfer equipment

The most frequently used cycled items of transfer equipment used for the automation of assembly processes are circular cyclic transfer units, known in short as circular cyclic units or circular indexing tables. In this case, a distinction is made in terms of their drive energy and kinematic design. Pneumatic or electric motor drives are used in the fields of precision and electrical engineering.

With kinematic design, a distinction is made between three basic types (Figure 5.77):

- Ratchet drive
- Maltese-cross drive
- Cam drive.

Circular cyclic units with a ratchet drive are preferably pneumatic, Maltese-cross and cam-drive electric motor drive.

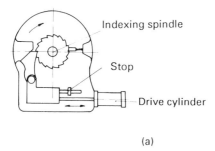

(a)

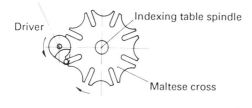

(b)

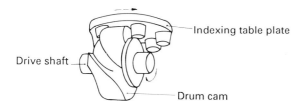

(c)

Figure 5.77 Possible drives for circular cyclic units: (a) ratchet drive; (b) Maltese-cross drive and (c) cam drive

152 Modules for the automation of assembly processes

The most common number of indexing points (number of stations) are 6, 8, 12, 16 and 24. At a uniform cycle time and with an increasing number of indexing points, the indexing angle and therefore the angular acceleration are reduced and the sequence of movements more favourable. Higher indexing accuracies can also be achieved with smaller indexing angles.

5.4.1.1.1 Pneumatically driven circular cyclic units
There is a wide range of pneumatically driven circular cyclic units with differing design features on the market. On some of these units, the drive cylinder is positioned on the outside of the unit housing so that it protrudes into the effective working space around the indexing table. With other designs, the drive elements are integrated in the housing so that the effective working space is fully utilized. Another design feature of other types is that, before indexing from one station to the next, the indexing table is raised by a few millimetres and is lowered upon completion of indexing. In this case, spur gearing is used for precise indexing.

This system has a high indexing accuracy but in many cases is not suitable for assembly purposes because of lifting of the table.

Figure 5.78 shows the design of a Bosch pneumatic circular cyclic unit. The mode of operation is as follows. A divider plate mounted on the axis of rotation of a circular table is indexed by the engagement and disengagement of a pawl by the action of a

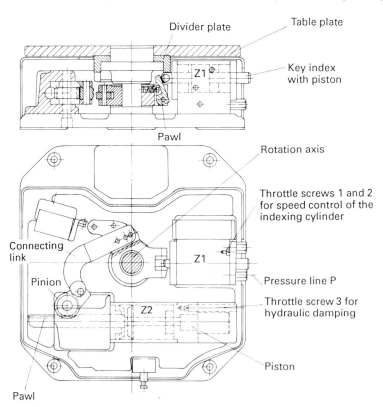

Figure 5.78 Design of a pneumatic circular cyclic unit ($Z1$ = piston for key index, $Z2$ = drive cylinder) (Bosch)

double-acting cylinder (Z2) via a toothed rack, pinion and connection lever. The divider plate and the table plate connected to it are fixed at the stopping position by a precisely located key index with a piston (Z1). The movement of the double-action piston (Z2) is damped by hydraulic end position damping. The damping can be adapted for the respective loading and indexing speed by throttle screw 3. The indexing speed can be altered and controlled via a throttle return flow valve by throttle screws 1 and 2.

5.4.1.1.2 Circular cyclic units with a Maltese-cross drive
With Maltese-cross drives, a distinction is made between internal and external Maltese crosses. This drive can be used for the direct or indirect indexing of a revolving unit. Figure 5.79 shows the design principle of a control mechanism with an external Maltese-cross arrangement.

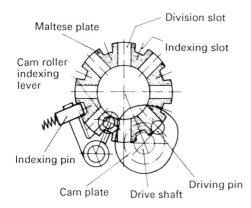

Figure 5.79 External Maltese-cross drive with an indexing attachment as a drive for circular cyclic units (Sortimat)

The mode of operation is as follows. The time for one rotation of the electric motor driven drive shaft is the cycle time. The plate attached to the drive shaft for holding the driving pin is in the form of a cam plate. The driving pin is moved in a circular path by the rotation of the drive shaft. During rotation it engages in the division slot of the Maltese plate and carries the plate until the driving pin disengages from the divider plate slot by the circular motion. At the same time the cam plate drives the indexing attachment. If the driving pin is disengaged from the Maltese plate slot, the taper indexing pin engages in the indexing slot Maltese plate in order to lock and therefore exactly position it in this position.

Circular indexing units for assembly purposes should have a short indexing and a long stationary time. The ratio of the indexing to stationary time varies with the number of indexing points on the Maltese plate. Figure 5.80 shows the ratios between the indexing and stationary times and for which with an increasing number of indexing points the ratio approximates to 1:1.

Figure 5.81 shows different design arrangements of circular cyclic units with a Maltese-cross drive. Figure (a) shows a type with an indirect drive. The drive shaft is driven by a gear motor via a suitable chain or tooth-belt drive. The time for one rotation of the drive shaft is equivalent to one cycle time of the circular cyclic unit. The plate for holding the drive pin is attached to the drive shaft. In order to control indexing during the stationary time, the outer contour of the plate is in the form of a cam (see Figure 5.79). The Maltese plate is rigidly connected to the flange for the support of the circular indexing table. A shaft located at the centre of the circular indexing table is driven by the drive shaft via toothed gearing at a ratio of 1:1. This

154 Modules for the automation of assembly processes

Divisions	Cycle time consisting of the indexing time (t_i) and stationary time (t_s)					t_i	t_s	Ratio t_i/t_s
	0°	90°	180°	270°	360°			
6						120°	240°	1:2.00
8						135°	225°	1:1.66
12						150°	210°	1:1.40
16						157.5	202.5	1:1.28
20						162°	198°	1:1.22
24						165°	195°	1:1.18

Figure 5.80 Variation of the ratio between the indexing and stationary times with an increasing number of indexing points

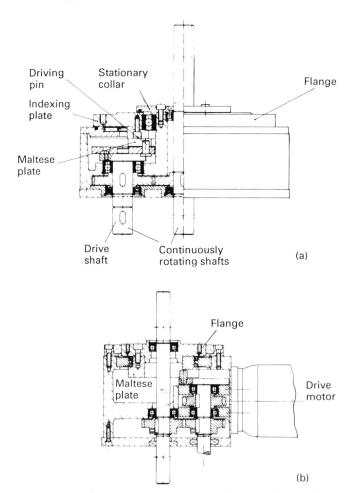

Figure 5.81 Design arrangement of circular cyclic units with a Maltese-cross drive: (a) with an indirect drive; (b) with a side drive (Sortimat)

continuously rotating shaft can be positioned on both sides of the circular cyclic unit. It runs synchronously with the cycle time by the transmission ratio of 1:1 and is used as a control shaft for movement sequences. Figure (b) shows a similar design, however, with a drive motor directly flange-mounted on the side which drives the shaft with the plates for the driver pin via a worm gear.

Units of this type can be used horizontally or vertically. With a direct Maltese-cross drive and as shown in Figure 5.81, the Maltese plate is mounted centric to the circular cyclic unit. With an indirect Maltese drive, a toothed wheel gear is connected between the Maltese plate and circular cyclic unit. These types are necessary if a high number of indexing points are required with short indexing and long stationary times. A schematic design arrangement of an indirect Maltese-cross drive is shown in Figure 5.82. The method of operation is as follows. The drive shaft a with plate b and driving pin f is electric motor driven and drives the four-piece Maltese plate c. The pinion d is rigidly connected to the Maltese plate. The rotary motion of the Maltese plate is transmitted via the pinion to gearwheel e which is rigidly connected to the circular cyclic unit table. The circumference of plate b which carries the driving pin f is in the form of a cam. The taper-form indexing pin e is operated via this cam by a connection linkage g. The pin is disengaged during the indexing period; during the stationary period it engates in the driven circular cyclic unit gearwheel for indexing. The number of stations of the circular cyclic unit is determined by the transmission ratio between the pinion c (pitch circle diameter $D1$) and the driven gearwheel d (pitch circle diameter $D2$).

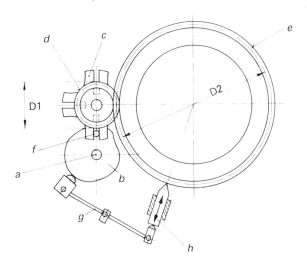

Figure 5.82 Indirect Maltese-cross drive of a circular cyclic unit (Menziken) (a=drive shaft, b=cam plate, c=Maltese plate, d=pinion, e=gearwheel, f=driving pin, g=lever linkage, h=indexing pin)

5.4.1.1.3 Cam drives for circular cyclic units
Cam drives have the advantage that the ratio of the indexing to the stationary time can be determined by the design of the diameter of the cam drum and cam form independent of the number of indexing points of the circular cyclic unit.

Figure 5.83 shows schematically a solution for a cam drive. The method of drive is as follows. The cam drum b with the cam part c is mounted on the drive shaft a. The

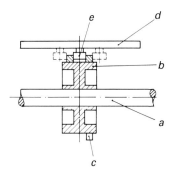
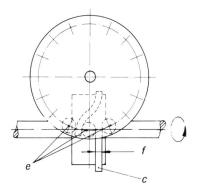

Figure 5.83 Cam drive for circular cyclic units (type Fischer, Brückner, Swanson and others) (a=drive shaft, b=cam drum, c=cam part, d=divider wheel, e=indexing roller, f=cam width)

divider wheel d is rigidly connected to the circular indexing table via a shaft. It has rollers e on its circumference corresponding to the number of indexing points. The spacing of the indexing rollers on the pitch circle diameter depends upon the width f of the cam part c and the minimum diameter of the rollers e. The cycle time of the circular cyclic unit is equivalent to the time for one rotation of the drive shaft with the indexing drum. For the most part, the cam part on the indexing drum runs, without a gradient in contact with the circumference of the indexing drum and, to a lesser extent, at a gradient which corresponds to the diameter of a divider wheel indexing roller. The divided wheel remains locked as long as the straight cam engages between two rollers, during rotation of the indexing drum. This period is the stationary time. The locking is released when the inclined cam part enters the next space between the rollers, and the cam then advances the indexing wheel forwards by one division. The ratio between indexing and stationary time is determined by the length ratio between the straight and the inclined cam bar.

If the inclined cam covers 120° of the indexing drum, the ratio between the indexing and stationary time is 1:2, and 1:3 with 90°. If the ratio between the indexing and stationary times is to be as large as possible, this is achieved by the largest possible indexing drum diameter. Figure 5.84 shows a cam-operated circular cyclic unit built in accordance with the diagram as shown in Figure 5.83. With this design, the cam drum drive shaft protrudes from the gearbox housing. On account of the synchronous running of the drive shaft and indexing cycle, the drive shaft can be used for controlling the movements of the other modules.

Figure 5.84 A cam-driven circular cyclic unit (type Swanson, USA)

A further solution possibility employing cam drives is shown in Figure 5.85. In this case, the indexing rollers are end-mounted on the divider wheel circumference. With this design, the cam drum indexing segment must have a concave enveloping contour in order to bring the rollers into contact with the cams across the whole surface line.

Figure 5.86 shows an actual sectionalized view of a cam drive. These drives are also suitable for the drive of longitudinal transfer equipment.

5.4.1.2 Cyclic longitudinal transfer equipment

The application of circular cyclic transfer equipment is limited in terms of assembly and machining directions and also with regard to the number of stations. On the other hand, depending upon their arrangement, longitudinal transfer equipment gives up to three assembly or machining directions and is hardly restricted with regard to the number of stations. Furthermore, they are also variable in terms of their length. A disadvantage is, however, that, by reason of their design, longitudinal transfer systems are, as a general rule, more expensive than circular cyclic units and that not all systems are suitable for precise positioning. Additional indexing must therefore be provided for the latter.

Depending upon their arrangement, longitudinal transfer systems can be classified into three groups:

- Overhead and underfloor longitudinal transfer systems
- Carousel longitudinal transfer systems
- Plate longitudinal transfer systems.

The drive of overhead, underfloor and also rotary systems is by electric motors and principally via the same cam drives as described for the drive of circular cyclic units. On the other hand, plate longitudinal systems can be pneumatically, hydraulically or electric-motor driven.

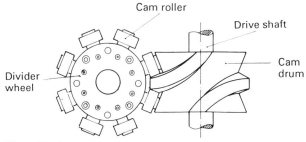

Figure 5.85 A cam drive for circular cyclic units (Ferguson)

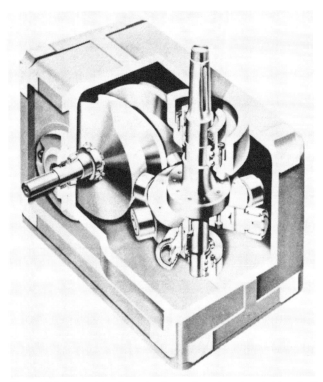

Figure 5.86 A cam drive for a circular cyclic unit and cycled longitudinal transfer units (Ferguson)

With overhead, underfloor or revolving systems, the workpiece carriers or assembly fixtures are either mounted on steel belts or chains or else chain links are used which are designed so that a workpiece carrier is supported by one chain link.

Continuous high-quality corrosion-resistant steel strips which have a high pitch accuracy, either with internally fitted carrier strips or carrier holes, are used in the steel strip design. The drive drums and tensioning drums have equally distributed, inserted, hardened and ground carrier slots on their circumferences; inserted carrier pins are used on the perforated strip types. The production accuracy of the steel strip determines the positioning accuracy of the equipment. Steel strip types are suitable for low workpiece carrier weights and a relatively small number of workpiece carriers. The positional accuracy decreases with increasing length and number of workpiece carriers. With the use of standard chains as a transport system, they are equipped with strips for holding the workpiece carriers. The advantage of this solution lies in the low cost of the elements by the application of standard chains and sprocket wheels as drive elements and tensioning drums. The achievable accuracy is, however, not generally adequate; the workpiece carriers must therefore be additionally positioned in the individual working stations.

A very high positioning accuracy is achieved with the chain-link system. A single chain link is also the workpiece carrier base plate. The positioning accuracy as required for assembly is achieved by precision production of the chain links. This accuracy is maintained even with a large number of chain links and relatively heavy workpiece carriers (1 kg to 2 kg).

5.4.1.2.1 Overhead and underfloor longitudinal transfer systems

Figure 5.87 shows schematically an overhead and underfloor longitudinal transfer system. The advantage of this arrangement lies in the fact that three assembly or machining directions a are available. A disadvantage with the overhead or underfloor system is that only the top side is usable. The return is below the working level during which the workpiece carriers return suspended. Consequently, with this arrangement, only about 40% of the workpiece carriers in circulation can be used. If the cam drum drive shaft which runs synchronously with the drive is designed to protrude out of the gearbox, it can be arranged along the transfer unit to control movements synchronously by a series of cams.

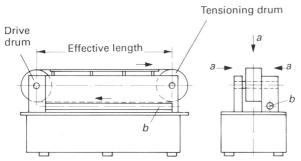

Figure 5.87 An overhead and underfloor longitudinal transfer system (a = assembly or machining direction, b = control shaft running synchronously with the drive)

5.4.1.2.2 Rotary longitudinal transfer systems

In contrast to an overhead and underfloor arrangement, with a rotary type the drive and tensioning drums are not arranged horizontally but vertically. All workpiece carriers on the forward or return runs can therefore be used for assembly and machining operations.

Figure 5.88 shows one such arrangement schematically. Depending upon the basic design, three assembly or machining directions (a and b) are available on both sides of the system. The arrangement of a control shaft (c) running parallel to the transfer unit is also possible with a rotary design using a cam drive.

5.4.1.2.3 Plate longitudinal transfer systems

The operation of a plate transfer system is shown schematically in Figure 5.89. In this arrangement, the plates are moved, in line, by one plate length, on guide rails. The workpiece carriers are mounted on the plates and, upon reaching the end of the installation, are moved transversely in the same manner on to a return track.

The plates are moved in a fixed cycle either by a cam drive via an oscillating lever or a pneumatic cylinder. The longitudinal dimension of the plates, x, is produced with high accuracy. A high positional accuracy is therefore achieved. During repairs, it is advantageous that individual plates can be removed and immediately placed since the elements are not rigidly connected to the transfer system.

5.4.2 Non-cycled transfer equipment

On non-cycled transfer equipment, the workpiece carriers or assembly fixtures have no rigid connection with the transfer equipment, but are moved by friction against

160   Modules for the automation of assembly processes

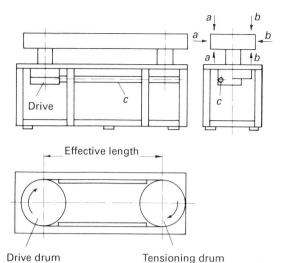

Figure 5.88 Rotary longitudinal transfer system (a, b = assembly and machining directions, c = cam shaft running synchronously with the drive shaft)

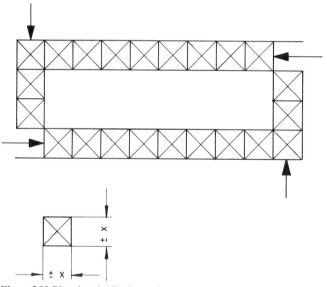

Figure 5.89 Plate longitudinal transfer system

variable stops. The workpiece carriers are preferably carried by a double-belt system in a modular longitudinal transfer arrangement. Assembly and movement directions are therefore possible from above, sideways and below. The latter is, for example, necessary if assembly forces have to be supported (see also Figure 3.14 and Sections 3.3.2.3 and 3.3.2.3.1). Centring units must be provided in order to increase the positioning accuracy of the workpiece carriers in contact with the stop. The cycle time of non-cycled transfer units in a double-belt arrangement should not be less than 3 s– 5 s.

Uncycled transfer units have the advantage that the number of workpiece carriers required is not determined by the spacing between individual stations. They can also transfer individual workpiece carriers over large distances. The possibility of accumulating several workpiece carriers between the individual stations presents the possibility of buffer formation irrespective of the cycle time and also the decoupling of the stations from a compulsory cycle. A further advantage is in the easy incorporation of manual work points with combined manual-automatic assembly lines.

The most commonly used non-cycled transfer systems is the FMS system by Bosch. Figure 5.90 shows the module of this system with which it is possible to construct various systems matched to a particular assembly operation. The workpiece carrier sizes are standardized in three sizes: 160×160, 240×240 and 320×400 mm.

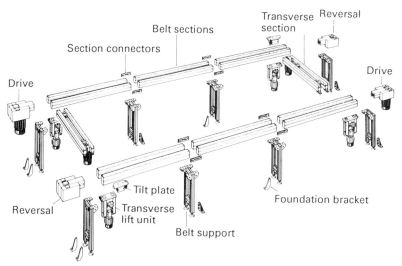

Figure 5.90 Modules for the construction of a non-cycled frame design transfer system (Bosch)

The transfer of the workpiece carriers from the longitudinal to transverse belt sections is shown in Figure 5.91. Approaching workpiece carriers on a belt section are stopped by a restrictor and fed individually to the transverse lift unit. The workpiece carrier moves on the transverse lift unit up to a stop. A lift motion is induced by the contact action and the workpiece carrier is raised from the double-belt system. Upon reaching the upper position, the belt drive engages the transverse lift unit in order to allow the workpiece carrier to move from the transverse lift unit on to the transverse section. The same procedure is repeated at the end of the transverse section with a second transverse lift unit for the transfer of the workpiece carrier on to the other belt section.

5.5 Screw-inserting units

In assembly, screw-inserting is the most commonly occurring assembly operation, so that great importance is attached to its automation. In the design of a product, the screw diameter and therefore also the tightening torque and the screw type are specified

162 Modules for the automation of assembly processes

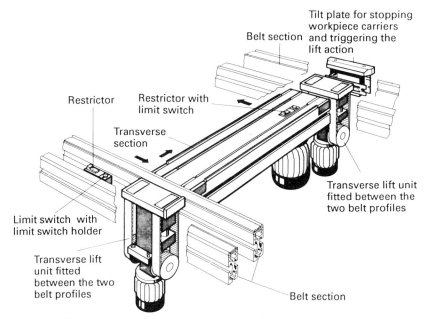

Figure 5.91 Transverse movement of workpiece carriers (Bosch)

by Product Development. The required tightening torque can be checked by specifying a further turning moment or a loosening torque. Figure 5.92 shows the curve of the screw-turning torque during the course of a screw-in operation. The required tightening torque is preceded by the contact torque. This occurs when the screw head makes contact during screwing-in. The contact moment and subsequently the required tightening torque can only be achieved if the screw can be turned without resistance until contact is achieved. Undesirable effects, e.g. a poor screw thread or female thread can generate torques which approximate to the contact or tightening torque. The screw head does not then make contact such that a force-fit connection can be formed. Therefore, not only the tightening torque but also the screw-in depth must be observed.

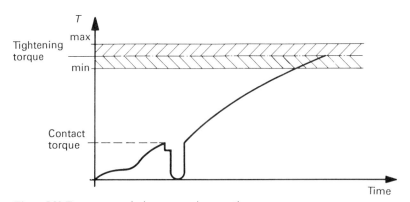

Figure 5.92 Torque curve during a screw-in operation

A screw unit must at least fulfil the following functions:

- Storage of unarranged screws
- Arrangement of the screws
- Placing the arranged screws in a magazine
- Distributing the screws
- Positioning of a screw in the assembly position
- Screwing-in of a screw
- Infinitely variable tightening torque.

So as to check the quality of the screw-in procedure, in addition to checking the tightening torque, measurement of the screw-in depth is also important.

The availability of an automatic screw-in unit is not only dependent upon the intrinsic reliability of the unit but also on the screw quality (see Section 2.2.6.2). For example, a burr in the screw also leads to faults in the aspect of arrangement and placing in magazines functions. Screw slots not located in the centre cause the screwdriver blade to contact and exert pressure on one side and consequently results in tilting of the screw during positioning. Irrespective of the screw feed head design, variations in the alignment between the screw shank and screw head can result in skew positioning of the screw in the feed head, depending on the design of the latter. The screw shank then carries out circular motion which results in assembly problems. Figure 5.93 illustrates one such situation.

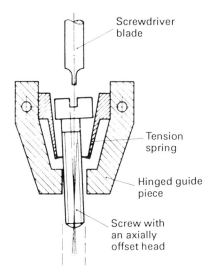

Figure 5.93 The effect of alignment accuracy between the screw head and shank when positioning a screw in a screw feed head

Cross-head screws increase the performance of screw fitting units since, in comparison with single slot screws, the screwdriver blade is engaged twice as fast. By reason of its conical shaped cross head and also the conical shaped screwdriver blade, the cross-head screw has the advantage that the engaged screwdriver blade has a centring effect on the screw.

Automatic screw-insertion machines are either pneumatically or electric-motor driven. With the screw feed on these machines, a distinction is made between drop tube and conveyor rail designs.

5.5.1 Drop-tube-type automatic screw-insertion machines

The application of automatic screw insertion with drop-tube feed of the screws is restricted to applications in which the ratio between the head diameter and shank length of the screws is at least 1:1.5. Experience shows that screws with a shorter shank length have a tendency to tilt or turn over in the drop tube. Figure 5.94 shows a screw feed head with the feed of the screws via a drop tube. This is either telescopically subdivided in order to follow the stroke movement of the spindle or it is a flexible hose type. The screw is arranged in the correct position by a vibrating spiral conveyor, separated on a discharge rail and gravity fed through the drop tube to the feed head. To ensure a more rapid and positive feed of the screws into the feed head, the drop action is quite often compressed-air-assisted.

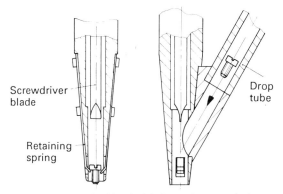

Figure 5.94 Screw feed head with drop tube screw feed

A further design is shown in Figure 5.95. In this arrangement, the sorted screws are fed into the feed head by a drop tube which can swing in and out. The drop tube swings into the feed head if the screw-insertion spindle is retracted into its upper position. A screw is separated and fed by gravity, assisted by compressed air. The drop tube swings away from the spindle by the action of the downwards movement of the screw-insertion spindle, and the screwing process is then carried out.

The feed head of these machines must be designed according to the screw size and screwing procedure. Figure 5.96 shows three different designs of feed heads and also various types of screwdriver blades. The feed head design (a) is suitable for screws whose shank length is at least four times larger than the shank diameter and consists of two parts of a taper spring-retaining tong whose internal taper is matched to the dimension of the screw to be handled with regard to the head and shank diameter and shank length. The spring-retaining tong is centred with the second part of the feed head which is a sleeve on the screw-insertion spindle. A fed screw is pressed through the retaining tong by the screwdriver blade. Types (b) and (c) are used for screws for which the ratio of the shank diameter to length is less than for screws which can be inserted with feed head (a). These feed heads have two hinge-action tong arms positioned on axes. When closed, they form a bore equal to the screw shank diameter. During feed the screw head comes into contact with the top edge of the hole formed by the closed tong arms. The hinged arms are opened by the downwards motion of the screwdriver blade during insertion of the screw and the screw head is pressed through. The different screwdriver blades are adapted to the types of screw to be inserted, e.g. hexagonal

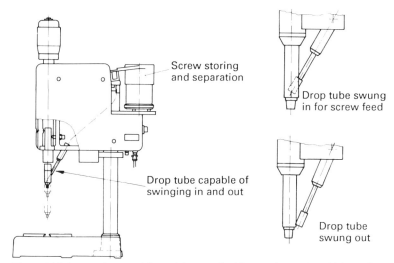

Figure 5.95 Automatic screwdriver with screw feed from a drop tube which can be swung in and out (Weber)

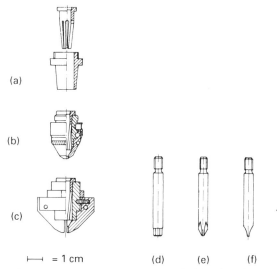

Figure 5.96 Design examples of screw feed heads and screwdriver blades suitable for automatic screw-insertion machines as shown in Figure 5.95 (Weber) (a) Feed heads for large-shank screws ($\phi \times L = $ min. 1:4). (b), (c) Feed heads for short-shank screws. (d), (e), (f) Screwdriver blades for hexagon-head, cross-head and slotted-head screws, respectively

head, cross-head or slotted screws. Screw-insertion units with a swing action drop tube are more suitable for rapid changing of the feed heads since, with a swing-action drop tube, the feed of the screws is not permanently linked to the screw-insertion spindle.

Figure 5.97 shows another type schematically with the screw feed through a drop tube. The feed head is fitted on the feed spindle. During the return motion at the end of its stroke (position 1) the spindle makes a helical rotary motion. The feed head moves to position 2 and is then aligned with the rigidly positioned drop tube.

166 Modules for the automation of assembly processes

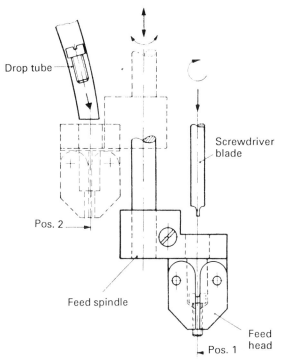

Figure 5.97 Swing-action feed head with a drop tube screw feed (Sortimat)

During screw selection, the screw is gravity-fed into the feed head. If a screw-insertion operation is started, the screw-insertion spindle moves into position 1 by the helical rotary motion and is aligned with the screwdriver blade. The blade is moved into the feed head and the screw-insertion operation commenced.

5.5.2 Automatic screw inserters with feed-rail feed

Automatic screw inserters with feed-rail feed are suitable for all screw types with a head; they are not suitable for inserting stud bolts. The screws are arranged in the required position by a vibratory spiral conveyor and fed to the screw-insertion spindle suspended from the head on an electromagnetically driven discharge rail. An advantage with this design is that in the event of feed difficulties access is possible at any time in the feed rail.

The feed head design and screw feed via a feed rail are shown schematically in Figure 5.98. The feed head a is in tong form, closed on one end and open on the other. The screws are fed by a feed rail c. One screw at a time is released by a sorting device. The cam-controlled swing motion operates a release slider b. One screw is forced into the feed head by the slider.

After retraction of the separating slider, the automatic screw-inserter spindle moves into the screw-insertion position, and the screwing operation is then undertaken.

Figure 5.99 shows a second possibility for the feed of screws in feed rails. In this arrangement, the screw-insertion unit feed head is in the form of a gripping tong; the mode of operation is shown in Figure 5.100.

Screw-inserting units 167

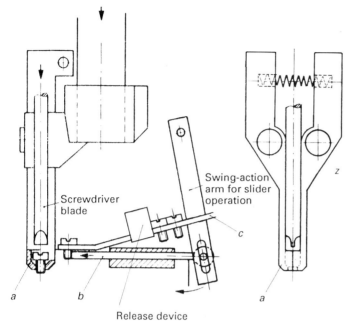

Figure 5.98 Screw feed by feed rail (Sortimat) (a = tong form feed head, b = separating slider, c = feed rail)

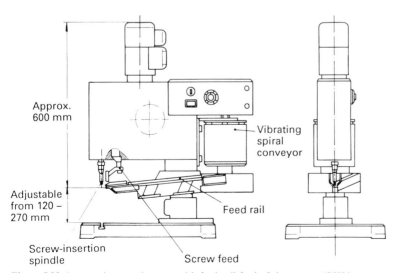

Figure 5.99 Automatic screw inserter with feed-rail feed of the screws (OKU)

168 Modules for the automation of assembly processes

(a) Held on the shank and individually selected

(b) Guided up to secure location of the thread in the workpiece

(c) Ready even during screw insertion

Figure 5.100 Operational sequence for a gripper feed head on a screw inserter as shown in Figure 5.99 (OKU)

The movement of the screw from the feed rail to the workpiece is controlled by the feed head in the form of a gripper tong.

5.6 Riveting units

Riveting is the alternative to screwing in cases where a detachable connection is not necessary or if detachable connections are not desired and also where the connection is not subjected to high tensile stresses.

Riveted connections are:

- Space-saving
- Generally cheaper then screw connections
- Preferable subjected to shear stresses.

In contrast to screw inserters, equipment for riveting is generally not fitted with an automatic feed device for the rivets. The reason is that, as a general rule, the rivets must be positioned as the first part so that the parts to be joined together can be centred over the rivet shanks, and the joining procedure then undertaken by re-forming.

Figure 5.101 shows a comparison of different riveting procedures. Pressing and rotating-mandrel riveting are the principal procedures used in the fields of precision engineering and electrical engineering.

5.6.1 Press-riveting

Figure 5.102 shows the operational sequence for a riveted joint formed by pressing. Figure (a) shows the joint structure and (b) the riveting procedure. With press-riveting, the parts to be joined must be form-locked at the rivet head since, during the pressing operation, the rivet shank is deformed in a manner so that the gap between the parts to be joined and the rivet shank is filled before forming of the rivet head.

Standardized units for press-riveting as available on the market are preferably pneumatically operated and work up to approximately 20 kN direct and above 20 kN

Riveting units 169

Feature	Hammering	Pressing	Rolling	Punch-rolling
Correct material rivet deformation	poor	satisfactory	good	good
Quality of riveting	good	very good	very good	very good
Surface protective coating on rivet	destroyed	destroyed	damaged	retained
Reproducibility of riveting process	satisfactory	good	good	good
Working speed	low	high	average	low
Axial force on rivet	high	high	average	low
Possibility of multiple riveting	limited	very good	limited	very good
Noise generation	Severe (up to 130 dB)	low	low	low

Figure 5.101 Comparison of various riveting processes

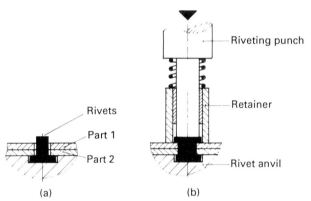

Figure 5.102 Operational sequence for a riveted joint by pressing: (a) joint structure; (b) riveting procedure

up to around 60 kN with a bell crank lever arrangement. Figure 5.103 shows one such unit. With the application of these press-riveting units in automated assembly equipment, care must be taken to ensure that the reaction force to the press force is controlled so that damage to the workpiece carrier or transfer equipment is avoided. When used at manual assembly points, for work safety reasons it is advantageous not to arrange the assembly fixture permanently in the press, but to locate it on a separate slide table type workpiece carrier.

When retracted and for loading the parts, the workpiece carrier is outside the area of the press. After loading, the slide table workpiece carrier is moved under the press ram and guarded so that unintentional access is precluded.

In addition to riveting, joining by re-forming includes processes such as notching and folding, etc. The same type of equipment is used. Figure 5.104 shows the operational sequence of a joining procedure by 'notching'.

The press is equipped with a notch-shearing tool instead of a riveting punch. Figure (a) shows the joint structure and (b) the notching procedure. The notch shearing tool cuts into the part and deforms the cut material with the outside edge of the tool so that the notched rivet is securely located.

170 Modules for the automation of assembly processes

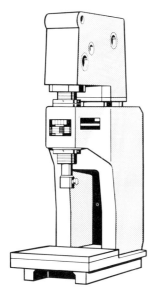

Figure 5.103 Pneumatic table press (Bosch)

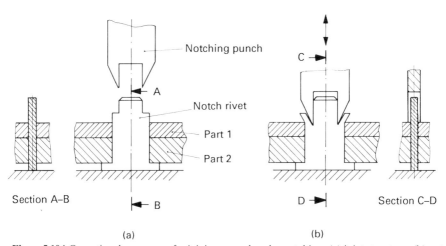

Figure 5.104 Operational sequence of a joining procedure by notching: (a) joint structure; (b) notching procedure

5.6.2 Rotating-mandrel riveting

In comparison with normal press-riveting, with rotating-mandrel riveting the riveting punch describes a rotating-type movement by reason of the inclination of the rivet spindle axis by a few degrees (see Figure 5.105). Since, during riveting, the riveting spindle rotates about the vertical axis, deformation of the rivet is limited locally by the contact line generated by this movement between the riveting mandrel and rivet. The yield point of the rivet material is exceeded locally so that the material flows into the desired final form of the rivet head in small stages.

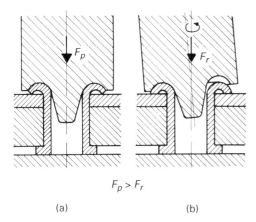

(a) (b)

Figure 5.105 Principle of (a) press and (b) rotating-mandrel riveting (Dr Herrmann)

Figure 5.106 shows a rotating-mandrel riveting machine and Figure 5.107 the design of a rotating-mandrel riveting tool. An important feature is that the riveting tool does not rotate about its axis but moves in a see-saw fashion on the rivet under pressure during the rolling movement. This results in advantageous material deformation and a fine finish of the rivet material surface. Consequently, the riveting tool is only subject to very slight wear. Furthermore, only a relatively low pressure is required between the rivet and tool on account of this limited deformation, so that, when working thin-walled hollow rivets, buckling can be avoided. During press-riveting, with hollow and also solid rivets, the rivet head formed by pressing exhibits a tendency to crack. This phenomenon is avoided with rotating-mandrel riveting. A further advantage is that the process has a low noise generation. With press-riveting, the riveting of quite often different rivets at various positions and heights at one point is not problematical since several rivets can be riveted at one time in a tool.

The rotating-mandrel riveting of several rivets in one product in sequence is time-consuming. The application of multi-spindle rotating-mandrel riveting machines is dependent upon the spacing of the rivets. Figure 5.108 shows the design of a multi-spindle rotating-mandrel riveting machine. Instead of one spindle, several satellite spindles are driven from a central spindle by gearwheels. Depending upon the riveting configuration, up to five spindles can be accommodated in this manner.

The disadvantages of this solution are:

- The whole multi-spindle punch-rolling rivet head is only suitable for a specific riveting configuration. Other arrangements require a new design and recalculation of the gearing.
- Small centre distances between the mandrels are difficult to achieve. Friction and lubrication problems are caused by the number of bearing points required.

The compensation of tolerances between the individual rotating-mandrel riveting spindles in terms of height for the riveting procedure is achieved by a fitted plate spring.

5.7 Welding units

Several special processes have been developed for welding and which have considerably furthered the wide application of this joining method in precision

172 Modules for the automation of assembly processes

Figure 5.106 Rotating-mandrel riveting machine (KMT)

engineering and electrical engineering. Resistance welding (spot welding) is particularly advantageous, in which a joint is formed within the shortest time by the simultaneous effect of heat and pressure without auxiliary materials. Laser welding has been widely used for some considerable time; it has the advantage that the welding procedure can be undertaken without contact.

Welding processes have the advantage of producing an unbreakable connection at low cost and are not linked to a geometrical form of the component. The following points must be considered during welding:

- Material pairing
- Wall thicknesses of the parts to be joined
- Grain structure changes at the weld point
- Heat effect on the surrounding points
- Slag formation.

The integration of welding machines into automated assembly processes requires that the workpiece carriers or assembly fixtures are designed in accordance with the welding technology.

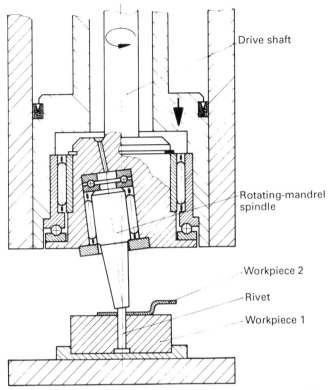

Figure 5.107 Design details of a rotating-mandrel riveting tool (Dr Herrmann)

5.7.1 Resistance welding

The principal components of a resistance welding machine are the welding head and the electrodes with the holder, the welding transformer, welding current and welding pressure control. The size of the welding head is dependent upon the welding operation; in order to achieve high-quality welding it must give low play, low friction and low mass control of the movable electrodes, good cooling of the electrodes and current-carrying conductor, and also easy replacement of the electrodes and high-wear parts without costly readjustment. As with the working tools, the electrodes are high-wear parts. For low-cost manufacture, they must be of simple form and easily exchangeable. Repeat reworking is advantageous. The selection of the electrode material and the surface depends upon the workpiece. The service life is increased by intensive cooling. The data for the dimensioning of the welding transformer, welding power with an extended connection period, the secondary voltage, short-circuit current, number of voltage steps and the external dimensions are dependent upon the respective applications. Semiconductor controls are used on mechanical welding equipment. Figure 5.109 shows a general arrangement of welding current control types for resistance welding and the related fields of application.

174 Modules for the automation of assembly processes

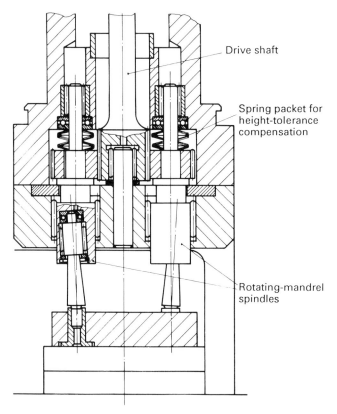

Figure 5.108 Multi-spindle rotating-mandrel riveting unit (Dr Herrmann)

The following pre-conditions must be fulfilled for precise switching of the current duration and amplitude:

- Mains synchronous switching on and off of the current
- Infinitely variable setting of the current amplitude
- Fine setting of the current time
- Fine selection of the secondary voltage
- Setting of the contact rating to transformer and machine.

Resistance welding machines with pneumatic control are principally used in the fields of precision and electrical engineering.

As a general rule they are capable of up to 200 strokes per minute and have an electrode closing force of 20 N to 20 000 N. The welding pressure control ensures that the electrodes make impact-free contact and can also follow the workpiece profile by a spring-loaded mechanism. Figure 5.110 shows the curve characteristic of a pneumatic control with impact damping and a two-stage pressure programme.

Figure 5.111 shows a resistance welding machine with a variable electrode force from 200 N to 3400 N [23].

5.7.2 Laser welding equipment

Principally, a laser consists of two parts, namely the active medium and the optical resonator, its principal being based on light amplification by stimulated emission. The

Welding units 175

Control type	Impulse form	Application
Alternating current		
Mains frequency 50 Hz		
Single-impulse, semi-wave		Spot welding
with multi-periods		Spot projection welding
with multi-periods and slope		Projection butt welding
Multi-impulse, semi-wave		Insulated wire welding
with multi-periods		Roll seam, projection welding
with multi-periods and current program		Insulated wire, hooked lug welding, hot pressing
with multi-periods and slope		Projection welding
Medium frequency 1 kHz		Spot welding
Single impulse, multi-period with multi-periods and slope		Spot butt gap welding
Low frequency		Spot projection welding
Direct current		
Capacitor impulse		Spot, projection, butt welding
Rectifier impulse		Spot welding
with slope		Gap welding

Figure 5.109 Summary of welding current control types (PECO)

term 'laser' is an acronym derived from 'Light Amplification by Stimulated Emission of Radiation'. The active medium can be solid, gaseous or fluid. The atoms or molecules of the active medium can absorb energy in a particular form which is later emitted as light radiation. A laser-active medium can, for example, absorb energy by absorbing the light emission. The laser-active atoms then change from a basic state I to a state of higher potential energy II or an excited condition. The energy difference is $h \times v_{12}$, where h is Planck's constant and v_{12} the frequency of the absorbed light wave. After a specific dwell time in the excited condition, the so-called life span, the exerted atoms revert to the basic state. The energy liberated is in the form of a light beam of frequency v_{12}. With the stimulated emission, the atoms are forced to revert to the basic state in an excited condition. This is achieved by the irradiation of a light source frequency v_{12} whose amplitude is amplified by the absorbed stimulated radiation. Normally, i.e. in an unexcited state, virtually all atoms are in the basic state. To achieve

176   Modules for the automation of assembly processes

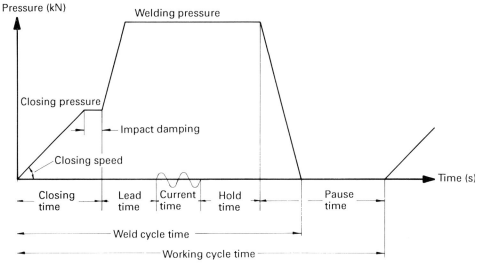

Figure 5.110 Electrode force curve with pneumatic control with impact damping and a two-stage pressure programme (PECO)

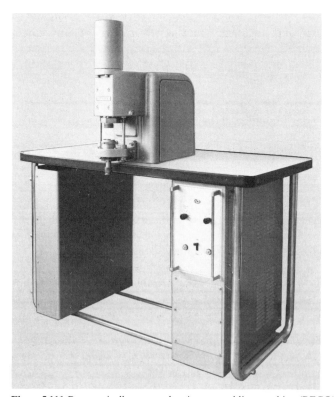

Figure 5.111 Pneumatically operated resistance welding machine (PECO)

amplification by stimulated emission, more atoms must be in an excited condition than in the basic condition, i.e. an occupation reversal or inversion must occur. In order to achieve an occupational inversion, a laser-active medium must have at least three energy conditions or energy levels. This is achieved by a light source of frequency v_{13}. The transition of the excited atoms from level III to level II is radiation-free. It is only during a transition from level II to the basic level that the liberated energy is re-emitted as a light beam at frequency v_{12}. Occupation inversion is achieved if the dwell time at level III is considerably shorter than at level II. An occupation inversion is achieved considerably more easily if the laser-active medium has four instead of three energy levels and for which the transitions II/III and IV/I are radiation-free and the dwell time in levels II and IV considerably shorter in comparison with that in level III.

The three-level laser is a ruby laser. Four-level lasers are, for example, glass and Yag lasers. An optical resonator is necessary in addition to the active mediums.

An optical resonator consists of two parallel mirrors between which a permanent light wave can form. The mode of operation of a resonator is shown in Figure 5.112. The resonator must contain a laser-active medium. An occupation inversion is achieved in this medium by an energy input. A proportion of the excited atoms reverts to the basic state before reaching the mean life period. If emissions travelling in the direction of the optical axis impinge on one of the two mirrors they are reflected and a permanent wave formed between the two mirrors of the resonator and whose amplitude is intensified by stimulated emission. Only a fraction of the light wave in the resonator can be separated – as a general rule around 20% to 50% by selecting one of the resonator mirrors with a reflection coefficient of <100%. For example, the active medium is activated by a xenon flash tube and in an elliptical reflector in whose focal line the flash tube or active medium is located. In technical language, this stimulation is known as 'pumping'.

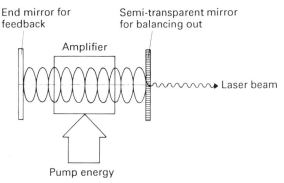

Figure 5.112 Mode of operation of a laser resonator (MI-Publication)

According to their function, laser welding units are designed on the modular construction principle. Various radiation waves which differ in terms of the maximum pulse energy and duration can be used together with the same optical and mechanical components.

The inside of the head with the laser bar, flash tubes and pump mirror is cooled by a deionized water flow. As high-wear parts, flash tubes must be replaced from time to time. The energy reservoirs, control electronics and cooling equipment are combined in a control cabinet. Figure 5.113 shows the schematic arrangement of a Siemens-type fixed-head laser. Figure 5.114 shows the Haas-type fixed-head laser.

178 Modules for the automation of assembly processes

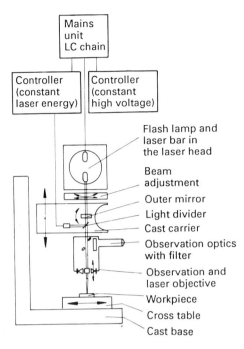

Figure 5.113 Schematic arrangement of a Siemens-type fixed-head laser (Brunst)

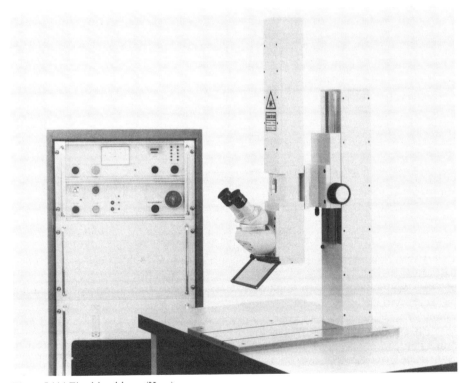

Figure 5.114 Fixed-head laser (Haas)

Welding with a fixed-head laser is spot-welding. If a seam is to be welded a number of points are located closely behind each other so that a continuous seam is formed. The advantage of modern equipment is that the number and geometrical arrangement of the points can be program-controlled as required.

From amongst gas lasers, the CO_2 laser with its high achievable output power is particularly important for welding applications. Figure 5.115 shows a schematic arrangement of one such item of equipment. The tube is folded once in order to achieve the length required for a continuous power output of 200 W. CO_2 lasers function at a wavelength of around 10.6 μm. It is noteworthy that light of this wavelength does not penetrate through optically transparent materials and also that CO_2 lasers, e.g. glass ampoules, can be sealed [25, 26].

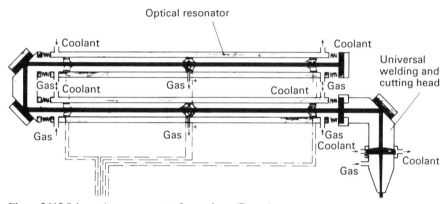

Figure 5.115 Schematic arrangement of a gas laser (Brunst)

5.8 Soldering equipment

Soldering is a process for the connection of metallic workpieces using a molten additional metal (solder) and for which the melting temperature of the solder must be less than that of the workpieces to be joined together.

Soldering is classified as follows by the working temperature into

- Soft soldering at working temperatures below 450°C
- Hard soldering at working temperatures above 450°C.

In precision and electrical engineering rosin-core solder in tubular form is predominantly used. This has the advantage that the tube is the solder alloy and is filled with the rosin flux. The application of a flux during soft soldering is therefore not necessary.

During soldering, the parts to be joined must be supplied with the energy for heating and also the solder.

Figure 5.116 shows a view of a pneumatically operated, automatic solder wire feed. Wire diameters of 0.7 mm to 1.5 mm can be used. The unit is fitted with two feed devices. The actual solder wire feed device is positioned by the adjusting unit. A movement of up to 50 mm is infinitely variable. The solder feed device has an infinitely variable wire feed from 0 mm to 12 mm per stroke. Figure 5.117 shows the design of an automatic soft soldering station. Two of the setting units shown in Figure 5.116 are

180 Modules for the automation of assembly processes

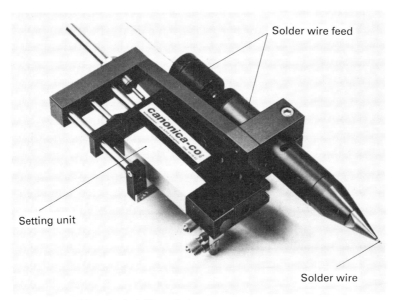

Figure 5.116 Solder wire feed (Canonica)

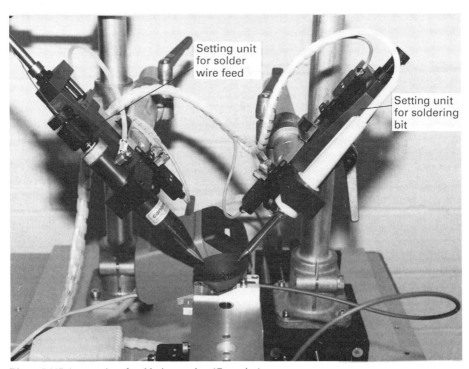

Figure 5.117 Automatic soft-soldering station (Canonica)

arranged on a column support. At one station, the soldering bit is brought into contact with the parts to be joined and the soldering wire fed at the other.

The heat energy required can be supplied by a resistance welding transformer. In this case, the parts to be joined are contacted by electrodes.

Silver solders are generally used when hard-soldering. In production operations using this soldering technique, integration of the soldering operation into the assembly operation is made possible by the application of high-frequency soldering units.

Pre-formed silver solders, for example in the form of rings, can as a general rule be fed automatically and can be handled as individual parts. The advantage of hard-soldering on a high-frequency basis is that the required energy can be supplied contactless by induction loops. Newly developed paste solders are rapidly replacing hard solders. They have the advantage that they can be automatically and accurately applied in the required volume by cylinder metering units. On the other hand, hard-soldering operations are difficult to integrate in automatic assembly processes under a protective gas atmosphere since, from the installation aspect, such units can hardly be connected with assembly equipment.

5.9 Bonding

Bonding is the jointing process which is currently making the greatest progress in application. In comparison with welding or soldering it has the advantage of generating none or only a low level of surface heating of the component. It can be used on large areas and produces joints with tensile strengths of up to 40 N/mm^2. Further advantages of bonding are:

- Good resistance to solvents and chemicals
- Bonding of different materials, e.g. glass to wood or any metals.

The production of bonded joints does, however, require certain preparations which differ considerably from other bonding techniques. Preparation commences with the design of the joint. Bonded joints should preferably be loaded in shear or compression and less in tension in order to prevent peeling of the joint.

For industrial applications, reaction bonding agents on an epoxy resin base are of particular importance. These are obtainable commercially as single- or double-component bonding agents in many forms and, by careful selection, give favourable adaptation to the required characteristic of the joint and also the production process. The setting of the bonding agent is by a chemical reaction the speed of which depends upon the type of material and ranges from a few seconds up to several hours. For bonding agents with a longer hardening time, the time can be significantly shortened by a suitable ambient temperature (e.g. 150°C).

A condition for permanent bonding is clean and grease-free workpiece surfaces. Roughening also increases the bond strength. The bond strength is also affected apart from the hardening temperature, by contact pressure during hardening. The application thickness of bonding agents is preferably around 0.1 mm. To a large degree, the bond quality is affected by application and metering. Figure 5.118 shows an automatic metering unit for single-component reactive bonding materials. The correct quantity is metered per cycle by a fine-adjustment metering piston. For a uniform

182 Modules for the automation of assembly processes

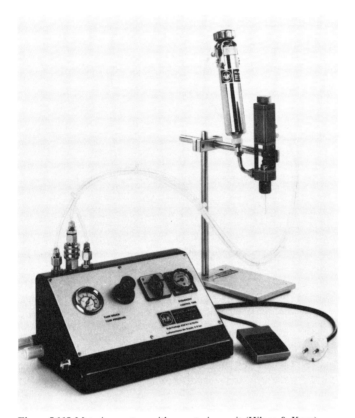

Figure 5.118 Metering system with a metering unit (Hilger & Kern)

application on the parts to be bonded together, either the workpiece or metering unit is moved so that the bonding surface is wetted.

In comparison to mechanical joining techniques, bonded joints have a considerable weight advantage.

Chapter 6

Design of assembly machines

6.1 Introduction

The design of assembly machines depends upon the complexity of the part to be assembled and the required production rate. A distinction is made between single- and multi-station assembly machines. Up to three parts can be assembled on single-station assembly machines and more than three parts on multi-station machines. The cycle time and therefore also the selection of the drive energy and kinematic design are determined by the required output.

The following functions must therefore be considered:

(a) Main movements, e.g. drives for transfer systems, X-Y movements of handling units, stroke movements of inspection units, etc. These principal movements can be by the following means:
   - Electric motor via switch, cam and level mechanisms
   - Pneumatically
   - Hydraulically.
(b) Secondary movements, e.g. for the clamping of parts, movement of grippers, separation of parts, centring, holding, peeling, etc., can also be carried out by the following means:
   - Mechanical, e.g. by cam and lever mechanisms
   - Electromechanically, e.g. by electromagnet
   - Pneumatically.
(c) Signal generation, detection, evaluation, e.g. interrogating whether existing parts are correctly positioned or if operations have been undertaken and specified parameters achieved.
   When the results of the interrogation are negative, secondary operations must be introduced and the results classified in terms of good and bad for sorting:
   - Electric
   - Electronic
   - Pneumatic
   - Fluid-operated
   components or combinations of them are available for the implementation of these functions.

Equipment for cycle times in excess of 3 s can be pneumatically or electric-motor-operated or also by a combination of both. On the other hand, the main movements of items of equipment with cycle times of less than 3 s must be electric-motor-operated.

184 Design of assembly machines

The use of pneumatic drives requires a follow-up control of the movement sequences which must be undertaken in sequence in a time series. Mechanically driven assembly machines permit mathematically accurate pre-determined movements which result in the overlapping of movement sequences and therefore also cycle shortening (see Section 5.3.1.1, Figure 5.48).

6.2 Single-station assembly machines

Single-station assembly machines are items of equipment on which all assembly operations can be undertaken in one position. They are suitable for the assembly of two to three parts and are restricted in their capacity in that the assembly operations must be largely undertaken in sequence. The finished assembly is only ejected and not arranged in a specific manner. These machines are preferably used if the base part can be loaded on to the machine. The feed and assembly of the second part should be by one unit, e.g. an automatic screw inserter, since the screw itself is automatically guided in the screw inserter. The same applies to the pressing of a part, during which the feed of the part to be pressed forms a part of the press unit. Typical single-station assembly machines are automatic screwing machines in the electrical engineering industry. By way of example, Figure 6.1 shows the insertion of a screw in a clamp. Figure 6.2 shows the appropriate arrangement of a single-station assembly machine.

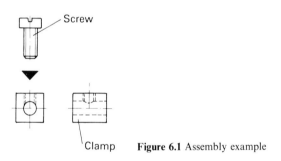

Figure 6.1 Assembly example

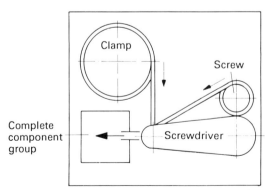

Figure 6.2 Layout of a single-station assembly machine for assembling the component group in accordance with Figure 6.1

Single-station assembly machines   185

It is comprised of standard commercially available units, namely a vibratory spiral conveyor for the correct positioning and feed of the clamps and an automatic screw-insertion station with correct arrangement and feed of the screws.

The design arrangement of the assembly fixture and sequence of operations is shown in Figure 6.3. Figure (a) shows the start position of the operation, (b) the assembly position and (c) the ejection of the completed assembly.

The operation is as follows. The clamps are arranged by a vibratory spiral conveyor and fed to the fixture correctly positioned via a discharge rail (1).

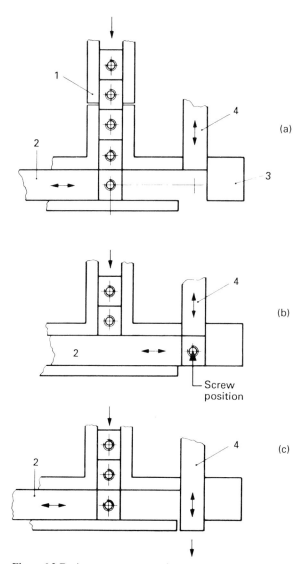

Figure 6.3 Design arrangement and operational sequence of a single-station assembly machine as shown in Figure 6.2: (a) start position; (b) assembly position; (c) ejection (1=discharge rail, 2=slider, 3=stop, 4=ejector)

186 Design of assembly machines

A clamp is separated and moved up against stop (3) in the screwing position by the sliding action of the slider (2). After insertion of the screw in the clamp, the completed assembly is ejected by the slider (4); at the same time slider (2) returns towards its start position so that the locking of the feed machine is released and another clamp positioned.

Figure 6.4 shows the sequence of movement in a time-displacement diagram. It is evident that the first four movements must occur sequentially. The unit functions pneumatically. On account of the short movements, the control of the movements can be by a programmed switch mechanism.

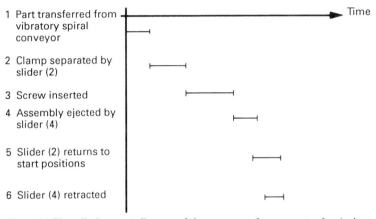

Figure 6.4 Time-displacement diagram of the sequence of movements of a single-station assembly machine as shown in Figure 6.2

Figure 6.5 shows an automatic screw-insertion machine with an automatic arrangement and feed of the screws via a drop tube, a vibratory spiral conveyor for the feed of the mating part and a pneumatic device for positioning of the part and ejection of the completed screw-fitted assembly.

6.3 Multi-station assembly machines

Multi-station assembly machines are items of equipment on which a number of operations must be undertaken separately via a number of single stations. As a basic structure for interlinking the individual stations, these machines require a transfer system for the workpiece carriers. The circular cyclic or longitudinal cyclic arrangements are formed depending upon the selection of the type of transfer system (see also Section 5.4).

The number of stations is dependent upon the number of single operations, e.g. feed, assemble, machine, test, adjust, stack, etc. The size and scope of the transfer equipment is determined by the possibilities of arrangement of the individual stations. Care must be taken to ensure that adequate free space is available between the individual stations so that, in the event of a fault, easy access is available for the implementation of remedial measures.

Stations for screw insertion, pressing, forming and welding, etc. generally require less space for their arrangement in relation to the transfer equipment than do part feed

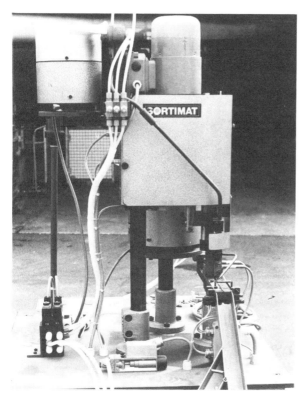

Figure 6.5 Single-station automatic assembly machine (Sortimat)

units. Figure 6.6 shows two different arrangements of feed equipment on a circular cyclic automatic machine. The arrangement of the feed unit is determined by the possibility for the positioning and feed of the part and has a varying space requirement. Figure 6.6(a) shows that on a feed unit for a part which can be positioned and fed such that its feed direction corresponds to the assembly direction, the feed rail can be aligned towards the centre of the circular indexing table. Arrangement (b) shows an arrangement in which the positioning and feed posibilities for the part are at 90° to arrangement (a). With such an arrangement the feed rail must be positioned tangentially to the circular indexing table in order to maintain the correct assembly position of the part. This arrangement covers several unusable stations on the circular indexing table. The same also applies to longitudinal transfer units. In order to avoid a tangential arrangement of the discharge rail on circular cycling machines or a parallel arrangement on longitudinal cycling machines, the unit for parts handling can be fitted in its Z-axis with a rotation device for turning the part through 90° between the pick-up and positioning points.

A decision between a circular cyclic and longitudinal cyclic arrangement is not only dependent upon the required number of stations and the space requirements, but it is based on the overall objective. With the design arrangement, care must be taken to ensure that the principal movements are kept as short as possible. The pick-up points for gripping workpieces must be positioned as near as possible to the assembly position so as to achieve the shortest possible X-Z movements on the handling equipment.

188 Design of assembly machines

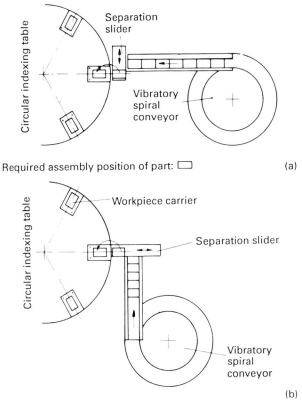

Required assembly position of part: ▢   (a)

Figure 6.6 Different arrangements of feed units. (a) Feed in direction of parts supply. (b) Feed transverse to direction of parts supply

Figure 6.7 shows schematically the optimum arrangement of a pick-up point to a workpiece fixture with the minimum possible movements of the Z-axis (a and c) and the X-axis (b).

The number of stations of an assembly machine principally determines the availability and therefore also the efficiency; quite often, the design of assembly machines is determined by design details of the workpiece carriers. These two criteria

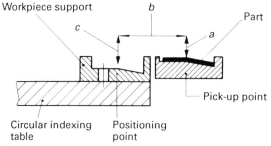

Figure 6.7 Optimum arrangement of the pick-up and positioning points for automatic handling: (a) Z-movement pick-up point; (b) X-movement pick-up positioning point and (c) Z movement positioning point

Multi-station assembly machines   189

are a principal part of the planning of automatic assembly equipment and are discussed further in Chapter 11.

6.3.1 Design of parts feed stations

The overall availability of an automatic assembly machine is principally determined by the individual reliability levels of the part feed stations. Accordingly, particular importance is attached to its design.

Figure 6.8 shows a typical design of a parts feed station consisting of a vibratory spiral conveyor for the positioning and feed of the individual parts, an electromagnetic discharge rail for storing the arranged parts and conveying to the separating station, a separating station (see Section 5.2.2.4) to remove the part from the build-up pressure of the workpiece line and move it to the gripping position, and a positioning unit for manipulation of the single part from the separating device and placing in the workpiece holder of the assembly machine transfer unit.

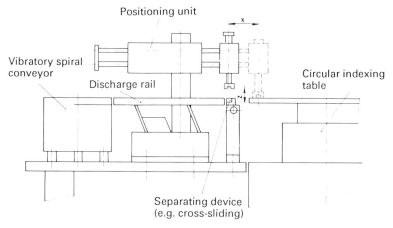

Figure 6.8 Design of a parts feed station

Its level of reliability is determined by the cleanliness of the parts and also the arrangement and accuracy of the parts feed station. Experience shows that the most frequent faults occur inside a vibratory spiral conveyor and during transfer from the spiral conveyor on to the discharge rail. The causes of these faults not only lie in the quality of the vibratory spiral conveyor, but principally in the unsatisfactory part quality, inadequate cleanliness of the parts and also in the inclusion of foreign bodies. Parts which enter the discharge rail correctly positioned are seldom the cause of any further faults during the whole working cycle.

In order to ensure that faults in the region of the vibratory spiral conveyor and at the transition from the conveyor on to the discharge rail do not affect the availability of the assembly machine, the length of the discharge rail must be such that a buffer of correctly arranged parts is formed on the discharge rail. The buffer capacity must be arranged so that faults in the region of the vibratory spiral conveyor can be promptly identified and also so that the discharge rail does not run empty. A 2 min reserve capacity is common.

To achieve this, an adequate feed section in the form of an intermediate buffer must be provided in the discharge rail by suitable signal recording and control.

190   Design of assembly machines

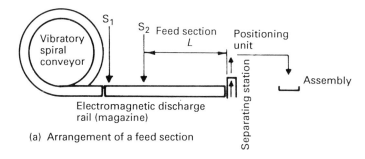

(a) Arrangement of a feed section

			Feed section from S_2						
			→	99.95	99.95	99.8	99.7		99.4
Total section →	99.35	99.85	99.9	99.95	99.95	99.8	99.7	98.5	
R_1 Quality of parts	R_2 Arrangement	R_3 Transfer to magazine	R_4 Magazine		R_5 Separation	R_6 Gripping transporting (positioning unit)	R_7 Assembly	R_T Total single station	R_B Total stations effective on the machine

R_i = individual reliability as a %
R_T = reliability of the total section
R_B = reliability after the feed section (buffer)

(b) Reliability calculation: $R_T = R_1 R_2 R_3 R_4 R_5 R_6 R_7$
$R_B = R_4 R_5 R_6 R_7$

Figure 6.9 Arrangement of a feed section and its effect upon reliability

Figure 6.9 shows the schematic arrangement of a feed section and the achievable overall reliability under the assumption of a particular parts quality. Two sensors must be fitted on the discharge rail. The vibratory spiral conveyor is matched to the machine capacity by sensor S_1. If the discharge rail is full, this condition is detected by sensor S_1 and the vibratory spiral conveyor switched off. A second sensor (S_2) is positioned so that a feed section in the form of a buffer capacity is formed between the sensor and separating station. The length of the discharge rail should hold a buffer capacity for two working minutes. If sensor S_2 records that parts do not approach this point, this condition is converted into an acoustic or visual warning signal. The machine operator is then notified that a fault has occurred in the region of the vibratory spiral conveyor. If this fault is rectified within the feed section buffer time, it has no effect on the overall reliability of the assembly machine. As shown in Figure 6.9, under the assumption of a certain reliability of the individual functions, the overall reliability of a station can be increased from 98.5% to 99.4% by integration of the feed section with signal detection and evaluation [2].

6.3.2 Checking stations

In order to avoid incorrect assembly in automated assembly operations the integration of checking stations is normal. These can be connected after the work stations for immediate checking of the process previously carried out. Positive results then trigger acceptance. Negative results can be variously evaluated as follows:

1. *Immediate switch-off principle*
   A checking station which functions on the immediate switch-off principle stops an assembly machine as soon as a fault is detected. The location and type of the fault can then be signalled. The cause of the fault is eliminated by the machine minder and the machine switched on again. This procedure has the disadvantage that the production equipment is stopped frequently.

2. *Switch-off of the machine following a pre-determined number of faults in sequence*
   If the checking station detects a fault, the machine is only switched off following a pre-determined number of faults occurring in sequence. For example, if the number is three, three faults can occur in sequence without stopping the machine. This has the effect that the down times are reduced, since in many cases and particularly in material feed technology such faults are self-eliminating after two or three cycles. A prerequisite for this is fault recording and their correlation to the workpiece carriers or workpieces concerned.

3. *Fault recording*
   In this case, the assembly station is not stopped upon identification of a fault. However, with fault recording, measures are taken to ensure that succeeding operations are not undertaken and that incorrectly or incompletely assembled units are separated in terms of 'good' and 'bad' upon discharge from the production equipment.

Checking stations differ in their design by two principles:

- Checking stations which are rigidly arranged on account of the position of the part to be checked.
- Checking stations which must perform a movement to fulfil their function.

Figure 6.10 shows the schematic arrangement of a checking station. The workpiece to be checked protrudes from the workpiece carrier such that the checking function can be undertaken by entry of the workpiece carrier into the checking station. With the example shown in the diagram, a check is made for the presence of the assembled workpiece by a light barrier.

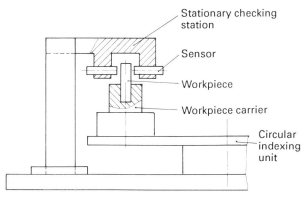

Figure 6.10 Rigid arrangement of a checking station

192   Design of assembly machines

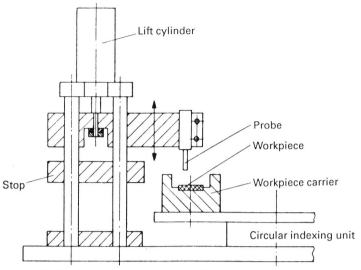

Figure 6.11 Checking station with a vertical lift movement

Figure 6.11 shows schematically a checking station for the availability checking of a workpiece which is located so low in a workpiece carrier that a checking station can only perform its function by a stroke movement of the probe element. In the example shown, the vertical movement of the probe, which is guided by two columns, is by a lift cylinder.

6.3.3 Design of pneumatically operated multi-station assembly machines

Extensive use of standardized commercially available units can be made in the design of pneumatically operated multi-station assembly machines. This means that a large proportion of the units can be reused upon cessation of an assembly operation. As already mentioned, the limits of applicability of pneumatically operated equipment are determined by the compressibility of the air, which has the effect that sequential movements of a unit must be performed in succession in a particular sequence. This necessitates an adequate level of control and monitoring functions. A requirement is that an adequate volume of compressed air is available at the required cleanliness. As a general rule, assembly equipment is operated at a pressure of 6 bar. Limitation of the operating noise and in particular of the exhaust air must be achieved by noise damping.

Figure 6.12 shows a schematic arrangement of a simple multi-station assembly machine with pneumatic units consisting of a circular cyclic unit, two part feed stations and a press. Since the pneumatically operated units are installed in sealed operationally functional units, they can be fitted at any point. In addition to the mechanical design concept, importance is attached to the pneumatic control. The time-related interrelationships between the parts operations must be determined by a program and they form the basis for the correct sequence of a program. The structure of program sequence plans is described by DIN 44 300 with the application of diagrams in accordance with DIN 66 001. A flow diagram must be constructed from the program sequence. The flow diagram is based on a displacement-step-time diagram. This is used to represent condition forms of the most important (principally drive elements) or all elements in a control sequence in relation to time and increment.

Multi-station assembly machines 193

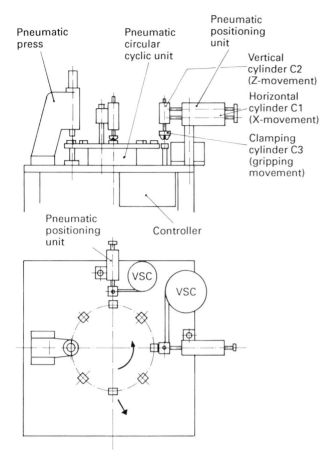

VSC = Vibratory spiral conveyor

Figure 6.12 Design example of a pneumatically operated circular-cyclic-type multi-station assembly machine

Figure 6.13 shows by way of example a displacement-time-increment diagram for the control of a pneumatically operated positioning unit. The following movements are to be performed after the start signal in a sequence, S1 to S4 being signal conditions in the end positions:

- Move horizontal cylinder C1 into the start position
- Move vertical cylinder for part gripping
- Operate clamping cylinder for part clamping
- Perform upwards vertical movement
- Perform horizontal movement
- Perform downwards vertical movement
- Open clamping cylinder for stacking a part
- Perform upwards vertical movement for return to start position.

The impulse-generating connections are shown in the displacement-time-increment diagram by vertical lines with an arrow direction, and mean that the signal for the first

194 Design of assembly machines

	Components					Valve	Start　　　　Cycle time　　　　Feedback
	Title	Ref.	State	Type	Ref.	Control output	Ref.	
Positioning unit	Horizontal cylinder	C1	1		S1	O1	V1	
			0		S2	O2		
	Vertical cylinder	C2	1		S3	O3	V2	
			0		S4	O4		
	Clamping cylinder	C3	1			O5	V3	Open　　　　Closed
			0			O6		

States: 1 = active
         0 = passive

Figure 6.13 Displacement-time-step diagram for a pneumatically operated positioning unit

vertical movement is given upon completion of the first horizontal movement. A signal for operation of the clamping cylinder is given upon completion of the vertical movement, and so on. The time t is the holding time in the stored program controller (PC, SPC) during which the clamping cylinder is given time to clamp.

Consequently, signal generators are not required for the clamping cylinder. Figure 6.14 shows the associated pneumatic circuit diagram for the positioning unit as per the displacement-time-step diagram. Circuit diagrams are used to illustrate a control with all items of equipment including lines and line connections. The pneumatic circuit diagram must be in accordance with specifications VDI 3226 and DIN 24 300.

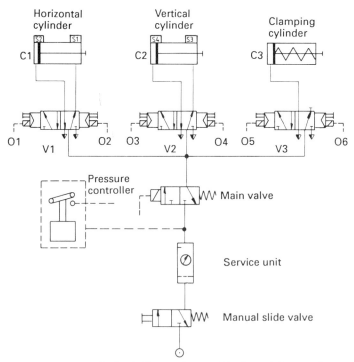

Figure 6.14 Pneumatic circuit diagram for a positioning unit in accordance with the displacement-time-step diagram in Figure 6.13

Servo-controls can be fluid, electrical or electronic. Manufacturers of pneumatically operated positioning units supply complete controls in modular form. Figure 6.15 shows an electronic stored program controller (SPC) with integrated electromagnetic valves and eight inputs and outputs. These controls are practically state-of-the-art because they can be reprogrammed without any alteration to the cable connection. Control can be switched from automatic to manual operation. When starting up assembly machines or with the elimination of faults, this has the advantage that every movement can be called up individually. The integration of electromagnetic valves into the controls reduces the circuit requirement during installation, because the circuit between the electronics and the magnetic valves is a part of the control, so therefore only the hose connections for the pneumatic feed and outlet points between the control and the individual cylinders are necessary.

Figure 6.15 An electronic stored program control unit (EGO)

A combination of electric motor based and pneumatic units is common. Here the transfer units are generally electromechanical and the handling and feed units, etc. pneumatically operated. With circular cyclic machines, the use of an electromechanical drive in the form of a Maltese-cross or cam arrangement is typical.

Figure 6.16 shows one such circular cyclic machine consisting of a circular cyclic unit with a mechanical Maltese-cross drive and pneumatically operated handling units.

6.3.4 Design of electric-motor-driven multi-station assembly machines

The same design guidelines which apply to pneumatically operated systems are also applicable in the design of electric-motor-driven multi-station assembly machines for the application of units with their own drive. Such designs are, however, rare, since, for example, the belt drive of positioning units can be replaced by an external drive with a synchronously running drive shaft. Depending upon the selection of the transfer unit, a distinction is drawn between systems with vertically and horizontally arranged drive shafts.

6.3.4.1 Design of assembly machines with a vertical drive shaft arrangement
Figure 6.17 shows schematically the basic design of a circular cyclic assembly machine with a vertical drive shaft. The circular cyclic unit a, circular indexing table b and drive shaft c form an in-line unit. The arrangement of the drive in the lower section of the machine base permits use of the drive shaft over its entire length and also the performance of the principal movements by plate cams which are attached to the drive

196 Design of assembly machines

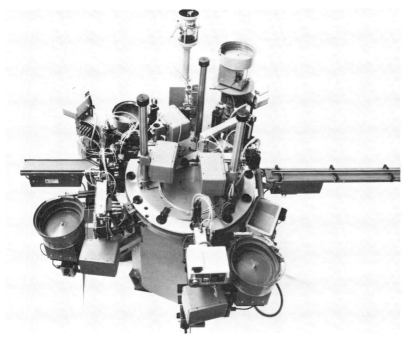

Figure 6.16 An automatic circular cyclic assembly machine with an electromechanical circular cyclic drive and pneumatically operated positioning units (Menziken)

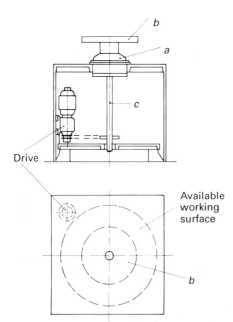

Figure 6.17 Basic design of a circular cyclic assembly machine with a vertical drive shaft (Sortimat). a = circular cyclic unit, b = circular indexing table, c = drive shaft

shaft. The whole circumference of the circular cyclic unit is available as a working surface due to the central arrangement of the drive shaft.

Figure 6.18 shows schematically the cam drive of a circular cyclic assembly machine with a vertical drive shaft and several drive cams (so-called dish cams). The workpiece carrier base plate *b2* is supported above a machine table plate *b1* by a circular cyclic unit *a* with a vertical drive shaft *c*. The drive shaft and therefore also the circular cyclic unit is driven by a chain drive via an infinitely variable speed geared motor *d*. The positioning unit cam drive is shown schematically to the right of the drive shaft and requires two dish cams. The vertical movement is generated by dish cam *e* via oscillatory lever *f* and the horizontal movement by dish cam *g* and the oscillatory lever system *h*.

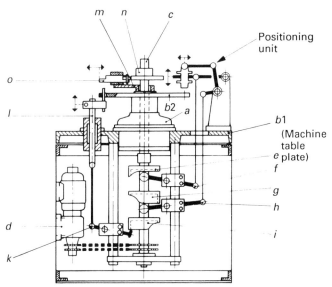

Figure 6.18 Design of a circular cyclic assembly machine (Sortimat). *a* = circular cyclic unit, *b1* = machine table plate, *b2* = workpiece carrier base plate, *c* = drive shaft, *d* = geared motor, *e* = dish cam, *f* = oscillatory lever, *g* = dish cam, *h* = oscillatory lever system, *i* = dish cam, *k* = oscillatory lever system, *l* = ejector device, *m* = stationary plate, *n* = cam plate, *o* = slider

The diagram to the left of the drive shaft shows the ejector device operated by dish cam *i* which ejects completed assemblies from the workpiece carrier by a vertical movement. The ejector device *l* is operated via dish cam *i* and oscillatory lever system *k*. The arrangement of the drive shaft *c* through the circular cyclic unit *a* and the workpiece carrier base plate *b2* makes the direct performance of movement possible above this plate. Mounted cam plates can operate oscillatory levers or sliders which are mounted on a stationary plate *m* positioned above the centre. In this example, the cam plate *n* moves slider *o* in a horizontal direction.

The use of a vertical drive shaft for driving horizontal secondary drive shafts is shown in Figure 6.19. Several secondary shafts can be driven synchronously to the main vertical drive shaft at a transmission ratio of 1:1 by bevel wheels. Cam plates are mounted on the secondary drive shafts for operation of the axes of the individual units by oscillatory lever mechanisms. By the possibility of arranging the secondary drive shafts in the lower area of the machine housing, dish cams can be fitted to the vertical

198 Design of assembly machines

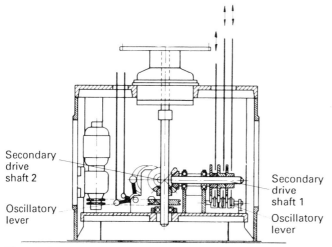

Figure 6.19 Arrangement of secondary drive shafts (Sortimat)

drive shaft in the space extending up to the machine table for the performance of other movements.

Figure 6.20 shows a further example for the performance of the main movements with a vertical drive shaft arrangement by cam plates or direct drive. The drive motor *a* and the toothed V-belt *b* drive the drive shaft *c* and thus the circular cyclic unit with the attached workpiece carriers is operated. With the application of cam plates, the movement generated must be transmitted by a suitable device. The oscillatory lever *f* which is mounted on a vertical shaft *e* is operated by cam plate *d*.

The pivot plate *h* is generated by the link *g* and the horizontal insertion of the linkage *g* transferred into a vertical motion by a link *i* for the operation of a vertically functioning unit *j*. A mechanically cam-operated positioning unit is driven by a second toothed V-belt *k* from the drive shaft.

The drive shaft of the positioning unit *m* is driven by a bevel wheel pair *l*, and the horizontal and vertical movements of the positioning unit are generated by the cam pair *o* and *p*.

6.3.4.2 Assembly machines with a horizontally arranged drive shaft

The use of circular cyclic units with a cam drive necessitates a horizontal arrangement of the drive shaft. This can be disadvantageous with the application of circular cyclic machines because not all the working space around the circular indexing table can be power-operated on account of the position of the drive shaft. If this is necessary, the arrangement of the secondary shafts which run synchronously to the main drive shaft can provide the required local accessibility to the units.

Figure 6.21 shows schematically the design of a circular cyclic machine with a cam-driven circular cyclic unit and also the arrangement of two secondary drive shafts.

The cam operation of the positioning unit by a horizontal main and secondary drive shaft is shown schematically in Figure 6.22. The mode of operation is as follows. The secondary drive shaft is driven by a chain drive *c* from the drive shaft *b* of the circular indexing unit *a*. The cam plate *e* rigidly connected to the secondary drive shaft *d* generates the vertical movement of the positioning unit *h* via the oscillatory lever mechanism *f* and link *g*. The horizontal movement of the positioning unit is generated

Multi-station assembly machines 199

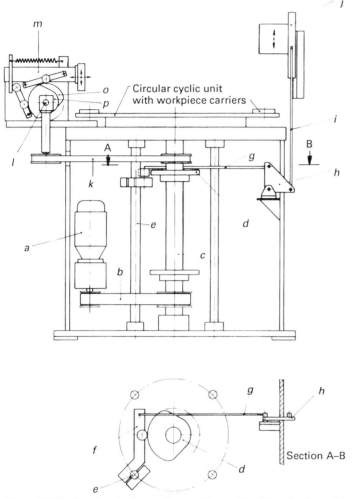

Figure 6.20 Design of a circular cyclic assembly machine (Menziken). a=drive motor, b=toothed V-belt, c=drive shaft, d=cam plate, e=shaft, f=oscillatory lever, g=link, h=pivot plate, i=link, j=unit, k=toothed V-belt, l=bevel wheel pair, m=positioning unit, o, p=cam sets

by the cam plate i on the drive shaft b. In this arrangement, the oscillatory lever j engages by a pin in a slot in the positioning unit so that the vertical movement is not obstructed.

The secondary drive shaft d is used as the oscillatory lever mounting, offset of the oscillatory lever design related to the cam plate i in relation to the contact point in the handling unit h being achieved by the sleeve-like design of the pivot bearing k.

The horizontal drive shaft arrangement is advantageous in longitudinal transfer systems. Figure 6.23 shows a schematic arrangement of a main and secondary drive shaft in a rotary arrangement of a longitudinal transfer unit.

The drive shafts run parallel to the direction of motion of the transfer unit so that the principal movements can be undertaken simply and coordinated to the point of work.

200 Design of assembly machines

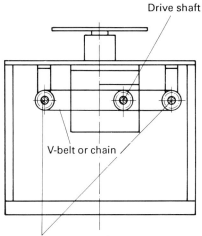

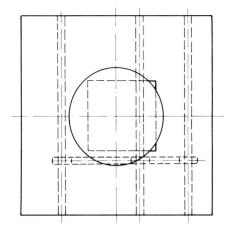

Figure 6.21 Design of a circular cyclic assembly machine with a horizontal main drive shaft and secondary drive shafts

Further axes must be provided parallel to the drive shafts for mounting of the oscillatory lever systems.

The mode of operation is as follows:

Vertical lift movements
For the performance of vertical movements, a plate cam on drive shaft a operates the oscillatory lever mechanism d which is mounted on shaft c.

Horizontal movements
A plate cam on secondary drive shaft b operates the oscillatory lever mechanism f which is mounted on shaft e; the vertical motion of lever g is transformed into a horizontal movement by a second oscillatory lever mechanism h.

6.3.4.3 Performance of several simultaneous principal movements by an oscillatory drive
Several principal movements to be performed simultaneously on assembly machines can be achieved with the application of drive elements which move to and fro

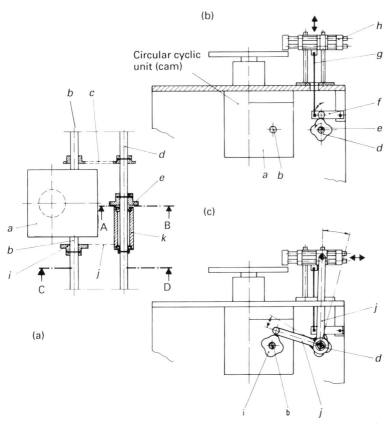

Figure 6.22 Cam drive of a positioning unit with a horizontally arranged main and secondary drive shaft: (a) plan view; (b) section A–B and (c) section C–D

(oscillating elements). Oscillatory drives can be in the form of oscillating plates on circular cyclic machines or oscillatory shafts in particular on longitudinal transfer machines.

Figure 6.24 shows for purposes of example the design arrangement of an oscillatory drive in the form of an oscillating plate positioned centrally beneath the circular indexing table. The oscillating plate is driven by a cam fitted on the drive shaft via an oscillating lever and a link. The forward stroke of the oscillating plate is automatically generated by the cam plate, and the return stroke by a pneumatic cylinder acting as a spring and which acts on the oscillatory lever during the return stroke of the oscillating plate in the form of an opposing force and maintains contact between the cam-follower roller and cam. The oscillating angle movement α is generated depending upon the cam plate form and the mechanical advantage of the oscillatory lever. Principal movements in the direction of the circular indexing table can be derived simply by the arrangement of the oscillatory plate beneath the circular indexing table. For example, an arrangement is shown in the lower half of Figure 6.24 in which an oscillating lever is operated by a cam on the oscillatory plate in order to achieve a swing-in type motion. A second example in the lower half of Figure 6.24 shows a similar drive by a carrier pin in the oscillatory plate which engages in a slot in the oscillatory lever.

202   Design of assembly machines

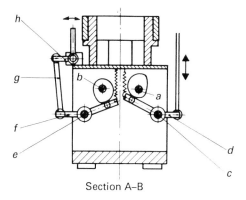

Section A–B

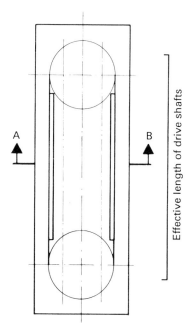

Figure 6.23 Horizontal arrangement of main and secondary drive shafts in longitudinal transfer systems

With many movement sequences on assembly machines two vertical movements must be made for one horizontal movement, which requires two separately driven oscillatory drives. Figure 6.25 shows a possible solution in the form of an oscillatory plate and a separately driven oscillatory ring. As shown by the example in Figure 6.24, the oscillatory plate is mounted concentrically beneath the circular indexing table. The oscillatory ring is mounted in the same plane as the oscillatory plate and also concentrically to it. It is mounted on several bearing blocks spaced around its circumference. Two cam plates mounted on the drive shaft operate the oscillatory plate and ring by an oscillatory lever mechanism and link. The oscillating plate performs one oscillatory movement during one working cycle and the oscillating ring two.

Multi-station assembly machines 203

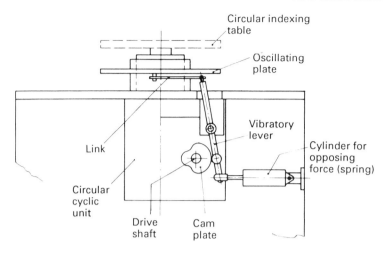

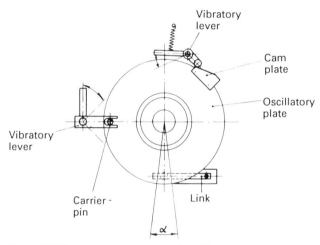

Figure 6.24 Design of an oscillatory drive with a circular plate

The oscillatory motion of the oscillating plate through angle α and also that of the oscillating ring through angle β are dependent upon the cam profiles and the mechanical advantages of the oscillating lever arrangement; they can be selected independently of one other in relation to both magnitude and phase angle.

As an application example, Figure 6.26 shows schematically the design arrangement of handling equipment with X-axis rotary and Z-axis translatory motion and which is driven by oscillatory drives.

Figure 6.27 shows the related time-displacement diagram and the sequence of motion. The oscillatory lever c is swivel-mounted on the housing l. The carrier pin b of the oscillatory drive a engages in a guide slot in the oscillatory lever c.

Its rotary motion is transmitted to the carrier lever e via the guide column d which is rigidly connected to the lift column j of the positioning unit.

204 Design of assembly machines

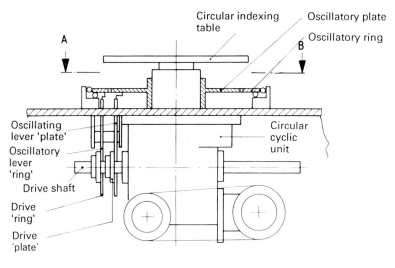

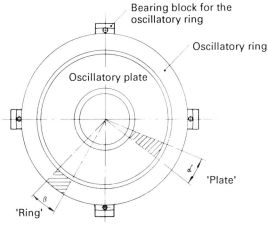

Section A–B

Figure 6.25 Drive arrangement for an oscillatory plate and ring

The arm k of the handling unit is mounted on the lift column. The oscillatory plate moves the oscillatory lever c; the lever moves lever e which is mounted on the lift column via the guide column so that the arm k performs the required swing motion.

The required vertical movements for the Z-displacements of the handling unit are generated by the oscillatory plate f. The centre shaft is moved by cam i by lever h via lever g, which is connected to the oscillatory plate. The cam raises and lowers the lift column j of the handling unit. At the same time, lifting is effected by the cam against the helical compression spring n and lowering of the lift column by an integral helical compression spring.

The same results can be obtained with longitudinal transfer units with oscillatory shafts. These are arranged parallel to the drive shaft and transfer unit. Figure 6.28 shows schematically the arrangement of an oscillatory shaft and the possible sequence of movement.

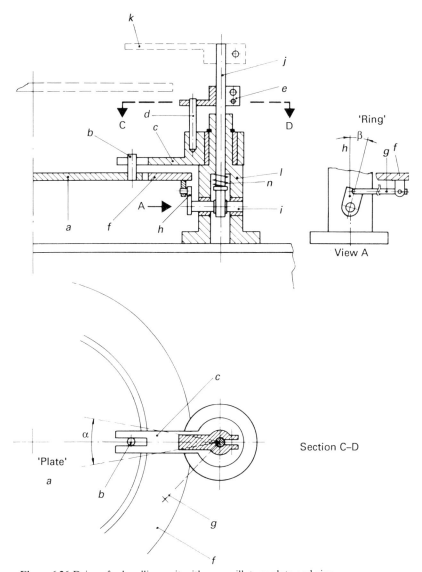

Figure 6.26 Drive of a handling unit with an oscillatory plate and ring

The oscillation angle α is dependent upon the design of the cam plate. Various displacements can be achieved by the lever length X.

6.3.5 Assembly machine systems

The unit construction principle is a design principle which represents the construction of a limited or unlimited number of different systems constructed from a range of standardized modules based on a programme or design specimen plan in a particular field of application [27]. A module is an assembly standardized in terms of its connection dimensions, design features and characteristics and which can be combined

206 Design of assembly machines

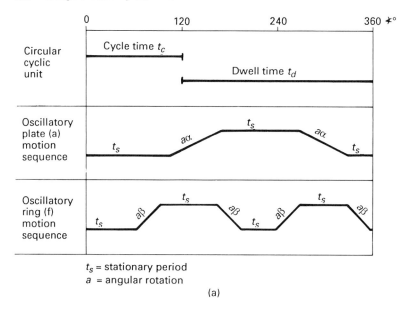

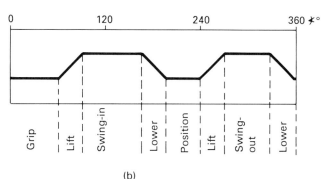

Figure 6.27 Time-displacement and movement sequence of a positioning unit as shown in Figure 6.26: (a) time-displacement diagram; (b) movement sequences

with other modules in different ways. Until recently this definition was only applicable to pneumatically operated modules. During the past few years, however, industry has also developed similar unit construction solutions for electric-motor-operated high-performance assembly machines which have resulted in a considerable reduction in the planning and design costs of automatic assembly systems.

Figure 6.29 shows a section of the basic module of an electric-motor-driven assembly machine. The basic element for the design of such circular cyclic machines are circular cyclic units which are standardized for 8, 12 and 16 stations. All necessary modules such as handling equipment, automatic screw inserters, press and inspection units and metering equipment for bonding agents and lubricants, etc. can be directly bolted to the basic module.

Figure 6.30 shows a schematic system and design of circular cyclic machines. The central drive elements are mounted on a machine column; these are the Maltese drive

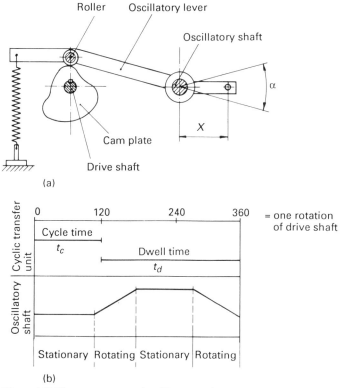

Figure 6.28 The arrangement of oscillatory shafts in longitudinal transfer systems: (a) schematic arrangement; (b) sequence of movement

for the circular indexing table, the bevel wheel drive for the drive shafts of the modules and the drive for the central lift shaft. The machine frame circumference is in the form of a polygon and, depending upon pitch number, has between 8 and 16 flanges for attaching the modules. The modules are attached depending upon whether or not they have their own drive or are driven from the central unit with different mounting housings.

The attachment housing is in the form of a bracket for a unit with its own drive, e.g. screw inserters or presses. With units which must be driven centrally, the attachment housing is a gearbox fitted with a drive shaft and bevel wheel for engaging in the central bevel wheel. Figure 6.30 shows schematically such cam-driven units. The vertical and horizontal movements of the carrier plate for the assembly tools such as gripper tongs, probes, punches, screw-insertion spindle, and metering equipment are generated by two plate cams and oscillatory lever systems with a link and in which the horizontal movement of the carrier plate for the assembly tools must be derived from a vertical movement. The solution in this respect is shown in Figure 6.31.

An additional oscillatory lever converts the vertical into a horizontal movement.

A vertical movement is made every cycle by the central lift shaft (see Figure 6.30). If a plate is fitted to the central lift shaft, this could, for example, accommodate an inspection station so that all the inspection procedures of the assembly machine could be operated by the central lift shaft.

208 Design of assembly machines

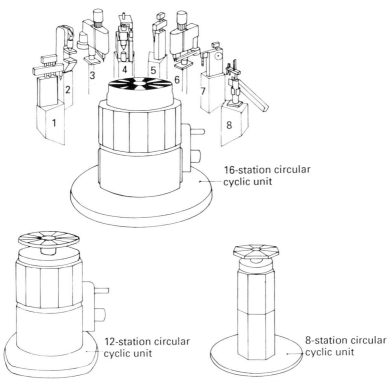

Figure 6.29 Basic modules for circular cyclic machines (OKU)

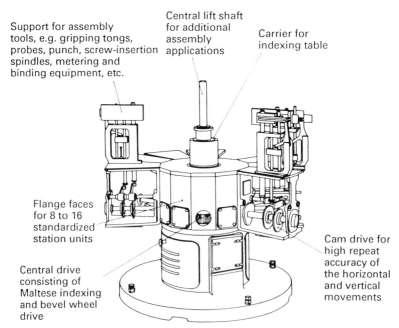

Figure 6.30 Design of circular cyclic machines (OKU)

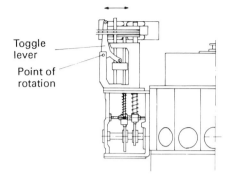

Figure 6.31 Transformation from a vertical to a horizontal movement (OKU)

On account of synchronization of all main drive shafts by a suitable design of the cam plate, these machines are suitable for a sinusoidal pattern of movement during which very short cycle times of one second and less can be achieved.

A further machine system is shown in Figure 6.32. The basic conception is that all necessary main movements are made centrally and via lift movements, and also beneath the circular indexing table. This has the advantage that handling equipment and inspection stations, etc. can be arranged from the centre above the circular indexing unit and the space on its circumference fully utilized for feed equipment such as vibratory spiral conveyors with discharge rails. For purposes of clarity, the time-displacement diagram has been added to the schematic diagram of the machine construction as shown in Figure 6.32.

The design and mode of operation of this assembly machine is as follows. The centre of the machine forms a cylindrical, tower-type structure. It contains the drive of the circular indexing table a and the dish cam drives of the lift plates b, c and d via a vertically arranged drive shaft.

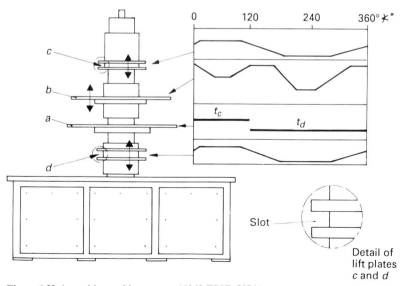

Figure 6.32 Assembly machine system (AMI-ERIE, USA)

Lift plate *b* makes two vertical movements of different magnitudes per working cycle. Lift plate pair *c* and *d* each make one movement per working cycle. All necessary items of equipment such as handling equipment and inspection stations are mounted as lift plate *b*. The vertical movement on assembly presses is performed by this arrangement via lift plate *b*. Movements beneath the circular indexing table arrangement or horizontally from the outside are generated by lift plate *d*. The lift plate has a circular groove in which the oscillatory lever can engage in order to convert the lift movements into vertical and horizontal movements by the oscillatory lever system. Lift plate *c* also has a circular groove in which lever linkages of units mounted on lift plate *b* can engage in order to convert the vertical lift movement into a horizontal movement.

Figure 6.33 shows handling equipment mounted on a lift plate *b* and operated by lift plate *c*. The method of operation is as follows. A positioning unit with its horizontal slide *e* is mounted on the lift plate. Diagram (a) shows the situation of the positioning unit when gripping a part from the gripping position. Lift plate *b* has made the shorter lift movement in order to bring the positioning unit into the gripping position. Lift plate *c* makes the same movement.

Figure (b) shows the position of the unit after grasping and raising of the lift plate *b*. Lift plate *c* makes the same movement. At the same time, the positioning unit with its gripper is moved over the workpiece carrier of the circular indexing table *a* in a horizontal direction towards the machine centre by the oscillating lever mechanism of deflection station *f*. Figure (c) shows the situation during assembly and placing the workpiece in the workpiece carrier. In this respect, lift plates *b* and *c* make a vertical movement downwards. This design permits the formation of handling and assembly operations with the shortest handling movements and also cycle times in the order of one second.

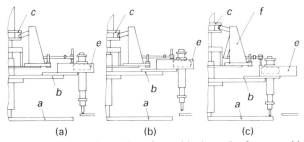

Figure 6.33 Mode of operation of a positioning unit of an assembly system as shown in Figure 6.32 (AMI-ERIE, USA): (a) grasping; (b) lifting; (c) assembly

6.4 Combining assembly machines to form assembly lines

The combining of assembly machines is necessary:

- If the complexity of a product or parts assembly is so high that the operation can no longer be undertaken by one unit, or
- If for production reasons, operational distribution over several machines is necessary.

The number of operations which can be undertaken on one machine is dependent upon the size of the machine (number of stations) and the required production capacity. The availability decreases with an increasing number of working stations and shorter cycle times, which is a further reason for the distribution of operations over several machines. Other reasons can exist for the fact that different assembly directions or workpiece carriers make different designs of assembly machines necessary. A differential is made between rigid and loose combinations.

A rigid combination is achieved if the workpieces or workpiece carriers are directly transferred from one machine to another by a handling unit. As far as possible, rigid combinations should be avoided.

With loose combinations, the interconnection of transport systems is necessary. The transport system provides a decoupling between the machines and with suitable dimensioning a buffer formation. Decoupling and buffer functions have the important advantage that, depending upon the size of the buffer, faults on individual machines do not result in immediate shut-down of the complete line.

Stoppages on assembly lines can be segregated into short- and long-term stoppages. Short-term stoppages are, for example, caused by deformed parts in the feed equipment. Their duration is of the order of a few seconds to a minute. Long-term stoppages are, as a general rule, caused by the equipment in the mechanical or control mechanisms and can last up to several hours. Lengthy operating experience with such systems has shown that hundreds of short-duration stoppages occur for one long-term stoppage. Depending upon the fault signalling and deployment of trained maintenance personnel, short-duration stoppages can be rectified within 1 min, which explains the 2 min buffer capacity as previously mentioned.

The type of transport system is determined by the following criteria:

- Can the product or assembly be transported so that it does not lose its arranged state without the assistance of a workpiece carrier, or
- Must the assembly or product be transported in a workpiece carrier in order to retain its arranged state?

Figure 6.34 shows two examples. Figure (a) shows an assembly with a rectangular base housing. This assembly can be transported in a workpiece arrangement without a workpiece carrier without losing its arranged state. Figure (b) shows a spindle assembly which would lose its arranged state without a workpiece carrier so that rearrangement would also be necessary on the following machine. The rearrangement of pre-assembled assemblies in terms of form and accuracy requirements is not always possible and should fundamentally be avoided since rearrangement incurs additional cost and possible fault sources for the work sequence.

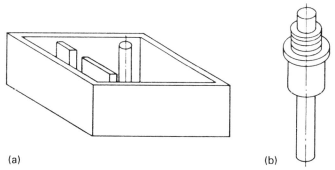

Figure 6.34 Assembly examples for products with different transport characteristics: (a) product with a base housing; (b) spindle assembly

212 Design of assembly machines

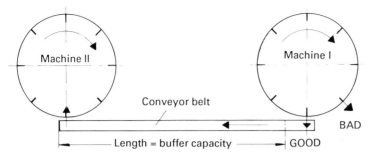

Figure 6.35 Combining of two circular cyclic assembly machines by a conveyor belt

Figure 6.35 shows a method for combining two automatic circular assembly machines for the transport of an assembly whose category is shown by Figure 6.34(a). Conveyor belts or electromagnetically driven rails (conveyor channels) can be used as transport systems.

If workpiece carriers are necessary during transport for maintenance of the ordered position they can be of a simple design. They have no function in assembly operations but simply hold the workpiece or assembly so that the ordered position is not lost. Figure 6.36 shows one such simple workpiece carrier suitable for the assembly as shown in Figure 6.34(b). The use of workpiece carriers necessitates transport systems with a forward and return arrangement.

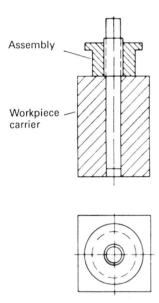

Figure 6.36 Workpiece carrier for the transport of an assembly

Figure 6.37 shows schematically one such arrangement for the combination of a circular cyclic assembly machine with a longitudinal cyclic machine. The mode of operation is as follows.

The pre-assembled assembly on machine I is placed at position a in workpiece carrier I by a handling unit. The workpiece carrier is moved from this point by a pneumatic cylinder on to a conveyor belt and transported by machine II. The buffer capacity is determined by the length of this conveyor belt. A handling unit of machine

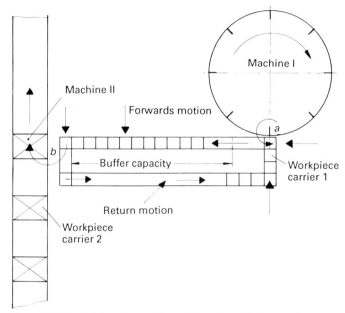

Figure 6.37 Combining of assembly machines by workpiece carrier systems

II removes the assembly from the workpiece carrier at point *b* at the end of the transport system and places it in workpiece carrier II of machine 2. The now empty workpiece carrier I is pushed transversely on to the transport system return belt by a pneumatic cylinder so that it can be returned to the loading station at the end point by a transverse movement.

The design of the transport systems is also determined by the cycle time, size of the assembled part and the required buffer capacity. With an assumed cycle time of 3 s and a buffer capacity of 2 min it must be possible to hold 40 workpieces in the transport system. With an assumed workpiece or workpiece carrier length of 40 mm this gives a conveyor-belt length of 1600 mm. This length is feasible for the combinations as shown in Figure 6.35 and Figure 6.37. With smaller parts, such sections can be shortened or else the buffer capacity increased. This is preferable if permitted by the availability of space. Longer cycle times permit a shortening of the transport systems with a constant buffer capacity or an increase in the buffer capacity with a constant transport system length.

With larger assembled parts or workpiece carriers, e.g. with a side length of 150 mm, on account of space considerations the required buffer capacity cannot be realized by the conveyor-belt length alone, so the interconnection of tiered buffers is necessary. Only very few products can be stacked within each other, so the use of stackable workpiece carriers is necessary. To increase the availability of an assembly line, it is not only necessary to buffer the workpiece carriers with the parts to be assembled on the feed belt, but also to buffer the empty workpiece carriers in the return conveyor belts. This ensures that, with a short-duration fault on the next machine, during which no workpieces can be removed from the transport system, the preceding machine can no longer operate because of the absence of empty workpiece carriers. Figure 6.38 shows one such workpiece buffer. Five workpiece carriers in a block on the conveyor belt are

214   Design of assembly machines

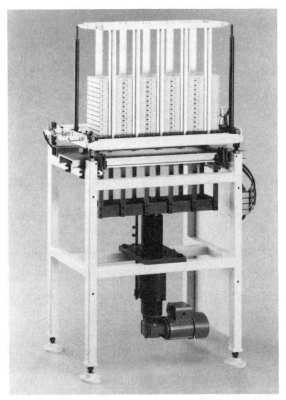

Figure 6.38 Workpiece carrier multi-tier buffer (Montech)

buffered in one tier and in the event of a deficient feed are removed from the tiers. Figure 6.39 shows a non-rigidly combined assembly system with three circular cyclic machines, with which the feed and return belts between the machines and workpiece carriers are integrated as shown in Figure 6.38.

6.5 Integration of manual work points in automated assembly lines

The integration of manual work points is necessary if:

- On account of their design or sensitivity the parts cannot be arranged and fed automatically and these operations must be undertaken manually, or
- If highly complex assembly operations are involved such that they cannot be economically performed automatically.

Different methods of construction for the integration of manual work points must be provided for the above-mentioned differences.

6.5.1 Manual work points for parts provision

On assembly machines with cycle times shorter than 5 s, the manual arrangement and provision of parts results in monotonous activity in pace with the assembly machine.

Integration of manual work points in automated assembly lines 215

Figure 6.39 Combination of three circular cyclic automatic assembly machines by workpiece carrier conveyor belts and tiered buffers (Montech)

Work science experience shows that this is no longer acceptable to the operatives. In order to achieve a limited decoupling from the fixed cycle, suitable items of equipment must be located upstream of the automatic stations.

Figure 6.40 shows three different solutions schematically. As shown in diagram (a), a second indexing table with the maximum possible number of indexing positions is connected in front of a circular indexing assembly machine at the manual work point. The indexing table functions in sequence with the circular indexing unit of the assembly machine. The part to be handled is placed manually by the operative in the indexing table workpiece holders. The part is removed from the indexing table by a positioning unit of the assembly machine and placed in the workpiece holder. The arrangement of a number of hoppers in this pre-positioned indexing table provides a larger working area for positioning the parts and also gives a limited decoupling effect from the cycle of the assembly machine. It gives a low individual freedom in the work rhythm since a certain buffer effect is formed between manual placing and automatic removal by the large number of workpiece carriers.

If the part for manual handling is designed such that it can be transported to a removal point with a positioning unit by a conveyor belt without losing its arranged position, decoupling from the cycle is provided by a conveyor belt arranged in front of the assembly machine. In this case also, the operative is given a certain degree of individual freedom by the buffer effect. Diagram (b) shows one such solution and the buffer effect which can be achieved between manual positioning and automatic removal.

A further possible solution is shown by diagram (c). With this solution stackable parts are arranged in a duct magazine by the operative. In order to achieve a buffer effect, two magazines are arranged on a rotational unit. A full duct magazine is positioned towards the assembly machine. The parts are removed from the bottom of the magazine in an ejector and placed in the workpiece holder by the positioning unit

216   Design of assembly machines

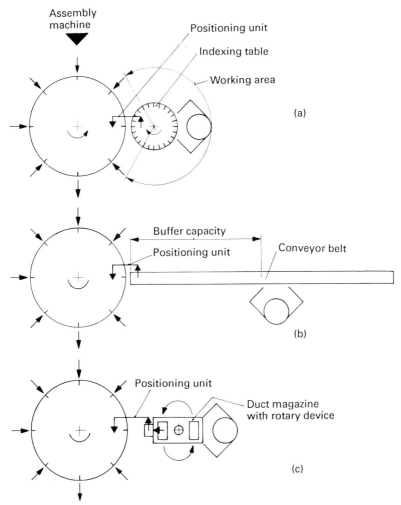

Figure 6.40 Units for the manual provision of parts and for the decoupling of operatives from the cycle of an automatic assembly machine: (a) indexing table; (b) conveyor belt; (c) duct magazine

of the assembly machine. The second duct magazine is available for manual filling. If the first magazine is empty and the second full, the rotary device is activated, the second duct magazine moved into the feed and the first into the filling position.

6.5.2 Manual assembly work points

On account of the level of difficulty of certain assembly operations, e.g. for interlocking or highly bendable parts, they cannot be undertaken at the cycle times of automated stations. This makes the performance of manual assembly operations outside of assembly machines necessary. The manual work points as necessary can be most appropriately integrated in the transport systems for combining the machines. Figure 6.41 shows one such example for workpieces which can be transported by a conveyor

Integration of manual work points in automated assembly lines 217

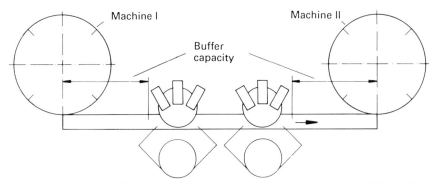

Figure 6.41 Integration of manual work points between two circular cyclic assembly machines

belt without losing their arranged state. Adequately long transport systems should be used in order to provide adequate working space for the manual work points and a sufficiently large buffer capacity. With an assumed cycle time of 3 s and a time of 6 s for manual assembly, two manual work points must be provided. A recent workpiece is therefore processed on each of these two work points. If the order of assembly permits that several manual assembly operations can be combined, this results in an increase in the work content at manual assembly points and, at least to the same degree, relieves the operatives of monotonous work.

Figure 6.42 shows the layout of a solution for such workpieces which can only be transported with the aid of a workpiece carrier. With an assumed assembly machine cycle time of 3 s and an increase in the work content of the manual assembly points to 9 s, three parallel work points must be included in the transport system. A rectangular workpiece carrier support system including three parallel work points fulfils this requirement. With the assumed work content increase to 9 s, the supply of ten workpiece carriers per workpoint means a decoupling of 1.5 min and also an individual performance requirement for each operative during a shift (see Section 3.3.2.3, Figure 3.18, and Section 5.4.2, Figures 5.90 and 5.91).

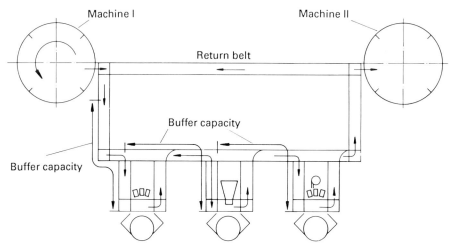

Figure 6.42 Integration of three manual parallel work points with the combination of circular cyclic assembly machines by a double-belt workpiece carrier transport system

218 Design of assembly machines

6.6 Uncycled assembly lines including manual work points

Products with a high work content and also a manual assembly activity can preferably be assembled on uncycled longitudinal transfer assembly lines. This line can be most advantageously constructed using the double-belt system as described in Section 5.4.2 with the application of workpiece carriers. The workpiece carriers are transported between work stations by friction on a double-belt system so that the arrangement both of the automatic and manual work points can be at different distances. A maximum possible distance between work stations permits the movement of workpiece carriers between the stations such that buffers can form independently of the cycle time. The individual automatic stations therefore represent individual units. The working cycle of a station commences with the approach of a workpiece carrier and its positioning. The work station is switched off following completion of the assembly operations and the workpiece carrier released for further transport.

Manual work points can be arranged directly on the double-belt system and particularly so if a specific work content is allotted to each work station. Depending upon the work content, a certain amount of decoupling from the fixed cycle is achieved by the buffer formation to and from the manual assembly stations. If a work content can be created on the manual assembly stations by the combination of several operations and it exceeds the cycle time of the overall installation by a factor of 1:2 or 1:3, the formation of several parallel work points with a high work content can be achieved by the possible distribution of the workpiece carriers (see Section 6.5.2). A particularly effective decoupling from the working cycle of the installation is achieved in this manner. Figure 6.43 shows a scheme for an assembly line realised by a rectangular double-belt arrangement and consisting of seven automatic stations and three manual parallel work points.

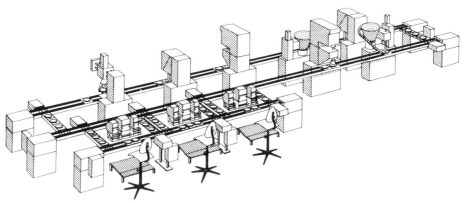

Figure 6.43 An uncycled longitudinal transfer assembly system with 7 automatic stations and 3 manual parallel work points (Bosch)

6.7 Availability of assembly systems

Assembly systems are always complex technical items of equipment which consist of a large number of individual units of different technologies, a fact which, in its entirety will always result in breakdowns. The assembly operations to be undertaken are also affected by the quality of the parts or materials to be processed so that stoppages cannot be totally avoided but can, however, be held within limits by appropriate precautions.

In order to achieve this objective with the minimum possible personnel deployment, in addition to considering the purely functional and technical aspects of the various assembly systems, it is also necessary to investigate their operational characteristics and to achieve the longest possible undisrupted running time by conceptual precautions. A requirement for the efficiency of capital-intensive plants is also achieved in this manner [30].

6.7.1 Parameters of the operational characteristic

The operational characteristic of an assembly plant is subject to periodic changes. In this respect the following parameters must be considered:

- Reliability
- Availability.

Reliability is defined as the functional reliability of an assembly plant. Reliability is the probability that a unit will not fail during a specified period of time under defined functional and environmental conditions. Reliability has a very large effect on the availability of a plant because it determines within a specified period of time over how many cycles the plant will remain functional and during how many cycles faults will occur. The fault-free service life between two faults can be calculated on this basis. Furthermore, the availability is dependent upon the mean duration of fault rectification [30].

Three availability terms are known from the literature: the initial availability, operational availability and the static availability. The static availability is applied to the operational characteristic of assembly plants. Its definition is [31]: 'The static availability A_{STA} (also known as the interval availability) defines the proportion of a service period during which the unit under consideration is functional.'

It is calculated by:

$$A_{STA} = \frac{\bar{T}_0}{\bar{T}_0 + \bar{T}_B} \qquad (6.1)$$

$\bar{T}_0$ = mean fault-free service life
$\bar{T}_B$ = mean breakdown duration

For the purposes of maintenance consideration units, the mean fault-free service life is equivalent to the arithmetic mean of the individual service life periods between two breakdowns of a plant. For repairable units the mean duration of breakdown is equivalent to the arithmetic mean value of the individual periods of breakdown. The breakdown period is the time between the occurrence of a fault and the time at which the repair procedure is completed [30].

6.7.2 Utilization

Quite often availability and utilization are confused with each other. Whilst availability is a purely technical parameter, stoppages resulting from organization measures are also included in the calculation of the utilization, e.g. lack of work, etc. As shown in Figure 6.44, the two parameters therefore have different reference time

220 Design of assembly machines

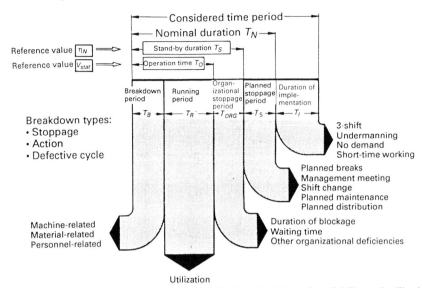

Figure 6.44 Summary of the time components for the calculation of availability and utilization (Wiendahl, Ziersch; IFA, Hanover University)

periods [30]. The utilization U_T is given by the following equation:

$$U_T = \frac{\bar{T}_0}{\bar{T}_0 + \bar{T}_B + \bar{T}_{ORG}} \tag{6.2}$$

where $\bar{T}_0$ = running time during the considered time period
$\bar{T}_B$ = stoppage time during the considered time period
$\bar{T}_{ORG}$ = organizational stoppage time in the considered time period.

Accordingly, the utilization can reach the value of the static (technical) availability.

In the following, the availability is considered in further detail because it is affected greatly by the technical concept of the equipment.

6.7.3 Factors of influence on the availability of assembly systems

The overall availability of an assembly system is principally affected by the following criteria.

6.7.3.1 Number of stations
In the case that the stations in a system are arranged behind each other without a buffer formation (e.g. circular cyclic automatic machines), the overall availability T_{tot} of the system is given as the product of the availability of the individual stations:

$$T_{tot} = T_1 T_2 T_3 \ldots T_n \tag{6.3}$$

Two factors are therefore decisive for the overall reliability:

- The reliability of the individual stations
- The number of stations.

On the one hand, the reliability of the individual stations is determined by the technical design and the reliability of their components and, on the other hand, by the quality of the parts to be processed. With a rigid combination of, for example, eight assembly stations in an assembly system with an individual reliability of 98%, the overall reliability T_{tot} is calculated by $T_{tot} = 0.98^8 = 0.85 = 85\%$. Therefore 15% of the possible cycles can result in breakdowns. This means that, with a rigid combination, the minimum possible number of individual stations should be included in an assembly system. For example, considerably higher availability is achieved by the distribution of 12 work stations in two non-rigidly combined assembly machines than with the integration of all stations into one assembly system.

6.7.3.2 Time required for the rectification of breakdowns
As shown by Equation 6.1, the time required for the rectification of causes of breakdowns is an important factor for the availability. The smaller the total time for individual breakdowns, the higher the availability.

The following precautionary measures are recommended so that, in the event of breakdowns, they can be rectified as quickly as possible by operatives and have no or only a slight effect on the availability:

- Easy local access to the fault areas and rapid accessibility to the fault points (e.g. snap-fit systems for cover strips on feed rails)
- Formation of an intermediate buffer capacity between fault points and the work point (see Sections 6.3.1 and 6.4)
- Optical and acoustic signalling of the fault point in order to minimize the search time by the operative.

Figure 6.45 illustrates how the number of stations and individual reliability of a station affect the availability with a cycle time of 3 s and a mean time of 10 s for the elimination

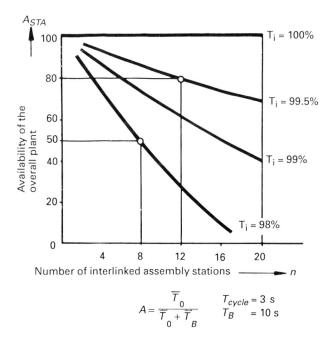

Figure 6.45 Availability of rigidly connected assembly stations in relation to the individual reliabilities of the stations (Warnecke; IPA Stuttgart)

of a fault. It can be assumed that, for example, an assembly system with 12 rigidly combined stations with a single reliability of 99.5% per station and a mean breakdown time of 10 s achieves an availability of approximately 80%.

The calculation is as follows:

$T_{tot} = T_i^{12} = 0.995^{12} = 0.94$
$P_{ACTUAL} = P_{spec} \cdot T_{tot} = 1200 \text{ pc/h} \times 0.94 = 1128 \text{ pc/h} = > 72 \text{ missing cycles/h}$
$T_0 = 3600 \text{ s/h} \cdot 0.94 = 3384 \text{ s/h}$
$T_B = 72 \text{ missing cycles/h} \cdot 10/\text{missing cycle} = 720 \text{ s/h}$

As per (6.1),

$$A_{STA} = \frac{3384 \text{ s/h}}{3384 \text{ s/h} + 720 \text{ s/h}} \times 100\% = 82\%$$

If on the other hand, every single station has a reliability of 99%, an overall availability of only around 65% is achieved. With eight stations and a single reliability of 98% this decreases to 50%.

6.7.3.3 Cycle time

The cycle time of an assembly plant has a marked effect on its availability. The calculated availability of rigidly connected assembly stations for various cycle times is shown in Figure 6.46 for various cycle times in relation to the number of stations and a mean down time of 10 s. It is shown that the availability decreases steeply with diminishing cycle times in spite of an extremely high individual reliability of the individual stations of 99.5%. This is attributable to the fact that the mean down time remains constant, which, in accordance with Equation 6.1, results in a decreasing availability with a decreasing mean fault-free running time. For example, the

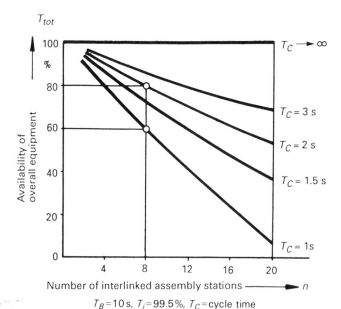

$T_B = 10$ s, $T_i = 99.5\%$, $T_C =$ cycle time

Figure 6.46 Availability of rigidly interlinked assembly stations in relation to the cycle time

availability with an eight-station machine and a cycle time of 1 s is only approximately 60%; it is around 75% with a 2 s cycle time. During the planning stage of automatic assembly machines, detailed investigations should be undertaken to determine and also calculate whether or not a somewhat longer cycle time with a significantly higher availability would result in a higher productivity than optimally short cycle times.

6.7.3.4 Quality of individual parts
The supply quality of the individual parts to be processed on automatic assembly machines has a considerable effect on the availability of such machines. In order to make the assembly of random parts possible taken from a hopper they must be completely interchangeable. Consequently, it is necessary that the normal scatter of the specified dimensions of the individual parts is restricted about the mean value (see Section 2.2.6.1, Figure 2.37). The effect of this restriction is that, in comparison with manual assembly, more stringent acceptance conditions are necessary by the specification of the lowest possible AQL values [1, 2, 10]. Furthermore, the individual parts must be extensively burr-free; deliveries must not contain foreign bodies. To achieve this, a number of organizational and technical measures are necessary in the pre-production area. The additional costs incurred in the production of the parts must be taken into account with an economic consideration of the automation of an assembly process.

The effect which the part quality level has on the production rate of an automatic assembly plant is shown by Figure 6.47.

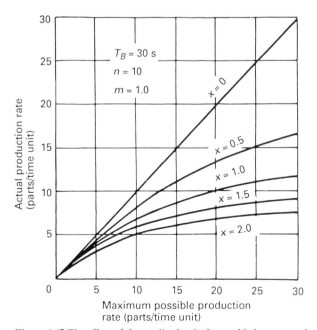

Figure 6.47 The effect of the quality level of assembled parts on the production rate of an assembly system (Boothroyd) (x=ratio of defective:correct parts, m=number of defective parts which cause equipment faults, n=number of automatic work stations)

6.7.4 Summary

Figure 6.48 shows the reference points for an availability improvement of assembly systems; these are a summary of the subjects as discussed in the individual Sections, such as assembly-oriented product design, assembly-extended ABC analysis, modules for assembly automation, and assembly processes and plant design, with regard to their effect on the availability of assembly plants.

Various strategies for an increase in availability are detailed in Section 11.10. Procedures and analytical methods for increasing the availability of automatic assembly plants are discussed in further detail in Chapter 13, 'The operation of automated assembly systems'.

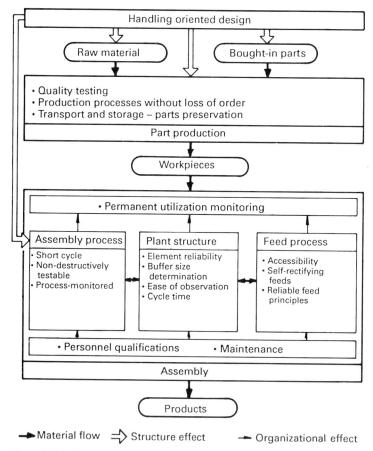

Figure 6.48 Reference points for availability improvement of assembly systems (Wiendahl; Hanover University)

Chapter 7
Design of flexible-assembly systems

7.1 Introduction

Strictly speaking, only manual assembly is flexible. A person performs work with his hands on the basis of his skill and also his sensory organs and intelligence with the application of tools and assembly processes. Automated assembly can only simulate these characteristics to a limited degree. The human hand and arm have more than 50 degrees of freedom. In a similar application, robots have four to six degrees of freedom.

Up to the present, flexible-assembly systems have only been used to a limited degree; they are, however, gaining in importance in the medium production area, which today is still exclusively undertaken manually.

A flexible-assembly system implies an assembly device in which different product variants of a particular product can be assembled in an arbitrary manner. This is principally achieved by the integration of programmable handling, joining and testing equipment. Flexible-assembly systems are therefore in competition with manual assembly. Figure 7.1 shows the percentage distribution of the most common assembly functions in precision and electrical engineering.

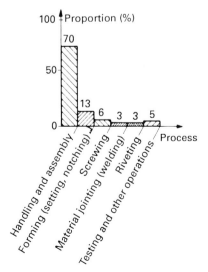

Figure 7.1 Percentage distributions of individual functions of assembly processes in precision and electrical engineering

The proportion of approximately 70% for handling and assembly is split into approximately half for actual handling and half for provision and ordering of the individual parts. Since handling can only be undertaken using flexible equipment, e.g. assembly robots, of the total requirement only around 35% of the necessary assembly processes are flexible by using programmable equipment. On the other hand, the other processes are not flexible since they must be undertaken using rigid type-related equipment.

Flexible-assembly systems must have the following characteristics:

- Complete assembly of a product family. Depending on the order details, the variants of these product families must be assembled in an arbitrary order by programming.
- It must be possible to retool flexible-assembly systems from one product to another, in the ideal case between two cycles.
- The standard modules used on a flexible-assembly system must, to a large degree, be able to be reused for other assembly operations.

Within the context of space cost estimations, the reusability of the standardized modules makes possible a subdivision into standardized components, whose complete period of utilization is fixed, and those whose period of utilization terminates with the planned production time of the product since, in this case, the modules are product-related [3, 12, 34].

In addition to the product complexity and production volume, the product design and peripheral costs are important in the design of flexible-assembly systems. By reason of their design, only a few products are suitable for fully automatic assembly using flexible-assembly systems. The integration of manual assembly points is necessary with a large number of products.

In order to achieve an optimum combination of manual and automatic activities, the manual work points must be incorporated into the assembly systems in accordance with scientific principles of work. The peripheral costs for the application of an assembly robot increase with an increasing number of single parts to be handled and also the required level of automation. Optimization calculations relating to the level of automation are therefore a fundamental requirement for the economic application of flexible-assembly systems.

With the application of an assembly robot, the arrangement of the working space of a flexible-assembly system is just as important as the arrangement of a manual assembly point. It has proved to be very helpful to analyse and optimize the operational sequence of an assembly robot by a detailed analysis of primary and secondary assembly operations (see Sections 4.2 and 4.3.2).

7.2 Primary–secondary fine analysis with the application of assembly robots

The five basic movements in accordance with MTM (see Figure 4.5) and their movement sequence are also fully applicable to automatic assembly with programmable handling equipment. Fine analysis for primary and secondary assembly (PAP and SAP) is applied for the selection of automatic assembly processes.

In contrast to manual assembly, in this case the requirement applies that the individual parts to be handled must be fed to the robot in an arranged manner.

Primary–secondary fine analysis with the application of assembly robots 227

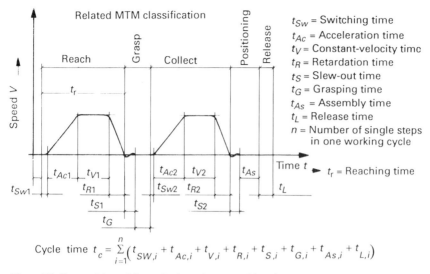

Figure 7.2 Composition of the cycle time of an assembly robot

The cycle time of the assembly robot is vital in relation to whether or not an economic method of working commensurate with the capital investment is achieved. Figure 7.2 shows the sequence of individual steps for one cycle of an assembly robot during which a part is moved from position 1 to a position 2, assembled at that point; the relationship to the basic movements in accordance with MTM is also shown.

It is shown that eight different types of time occur during the five basic movements, reach, grasp, collect, assemble, release, of which the cycle time t_C is composed:

$$\text{Cycle time } t_C = \sum_{i=1}^{n} (t_{SW,i} + t_{AC,i} + t_{V,i} + t_{R,i} + t_{S,i} + t_{G,i} + t_{AS,i} + t_{L,i}) \tag{7.1}$$

The cycle time components can be separated into affectable and non-affectable time sections. The switching time t_{SW}, the acceleration time t_{AC} and the retardation time t_R are fixed values depending upon the type of robot. The constant-velocity time t_V is dependent upon the length of the distance to be travelled. The slew-out time t_S is principally dependent upon the magnitude of the mass to be handled (part and gripper). The gripping time t_G is dependent upon the degree of correct positioning of the parts. The assembly time t_{AS} is dependent upon the product design, point of assembly, assembly direction, assembly space availability and assembly tolerance. The same basic principles as applicable to fine analysis for manual assembly also apply with the application of a primary–secondary fine analysis for assembly robots, i.e. that the necessary minimum requirement for movement operations can be considered as being a so-called primary assembly operation, any other requirement being a secondary assembly operation.

The limits for the minimum requirements which are to be determined are variable and predominantly dependent upon the product size and design.

7.2.1 Reaching

Reaching is defined as the movement which the robot's gripper must make to grasp a new part after releasing a part from the previous assembly position. The magnitude of

228 Design of flexible-assembly systems

the whole movement sequence is dependent upon whether or not the shortest path can be traversed from the point of release to gripping; the sequence comprises horizontal and vertical movements. As shown in Figure 7.2 the time t_r for a basic movement 'reach' is resolved into the following steps by a primary and secondary time analysis:

$$t_r = t_{SW1} + t_{AC1} + t_{V1} + t_{R1}$$

where t_{SW} = switching time
t_{AC} = acceleration time
t_V = constant-velocity time
t_R = retardation time.

With a given robot design, only the constant-velocity time t_V in the time requirement for 'reaching', t_r, can be affected by the design of the track section and work point arrangement. The time requirement for the switching acceleration and retardation time are fixed values related to the robot design. Depending upon the robot type and also the arrangement of very short distances between gripping and assembly, the constant-velocity time is never reached. On the other hand, with long distances, the constant-velocity time is important.

The requirement for the primary operation is determined by the limitation of the distance requirement between gripping and assembly or similarly by the definition as given in Figure 4.6. Any other requirement is of a secondary nature [13].

7.2.2 Gripping

The time requirement for the basic movement 'gripping' as shown in Figure 7.2 is comprised of the slew-out time t_{S1} and the gripping time t_G. The slew-out time is basically a secondary operation. The direct gripping time, i.e. the closing of the gripper in order to bring a part under control is, on the other hand, a primary operation.

All necessary movements of the gripper before actually gripping, such as rotation of the gripper about its axis, positioning of the gripper finger to a part-related distance, or slewing movements during revolving gripping, are, in so far as they cannot be undertaken as parallel movements during the basic movement 'reaching', secondary operations. If a gripper change is necessary, the whole of the movement sequence as necessary is a secondary operation [13].

7.2.3 Collection

Collection is the basic movement by which, and after gripping, the part is brought into the assembly area.

As in the basic movement 'reaching', this sequence of movements contains the same elements, though in the reverse order, and is also affected by the track system.

The parameters relating to the basic movement 'reach' are also applicable for the determination or correlation of the primary–secondary requirement for the basic movement 'collect' (see Section 7.2.1).

7.2.4 Assembly

The requirement for the basic movement 'assemble' as shown in Figure 7.2 is comprised of the slew-out time t_{S2} and the assembly time t_{AS}. Slewing out is a secondary requirement; assembly is a primary requirement.

If auxiliary tools such as screwdrivers, press-in tools, etc. are required and if they can be manipulated by the assembly robot, the specified limits of the basic movements are applicable to the 'reaching' or 'collection' movements for the tool, i.e. the movements required to deliver and return the tool which exceed specified limits are classified as a secondary requirement. This rule also applies if the assembled unit must be moved from the assembly point to an auxiliary tool [13].

7.2.5 Release

Of all the basic movements, 'release' is the one with the minimum requirement. For this reason, 'release' is classified under the term 'primary operation'. If, however, 'release' is followed by a dwell time, the dwell time is classified as a secondary operation.

7.3 Working space

The cycle time of an assembly robot is principally determined by the length of the reaching and moving distances between the positions of the parts or tools to be moved to the assembly or machining position. Figure 7.3 shows that with the provision of the individual parts to be handled by feed rails (gravity or electromagnetically operated) a constant gripping position is achieved. If, on the other hand, the parts to be handled are provided on pallettes, the reaching and moving distances vary with increasing emptying of the pallettes [34].

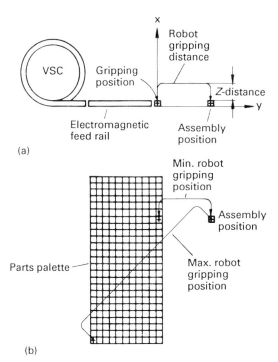

Figure 7.3 The effect of the method of parts provision on the cycle time of a robot: (a) parts provision with a feed device (constant gripping position); (b) parts provision by a pallette (variable gripping position)

7.4 Gripper

Gripper changing should be avoided as far as possible because gripper change times are a purely secondary requirement and therefore have a negative effect on the achievable efficiency. If different numbers of parts have to be gripped, this can be achieved by multiple grippers depending on the assembly positions. Figure 7.4 shows an example in this respect. This is a parallel gripper which can basically cover a parallel gripping distance of 0 to 10 mm.

Gripping fingers are arranged on the basic gripper, by means of which three different dimensional ranges can be gripped. The gripper fingers arranged in the centre function within a range of 0.5 mm to 10.5 mm so that, for example, parts with 0.8 mm, 5 mm and 10 mm gripping widths can be held. A second gripping tong arranged on the side functions within a range of 45 mm to 55 mm and, for example, grips a workpiece with a gripping width of 50 mm. In this example, the third gripper arrangement opposite is for a gripping range of 10 mm to 20 mm and is used for gripping a tool with a gripping area of 15 mm [35].

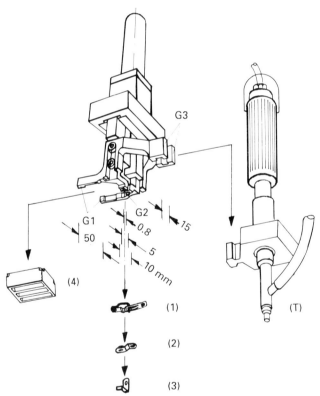

Figure 7.4 Multi-purpose gripper for handling four different workpieces and a tool (1–4 = workpieces, T = screw-insertion tool, G1–G3 = gripper jaw pairs)

7.5 The design of flexible single-station assembly cells

Experience shows that, in the design of flexible-assembly systems, the costs for the basic equipment such as assembly robots, controls, working table and safety equipment are only a part of the necessary investment required to make a flexible-assembly system ready for production.

The so-called peripheral costs consisting of the feed elements required to present the individual parts in the correct position, the assembly fixtures, transport systems, gripper systems, etc. can on the other hand and depending upon the level of complexity and automation, exceed basic costs several times over. In many cases, semi-automation is a more economic situation than full automation. Semi-automation implies the inclusion of manual activities in a specifically defined production unit.

7.5.1 Semi-automatic flexible-assembly cells

Figure 7.5 shows a solution in this respect in the form of a flexible semi-automatic assembly cell. A SCARA-type assembly robot is used. A square arrangement workpiece carrier system is placed between the robot work point and the work point for manual assembly as a means of decoupling between man and machine.

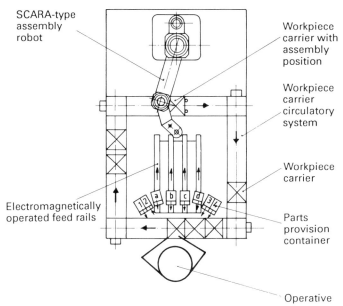

Figure 7.5 Arrangement of a semi-automatic flexible-assembly cell (1–4 = manually assembled workpieces, a–d = manually arranged, mechanically assembled workpieces)

In this example, four parts (1 to 4) are manually assembled by the operative and four different parts (a to d), which occur several times on the assembly, are manually arranged by the same operative and placed in electromagnetic feed rails. The parts are fed by the rails in an arranged form to the robot gripping positions. The parts are grasped individually by the robot and automatically assembled. At the same time, the arrangement provides an absolutely positive space separation of the robot from operatives since, even in its outmost extended position, the robot does not extend into the grasping area of the operatives. The workpiece carrier circulatory system and the electromagnetically operated feed rails serve as an intermediate buffer between the automatic and manual activity and therefore also for the decoupling of both. To achieve a uniform loading of both work points, the work content must be approximately equal for the manual and automatic activities.

The flexibility of the assembly cell can be defined as follows:

- Products consisting of eight different parts are assembled in different arrangements by programming in variants.
- Assembly robot workpiece carrier system and drives for electromagnetically operated feed rails are standard units which, following phase-out of the product, e.g. after four years, can then be used for other purposes.

Figure 7.6 shows an alternative solution for a semi-automatic flexible-assembly cell in which the parts to be handled are supplied by the robot via vibratory spiral conveyors and discharge rails and are automatically fed to sorting stations, with additional operations being undertaken manually.

The system is also constructed with a square arrangement workpiece carrier circulatory system so that the working space of the manual work point falls within the working area of the robot buffering and therefore a decoupling between the manual and automatic station is achieved by the circulating workpiece carriers.

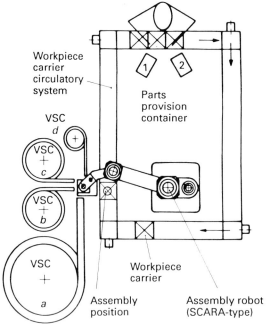

Figure 7.6 Flexible semi-automatic assembly cell (1, 2 = manually assembled workpiece, a–d = mechanically assembled workpiece, VSC = vibratory spiral conveyor)

7.5.2 Automatic flexible-assembly cells

As with other assembly machines the problem of the fully automatic assembly cell lies in the area of parts provision. The vibratory spiral conveyor is the currently most widely used arrangement and feed unit. The achievable reliability of this feed technique is largely determined by the quality of the parts, foreign content, level of contamination, degree of difficulty in arrangement and the quality of the feed unit itself. In order to maintain the overall availability of a flexible-assembly cell as high as possible, not more than four or five automatic feed stations should be arranged per robot. If more parts must be fed automatically, in order to increase the overall availability of such a system, distribution of several flexible-assembly cells is necessary with flexible interlinking and appropriate intermediate buffering.

Since, up to the present, the programming of automatic feed stations has not been economic, flexibility must be achieved by rapid arrangement and a high reusability factor by the application of standardized modules. Figure 7.7 shows an example of a flexible-assembly cell with the automatic feed of three different parts. With this solution, the product-specific periphery is arranged on a common-angle table and is connected to the base unit on face a. The same applies to the product-specific assembly fixture which is arranged on face b. The setting time is reduced to a few minutes by this peripheral arrangement. Interfaces for connection to compressed air and control should be provided in the standard equipment and also the angle table for equipment connection. The angle table must also be standardized for adaptation to the standard unit.

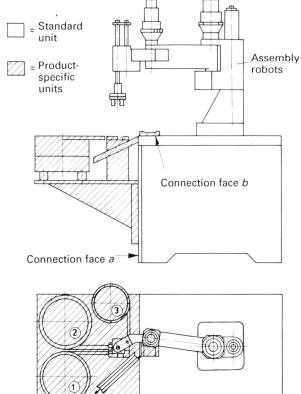

Figure 7.7 Flexible-assembly cell

The construction of a flexible-assembly cell including all necessary equipment for riveting, screw insertion, welding, pressing, etc. can be realized by a box arrangement as shown in Figure 7.8.

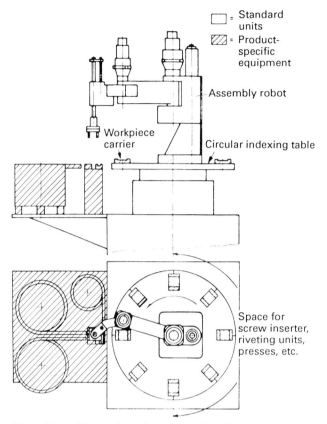

Figure 7.8 Flexible circular indexing assembly cell

The circular cyclic unit must have a fixed centre so that the assembly robot can be placed at this point. This arrangement has the advantage that peripheral equipment can be connected around the total circumference of the circular cyclic unit. The robot then functions from outside inwards. This arrangement is also necessary if a gripper change is necessary, on account of different handling operations. Every operation is divided several times by the circular table arrangement in accordance with the pitch of the circular indexing table so that the gripper change times are divided by the pitch number of the circular table arrangement. In this case, flexibility is also achieved by the programmable assembly operation of the robot by short retooling times with the connection of specially designed, product-specific peripheral equipment positioned around the indexing table arrangement.

Figure 7.9 shows a further example of a flexible-assembly cell. The portal-type assembly robot is the basic item of equipment of the flexible-assembly cell. The peripheral equipment in the working area, such as parts provision and workpiece transport, is not mounted directly on the machine base plate but on an intermediate plate. If necessary, the total peripheral equipment can be removed from the working area by the attachment shown in the diagram as a replaceable carrier plate and a new plate inserted for a different product. Short setting times are achieved in this manner [34, 35].

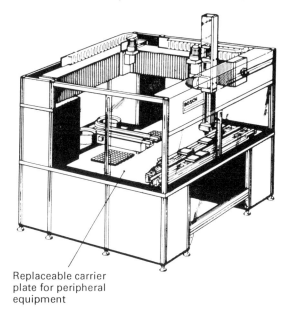

Replaceable carrier
plate for peripheral
equipment

Figure 7.9 Portal-design flexible-assembly cell (Bosch)

7.6 Assembly lines with flexible-assembly cells, interconnected by manual work points

The integration of manual work points is indispensible with increasing product complexity. The following rules have proved themselves for the optimization of manual and also automatic assembly operations (see also Section 4.4.6):

- The cost for secondary assembly must be kept low by short collection and reaching distances. Intermediate placing and regrasping should be avoided.
- As many repeat operations as possible must be provided in order to achieve the so-called conveyor-belt effect with manual assembly and with automatic assembly to keep the expense for gripper change as low as possible.
- With line production, the work points are decoupled cyclically by an adequately large buffer in order to adapt to the personal performance at manual work points and, with automatic stations, so that short-duration faults do not affect following work points.

7.6.1 Solution examples

Based on two (theoretical) examples, the possibility of linking manual and automatic assembly in an assembly line will be investigated in relation to the workpiece size, the best possible single work point layout and also decoupling from work point to work point.

7.6.1.1 Example 1
Figure 7.10 shows schematically the arrangement of an assembly line for a hypothetical product of size approx. 50 mm × 50 mm × 30 mm consisting of 11 single parts

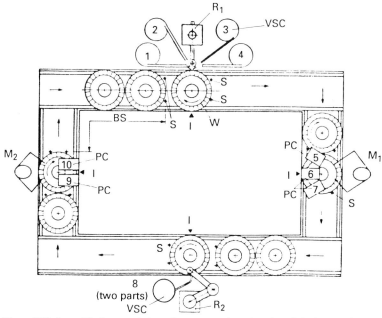

Figure 7.10 Assembly line consisting of two manual work points (M_1, M_2) and two assembly robots (R_1, R_2) (I = indexing unit, VSC = vibratory spiral conveyor, W = workpiece carrier, PC = parts container, S = stopping point, BS = buffer section, 1 ... 11 = single parts)

(including one screw fitted twice). Plate-type workpiece carriers (W) are used in the installation, each with 24 workpiece holders. In comparison, to coordinate design workpiece carriers, plate-type workpiece carriers cycled in the working area have the advantage of a uniform assembly position, i.e. the collection and reaching distances are equal for all 24 workpiece carrier holders [2, 13, 34, 35].

The workpiece carriers are fed to the four work points by the rectangular conveyor-belt system, which consists of two manual workpoints M_1 and M_2, a robot assembly station R_1, and a robot screw-insertion station R_2. The four assembly points are equipped with indexing units I for the workpiece carriers. The conveyor-belt system transports the workpiece carriers over the indexing units. The workpiece carriers are stopped over the indexing unit by the stop units S. The indexing unit moves from below with a short lift and engages in a plate-type workpiece carrier which in turn is raised by the conveyor belt and can be cycled by the indexing unit. To avoid a secondary activity, the work content of the product should be equally distributed over the four assembly points. For example, four single parts 1 to 4 are supplied at robot station R_1, by a vibratory spiral conveyor VSC, and are fitted by the industrial robot. Each assembly operation occurs 24 times in sequence. After assembling 4×24 parts (96 operations) the workpiece carrier is lowered on to the conveyor-belt system and transported to the next station, by which means a buffer reserve can form. The manual assembly points and robot stations were optimized by the primary–secondary fine-analysis procedure.

With a requirement of 2.5 s per automatic operation, the calculated work content per workpiece carrier at robot station R_1 is $4 \times 24 \times 2.5$ s = 240 s = 4 min. The buffer sections BS are arranged so that a minimum of three workpiece carriers can accumulate between the individual work stations. With a work content of 4 min per

Assembly lines with flexible-assembly cells, interconnected by manual work points   237

workpiece carrier, the buffer capacity is 12 min. This is sufficient to decouple the manual work points in the line from the cycle and to rectify short-duration faults on automatic stations.

7.6.1.2 Example 2

With the assembly of larger workpieces (in the following example of dimensions 100 mm × 100 mm × 80 mm) the diameter of the plate-type workpiece carrier would be too large. In this case, the single workpiece carrier principle is applicable. However, in order to achieve the same advantageous work point arrangements as in the previous example, the workpiece carriers are fed to the assembly points from the workpiece carrier circulatory belt on to a circular indexing table. Figure 7.11 shows the section of an assembly line with circulating workpiece carriers with the application of circular indexing tables at the assembly points (see Section 3.3.2.3, Figures 3.19 to 3.22). In principle, the work sequence and distribution of the work contents is as in the example shown in Figure 7.10. The pitch number of the circular indexing tables is dependent upon the workpiece carrier size (24 in the example). The arrangement of a manual assembly point M_1 and a robot assembly point R_1, is shown in Figure 7.11. Both stations have four handling and assembly operations.

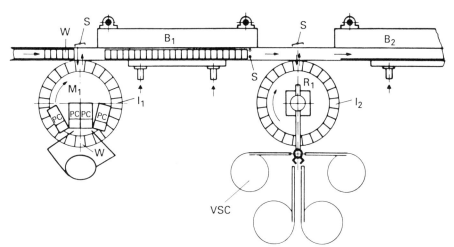

Figure 7.11 Flexible-assembly line consisting of a manual work point (M_1) and an assembly robot (R_1), multi-level buffers (B_1, B_2), circular indexing tables (I_1, I_2), stop units (S), parts containers (PC), vibratory spiral conveyors (VSC) and single workpiece carriers (W)

The manual parts dispensers at the manual work points are positioned so that in terms of a primary–secondary fine analysis there are no secondary operations for 'reaching' and 'collecting'. The workpiece carriers arrive from the conveyor belt and are pushed into the holders of the circular indexing table arrangement from which a workpiece carrier has been previously removed. With 24 workpiece carriers on the circular indexing table, 24 operations occur in sequence, so that with the arrangement of four individual parts, 96 operations are necessary until the workpiece carriers leave the circular indexing table again. The same mode of operation applies to an automated work point with the assembly robot.

Tiered buffers B are arranged between the individual stations which hold the workpiece carriers if work is undertaken with cycle differences between the stations.

The capacity of the multi-level buffer differs in accordance with the dimensions of the product to be assembled. They are, however, designed so that, as a general rule, ten levels can be occupied and in which the number of workpiece carriers of a circular indexing table arrangement can be held. With a requirement of 2.5 s per operation, a circular indexing table loading of 24 workpiece carriers and an allotment of four operations has a work content of $4 \times 24 \times 2.5 \text{ s} = 240 \text{ s} = 4$ min. An intermediate buffer with ten levels therefore has a total buffer capacity of 40 min.

7.6.2 Summary

With the interlinking of manual assembly points with flexible automatic assembly cells, the circular arrangement of workpiece carriers has the effect that the requirement for secondary operations is kept low by the best possible work point arrangement. If work is undertaken using multiple workpiece carriers in a coordinated arrangement, the reaching and collection distances would alter in the course of work during every operation and reach magnitudes which include secondary operations. A good level of decoupling is obtained between the stations by the buffer effect and therefore also an improvement in the overall reliability of the assembly line. The achievable flexibility lies in the programmability of the assembly of product variants and also, depending upon the application of standardized modules, in a huge reusability factor for new assembly operations. The assertions made in Section 6.7 for the availability of assembly machines are applicable to flexible-assembly equipment [13, 34, 35].

Chapter 8
Stored program controllers [46]

8.1 Introduction

With very many control functions which are characterized by a large number of inputs and outputs on the control unit for cost reasons, the application of stored program controllers (SPC) is preferable to conventional electromechanical-type controllers. With an SPC, the required connections of the sequence of control are written in the form of a program and not provided in the form of wiring or cabling. High processing stages, i.e. complex connections and/or sequences, can be achieved at low cost. A high processing speed is achieved by the electronic processing of information. The reaction time of the control unit to signals from the connected machine or equipment is therefore correspondingly short.

A wide range of stored program controllers are available on the market.

Within the context of this book, the design, operation and programming will be described by an example of an FPC 404 type Festo controller.

8.2 Design of stored program controllers

In terms of their design, SPC control units are information processing systems which have been specially designed around the particular requirements of control technology. Special modules record the incoming signals, and others transmit output signals to solenoid valves, relays and motors, etc. on the basis of the incoming signals and the commands undertaken in the processor within the scope of the program. One or several microprocessors are generally used as processors in a modern SPC. Visual display screens, programming units, printers or units for data exchange (e.g. with a central memory) can also be connected to an SPC.

An SPC comprises five principal component groups as described in the following subsections.

8.2.1 Input modules

These make possible a reliable detection and filtration of the signals emanating from the sensors.

8.2.2 Signal processing modules

These contain the processor with a memory for programs.

8.2.3 Output modules

These generate a direct control of the actuators such as solenoid valves on hydraulics and pneumatics, motors, power switches for electric motors, etc.

8.2.4 Network modules

These form the connection to printers, monitors, central processing units or other SPCs.

8.2.5 Program

The program determines the function of the equipment.

In SPCs, these units are either available in the form of 'cards' (memory, input cards, ...) separately or combined as modules (central unit modules, input/output modules).

All the single components can be combined to form complex control units. The type of programming of SPCs is oriented to the user's requirements.

The widely used simple program languages are:

- Instructions such as 'and', 'or', 'load constant', 'set output', are compiled in a list in the Instruction List (IL).
- The flow diagram (FD) is a graphical representation of the logic connections from the inputs and outputs and the incremental operation sequence of the machine to be undertaken incrementally.
- The contact plan (CP) was derived from the transposition of current flow plans from the electronics on to the SPC. Connections are shown graphically, and clearly by this technique sequence programmes must be constructed by internal markers and in accordance with the auxiliary relay on electronic controls.

8.3 Programming equipment

Programming equipment is used for the compilation of programmes for an SPC. The programmes are input, modified and stored by this equipment in the particular program language, checked on the monitor or printed out on the printer. The final program is then read into the control unit via a cable or by direct transmission on to the memory of the SPC.

During the initial operation phase the machine operating sequence is checked by the programming unit. For example, the signals and operating conditions on the inputs and outputs (I/O) can be called and displayed; outputs can be set and cancelled. A complete manual 'remote control' of the equipment is thus possible during initial operation or maintenance periods.

As a general rule, the programming unit is isolated from the control unit during the operating phase.

Details of the terminology, parameters and definition for SPCs are given in the respective VDI and DIN publications, such as:

VDI Guideline series VDI 2880 (December 1982, draft): Stored Program Control Equipment.

DIN 19237 (February 1980, tentative standard): Control Technology,

Terminology.

DIN 19 239 (May 1983): Control Technology, Stored Program Controllers, Programming.

DIN 40 719, Part 6 (March 1977): Circuit Documentation, Rules and Graphics Symbols for Function Plans.

DIN 40 700, Part 14 (July 1976): Circuit Symbols, Digital Information Processing.

8.4 Modules

Figure 8.1 shows a module with processor, memory for the user program, connection for eight input signals and operational modules for the determination of further functions:

- Data exchange with other electronic data processing systems such as the central computer and screen for signalling and printers, etc.
- Analogue adjuster which can be infinitely varied using a screwdriver, e.g. for pre-set time or numerical values.
- Additional memories for particularly comprehensive programmes and the connection of other FPC 404 controls via a glass-fibre cable.

Other plug-connected modules give further flexibility in relation to the number of inputs and outputs.

The control can be adapted to the required level of complexity by the addition of other central unit modules with a processor and memory (see Figure 8.2).

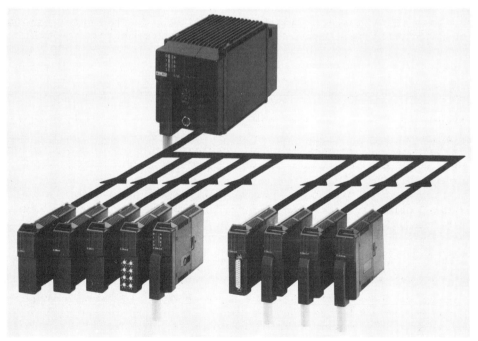

Figure 8.1 Festo FPC 404 with a central unit module and modules for maximum modularity and flexibility (Festo)

Figure 8.2 Configuration using 64 input/output, 8 analogue adjusters and 2 central units (Festo)

8.5 Operating system

The technically equipped modularity and flexibility is additionally supported by the operating system. Program parts are defined as an operating system which the manufacturer of the SPC has permanently installed and have the purpose of ensuring the function of the overall system. The operating system of the Festo FPC 404 permits programming in three different program languages which can function either completely independently of each other or, if required, coupled together. The selection of the program language depends upon the function to be undertaken and the expertise of the user.

Function type 1:
The programming of technical control functions, quite often in a combination of connection and sequence. The programming for this purpose is used in an instruction list (IL).

Function type 2:
Communication between controls. Input/output of text, signals, programming of control algorithms. The programming language Basic as used by personal computers has gained world-wide acceptance for this function. This has been extended and supplemented for technical control purposes.

Function type 3:
The processing of time-critical functions. Rapid-reaction programs are required for this purpose. These are written in a machine-compatible language (Z80 Assembler), are provided by the manufacturers as program modules and, if necessary, are installed by the users as the supplementary program structure.

8.6 Programming of SPCs

For purposes of comparison a control function is written below in various computer languages. This program only represents a section from the considerably more comprehensive complete program.

Figure 8.3 shows the objective. The workpieces are fed from a chute magazine to the stamping machine and clamped by cylinder A; the workpiece is then stamped by cylinder B. The parts are then ejected by cylinder C.

The sequence is as follows:

- Stage 0: Feed and clamp the workpiece
  If the start button is pressed and the start enable given (start position confirmed) the solenoid valve for the stroke or cylinder is operated.
- Stage 1: Raising of the punch
  If cylinder A has reached its end position – this is signalled by a corresponding signal on limit switch A1 – the solenoid valve for cylinder B is operated.
- Stage 2: Retraction of the punch
  If limit switch B1 on cylinder B is contacted, the solenoid valve for cylinder A is switched to 'retract'.
- Stage 3: Release of the workpiece
  If cylinder B has reached its stationary position and contacted limit switch B0, the solenoid valve for cylinder A is reset.

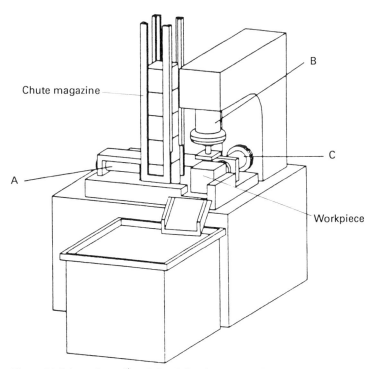

Figure 8.3 Schematic positional sketch for the stamping fixture (Festo). A, B and C = cylinders

- Stage 4: Ejection of the stamped workpiece
  If cylinder A has reached its end position and thereby operated the limit switch, then the solenoid valve for cylinder C gets the signal to 'eject'.
- Stage 5: Start position of the stamping device
  If cylinder C has travelled its full extent and contacted limit switch C1, then the solenoid valve for cylinder C is switched to 'return' and the sequence of events goes back to stage 0.

Figure 8.4 shows the computer print-out as it comes out of the printer after the program has been fed into the programming unit.

A comparison of the functions formulated above and the program print-out shows that the wording is designed in a similar way:

If ..., then ....

In this respect, the conditions (condition section) follow the 'if' and the operations to be undertaken (operation section) by the SPC after the 'then'. The condition and operation sections form a sentence.

All the sentences combined in a stage are processed virtually at the same time (combination program), and all individual stages are carried out in succession (sequence program).

The numbers given are the number of the inputs and outputs on the SPC

```
FESTO FPC-404/STD/D                                        PAGE 001
PUNCH  PROGRAM   0.00              VERSION 01   DATE*   22.02.84
                INSTRUCTION                  COMMENTS

0000  PROGRAM   PUNCH  0.0  V  01

0001  Step 0                              Feed
0002  If            E0                    Start button
0003       And      E1                    Limit switch A0
0004       And      E3                    Limit switch B0
0005       And      E5                    Limit switch C0
0006  Then Set      A 1.0                 Solenoid valve cylinder A

0007  Step 1                              Punch
0008  If            E2                    Limit switch A1
0009  Then Set      A1.1                  Solenoid valve cylinder B

0010  Step 2                              Punching complete
0011  If            E4                    Limit switch B1
0012  Then Cancel   A1.1                  Solenoid valve cylinder B

0013  Step 3                              Release
0014  If            E3                    Limit switch B0
0015  Then Cancel   A1.0                  Solenoid valve cylinder A

0016  Step 4                              Eject
0017  If            E1                    Limit switch A0
0018  Then Set      A1.2                  Solenoid valve cylinder C

0019  Step 5                              Retract
0020  If            E6                    Limit switch C1
0021  Then Cancel   A1.2                  Solenoid valve cylinder C
0022       SP after S0                    Program end
```

Figure 8.4 Instruction list of a program for the problem shown in Figure 8.3 (Festo)

(connection numbers), for example to which the start button (E0) or the solenoid valve for the feed cylinder A(A1.0) are connected.

Figure 8.5 shows the same program in the form of a flow diagram (FD). In this case, the stages are identified in square stage symbols. The following actions are shown in the right-hand adjacent field.

The contact plan was transposed on to the SPC as a further development of the circuit diagrams. At the same time, the form of the circuit diagram symbols was simplified so that they could be printed out on alpha-numerical printers. The sequence programs must be rearranged by the setting and cancellation of markers in accordance

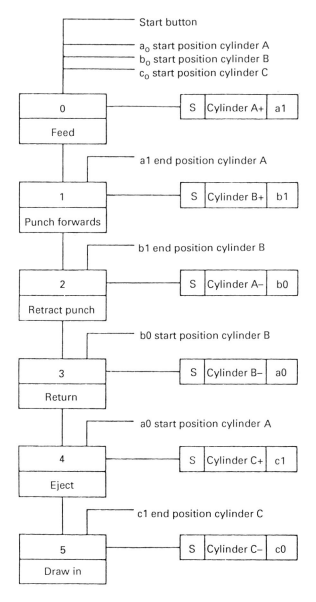

Figure 8.5 Flow diagram of a part program for the stamping device shown in Figure 8.3 (Festo)

with the functions of the circuits. With regard to the sequence, their program language is particularly suitable for combination programs (Figure 8.6).

8.7 Ease of maintenance

Since, with an SPC, only the connections from the SPC, input units (such as limit switches, initiators, switches) and output units (such as relays, motors and solenoid valves), and ignoring safety critical connections, no logic connections exist in permanently wired form, the overall construction of the equipment remains visually clear. In addition, automatic documentation of the program in the form of lists or

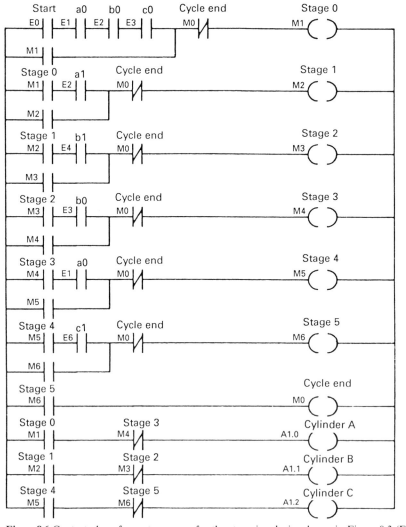

Figure 8.6 Contact plan of a part program for the stamping device shown in Figure 8.3 (Festo)

diagrams gives a high degree of ease of maintenance, since the actual state of a program and therefore also all the machine functions are obtainable in written form; all necessary information for the rectification of breakdowns is available to the maintenance personnel at any time. For the same reasons, in comparison with other technologies, the commissioning phase of equipment with an SPC is considerably shorter.

8.8 Availability

Practical experience indicates that approximately 95% of all faults or functional breakdowns occur outside the SPC. The SPC therefore contributes towards a low fault rate operation of a complete plant.

8.9 Data exchange

SPCs can be included in existing computer data processing systems and therefore in high-level company-specific organizations without difficulty. Figure 8.7 shows an example of a control hierarchy of flexible systems with the application of industrial robots.

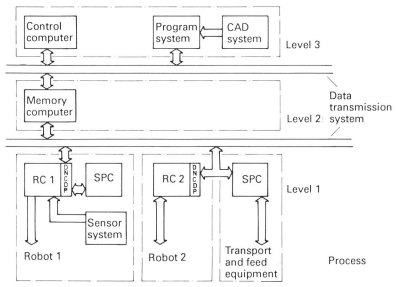

Figure 8.7 Control hierarchy of flexible systems with an industrial robot (Bosch)

Chapter 9
Practical examples

The designs of modules for automation and also the design of assembly machines and flexible-assembly systems is discussed in further detail below, based on practically achieved automation projects.

9.1 Assembly machines

9.1.1 Example 1: Rocker

9.1.1.1 Objective

The rocker as shown in Figure 9.1 as an assembly is a part of a temperature safety switch and consists of two single parts, namely a pin and a carrier bracket. Both parts are joined by resistance welding.

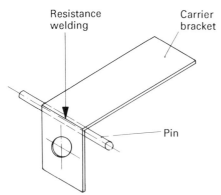

Figure 9.1 Assembly example 1: rocker

The required production rate is 10 000 parts per working day. Under the assumption of single shift working (480 min per day) and an availability of the production equipment of 80%, the theoretical cycle time for the production objective is:

$$t_C = \frac{480 \times 60}{10\,000} \times 0.8 = 2.3 \text{ s} \tag{9.1}$$

9.1.1.2 Operations to be performed
The following sequence of operations results from the product design:

1. Feed of the pin
2. Check if the pin was fed and correctly positioned
3. Feed of the carrier bracket and positioning over the pin
4. Check if the carrier bracket was fed and correctly positioned
5. Resistance weld
6. Eject, with sorting into assembled and non-assembled parts.

9.1.1.3 Criteria for method selection
At the required cycle time of 2.3 s it is advisable that all principal movements be mechanical. The required six operations make the application of a circular cyclic unit possible as a transfer unit. The positioning and ease of feed permit a fully automatic operational sequence.

9.1.1.4 Description of equipment
Figure 9.2 shows the actual layout of the assembly machine in a circular cyclic form. A circular cyclic unit with eight stations and fitted with a Maltese-cross drive is used as the transfer unit. This is fitted with a fixed plate in the centre above the circular indexing table arrangement for the location of the inspection stations I1 to I3. The parts to be assembled are positioned by the workpiece carriers fitted on the circular indexing table; they are designed so that the upper and lower electrodes of the welding press have unobstructed access.

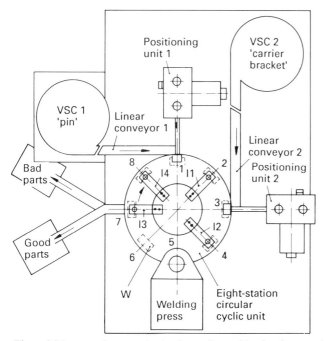

Figure 9.2 Layout of automatic circular cyclic machine for the assembly of a rocker as shown in Figure 9.1 (VSC = vibratory spiral conveyor, I = inspection station, W = workpiece carrier)

Both single parts are positioned by a vibratory spiral conveyor and fed to the sorting stations on discharge rails. After sorting, the parts are held by electric-motor-driven and cam-controlled positioning units and placed in workpiece holders. To increase the availability as shown in Figure 6.9, the discharge rails are equipped as feed sections with a suitable monitoring device. The workpiece carriers are equipped with a coding pin for recording the results of the inspection stations (see Figure 11.9, Section 11.5.1). With a negative inspection result, the coded pins are displaced and the subsequent operations disabled so that no further operations can take place (so-called idle cycle).

The operational sequence is as follows:

Station 1:
Feed of the pin and placing in the workpiece carrier by the positioning unit.
Station 2:
Station P1 checks if the pin was correctly positioned.
Station 3:
Feed of the carrier bracket and positioning by the positioning unit in the workpiece carrier.
Station 4:
Inspection station I2 checks if the carrier bracket was correctly positioned.
Station 5:
With a positive result from the previous inspection station I2, a resistance welding operation is enabled on the respective welding press. A negative inspection result causes an idle cycle on the welding press.
Station 6:
Vacant station.
Station 7:
Inspection station I3 checks if the welding operation has been undertaken. The rocker is then ejected into a discharge channel by a pneumatic device. The ejection channel has a flap plate which is set in accordance with the result of inspection station I3. With a positive result, the rocker is fed to a container for good parts and with a negative inspection result, the rocker is not ejected as an assembly but as a single part and fed to a container for bad parts.
Station 8:
Inspection station I4 checks if the workpiece carrier is free for reloading. If, for example, a pin still remains in the workpiece carrier, a new pin is not fed at station 1.

Figure 9.3 shows a photograph with a section of the described circular cyclic automatic machine. The resistance welding press is shown on the right-hand side; the discharge rail and the positioning unit for the feed of the carrier bracket are to be seen in the background.

9.1.2 Example 2: Valve plate

9.1.2.1 Objective
The assembly as shown in Figure 9.4 is considerably more complicated than that as shown in example 1. It consists of eight individual parts which are to be assembled together in sandwich form. The assembly process is by rotating-mandrel riveting. The required hourly production capacity is 575. In spite of the large number of parts which must be handled in an automatic assembly machine, an availability of 80% was assumed, based on the relatively low production level. The theoretical cycle time is 5 s.

Assembly machines 251

Figure 9.3 Section of a circular cyclic assembly machine for the assembly of a rocker as shown in Figure 9.1

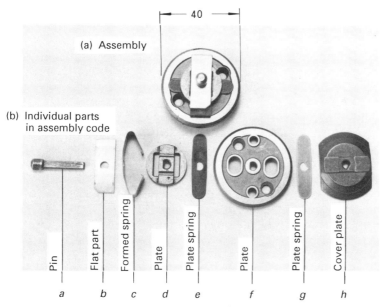

Figure 9.4 Assembly example 2: valve plate (Menziken)

Figure 9.5 shows the design of the pin part *a* of this assembly in detail. The shank has a flat face which prevents rotation of the other parts when in their correct fitting position. This requires alignment of all the parts to be fitted together relative to the position of this face on the pin part *a*.

252 Practical examples

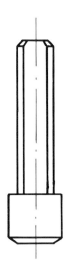

Figure 9.5 Design detail of pin part *a* shown in Figure 9.4 (Menziken)

9.1.2.2 Operations to be performed
The following sequence of operations results from the product design:

1. Feed of pin part *a* and positioning of the workpiece carrier, aligned to the pin flat face
2. Feed of flat part *b* and assembly by pin
3. Feed of formed spring *c* and assembly by pin
4. Feed of plate *d* and assembly by pin
5. Feed of plate spring *e* and assembly by pin
6. Check if parts *a* up to and including *e* have been fed and assembled
7. Feed of plate *f*, align to profile bore and assemble over pin
8. Feed of plate spring *g* and assembly over pin
9. Feed of cover plate *h*, align to outer contour and assemble over pin
10. Hold down assembly and complete rotating-mandrel riveting operation
11. Place fully assembled unit on conveyor belt in a pre-determined position
12. Check if the workpiece carrier is empty.

9.1.2.3 Criteria for method selection
The use of pneumatically operated positioning units is made possible by the 5 s cycle time. As shown by the sequence of operations, the assembly machine must be equipped with 12 stations so that a circular cyclic unit can be used. The plate *f* is ground on both faces and cannot be fed by a vibratory spiral conveyor because of the possibility of

Assembly machines 253

damage to the surfaces. This part is therefore provided manually on a conveyor belt. All other single parts can be arranged and fed by vibratory spiral conveyors.

9.1.2.4 Description of equipment

Figure 9.6 shows the actual layout of the assembly machine in the form of a circular cyclic automatic machine. The principal movements on the circular cyclic transfer unit are by an electric motor with an indirect Maltese-cross drive. The handling equipment is pneumatically operated. The circular cyclic unit has 12 stations and is constructed so that the centre of the circular cyclic unit has a stationary plate of lower type construction for holding the handling equipment. The handling equipment functions suspended from inside outwards. This gives good accessibility to the individual stations around the circular cyclic unit. To increase the availability, the discharge rails between the vibratory spiral conveyors and the sorting stations are designed so that a small buffer capacity is formed. The workpiece carriers are equipped with coded pins

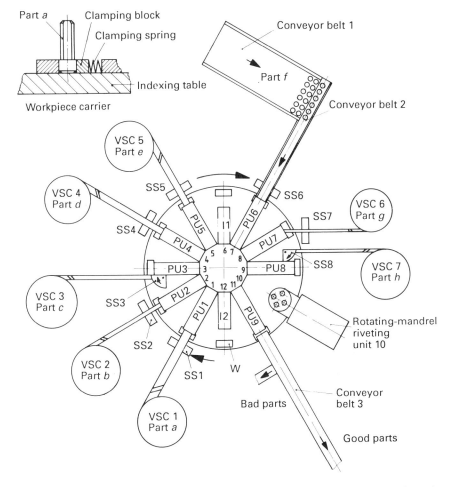

Figure 9.6 Layout of a circular cyclic automatic assembly machine for valve plates as shown in Figure 9.4 (Menziken) (PU = positioning unit, VSC = vibratory spiral conveyor, SS = separating station, W = workpiece carrier)

for recording the results of the inspection stations (see Figure 11.9). With a negative inspection result, the coded pins are displaced, so that the following stations are disabled so that no further operations are possible (so-called idle cycles). The coding pin is reset to its initial position by a pneumatic cylinder after unloading of the finally assembled unit and checking that all parts remain in the workpiece carrier.

The operational sequence is as follows:

Station 1:
The pin a is aligned relative to the shank face by vibratory spiral conveyor 1, and fed suspended via an electromagnetically operated discharge rail to the separating and turn-round station SS1. During separation, the pin is turned through 180° so that the shank points upwards. The positioning unit 1 grips the pin and places it in the workpiece carrier aligned relative to the shank face. The workpiece carrier is shown schematically in Figure 9.6 (top left). The pin is held in the workpiece carrier by its clamping device consisting of a clamping block and spring so that it cannot turn during further stages of the assembly process. A checking device for an availability check of the pin by sensors is integrated in the separating and turn-round station.

Station 2:
The flat spring b is arranged in the correct position by vibratory spiral conveyor 2, fed to separating station SS2 via a discharge rail and placed over the pin in the workpiece carrier by positioning unit 2. The availability check is integrated in the respective separation station SS2.

Station 3:
The formed spring c is arranged by vibratory spiral conveyor 3 and also fed to separating station SS3 via a discharge rail. During separation, the separating station makes a rotary movement through 90° in order to move the form spring from the correct gripping position into the fitting position. The formed spring is placed over the pin by positioning unit PU3. The availability check is made at the sorting station.

Station 4:
The plate d is arranged in the correct position by vibratory spiral conveyor 4 and fed to separating station SS4 via a discharge rail, grasped by the positioning unit PU4 and fitted over the pin.

Station 5:
Plate spring e is arranged in the correct position by vibratory spiral conveyor 5 and fed to separating station SS5 via a discharge rail. It is grasped by positioning unit PU5 and fitted over the pin. The availability check is also made at the separating station.

Station 6:
The availability of all previously fed single parts is rechecked by inspection station I1.

Station 7:
As already explained, on account of its sensitive surface, the plate f cannot be arranged and fed by a vibratory spiral conveyor. The part is therefore placed manually on a wide conveyor belt 1 and then transferred from this belt to a narrow transverse conveyor belt 2. Conveyor belt 1 is dimensioned so that it can hold a parts reserve for 20 to 30 min production. The parts move in a line on transverse conveyor belt 2 to separating station SS6. The drilled hole pattern of the plates is scanned, and they are then rotated until they are in the correct position. The plate is grasped by positioning unit PU6 and fitted over the pin.

Station 8:
The plate spring g is correctly arranged by vibratory spiral conveyor 6 and fed to separating station SS7 via a discharge rail. The discharge rail is arranged so that the

part is fed in a longitudinal direction and that the position of the profiled hole corresponds to the fitting position. The part is grasped by positioning unit 7 and fitted over the pin.

Station 9:

The cover plate *h* is correctly arranged by vibratory spiral conveyor 7 and fed to separating station SS8 via a discharge rail. This rail includes the availability check and during separation swings through 90° in order to bring the part into the fitting position. The cover plate is grasped by positioning unit 8 and fitted over the pin.

Station 10:

The assembly is held by a fixture and the assembly operation completed by a rotating-mandrel riveting unit.

Station 11:

The finally assembled assembly is removed from the workpiece holder by positioning unit 9 and placed correctly positioned on conveyor belt 3. A differential is made between good and incorrectly assembled units based on the inspection results of the individual stations. In the case of a negative signal, the assembly is removed transversely from the conveyor belt.

Station 12:

A check is made using inspection unit I2 to determine if the workpiece carrier is empty, and if necessary any residual parts from incompletely assembled assemblies are removed by compressed air.

The results of the integrated or direct inspection station are stored and, in the event of three successive faults, result in stoppage of the machine. The fault point is identified optically. This machine is shown in Figure 9.7. The bell-shaped guard which is arranged over the centre of the machine has been raised for purposes of clarity. On account of the relatively long cycle time of 5 s, the high level of integration of functions in a permanent arrangement does not adversely effect the availability, so the planned output is fully achieved.

9.1.3 Example 3: Spray nozzle–spray head

9.1.3.1 Objective

Spray parts for spray nozzles are required in high volumes. Figure 9.8 shows the design of a spray nozzle–spray head which is to be assembled fully automatically at a production rate of 150 to 170 parts per min. The spray head consists of five single parts of pre-assembled units. The assemblies' valve cones *b* and valve housings *c* are fully automatically assembled on separate assembly machines and can be fed correctly positioned as single parts. The hose part *e* is a highly flexible part and can only be handled automatically from the hose drum, so production of the hose, i.e. cutting to length, should be integrated in the assembly procedure.

9.1.3.2 Operations to be performed

The following operations are necessary for the assembly of the spray head:

1. Arrangement and feed of the valve plate *a*
2. Checking completed feed of the valve plates
3. Arrangement and feed of the pre-assembled valve cones *b* and assembly in the valve plate
4. Checking completed feed and fitting of the valve cones

Figure 9.7 Circular cyclic automatic assembly machine for valve plates as shown in Figure 9.4 (Menziken)

5. Arrangement and feed of the pre-assembled valve housing c and fitting over the valve cone in the valve plate
6. Crimping of the valve plate to form a form-locking connection between the valve plate with the valve cone and the valve housing
7. Checking the completed feed of the valve housing and completion of the assembly procedure by crimping
8. Arrangement and feed of the spray heads d and assembly on to the valve cone
9. Checking completed feed and assembly of the spray head
10. Withdraw hose from the hose drum, cut to length and fit on to the valve housing
11. Checking completed feed and fitting of the hose
12. Removal of the fully assembled assembly and sorting into good and bad parts.

9.1.3.3 Criteria for method selection
Under the assumption of an availability of $0.8 = 80\%$ and a mean output capacity, the required production rate of the equipment is:

$160 \div 0.8 = 200$ parts per minute

Assembly machines   257

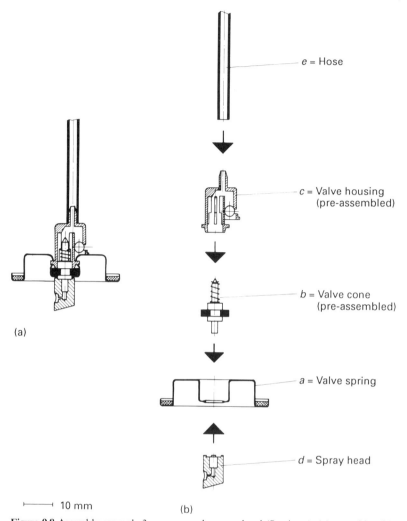

Figure 9.8 Assembly example 3: spray nozzle–spray head (Sortimat): (a) assembly; (b) parts

This gives a theoretical cycle time of 0.3 s. This output can only be achieved by a duplicate design of all single stations so that the cycle time for each station is 0.6 s.

A cycle time of 0.6 s can only be achieved by cam-operated main movements. A circular cyclic arrangement with 16 stations, with a Maltese-cross-driven circular cyclic unit and a central drive shaft for the drive of the cam-operated main assembly movements, is a necessity for such a short cycle time.

9.1.3.4 Description of equipment
Figure 9.9 shows the layout of the actual assembly machine as a duplex design circular cyclic automatic machine. All the main movements for the transfer unit, positioning units and inspection station are mechanically cam-operated. To increase the availability, the feed stations are equipped with feed sections (see Figure 6.9). The circular cyclic table is equipped with two workpiece holders per station. Each

258 Practical examples

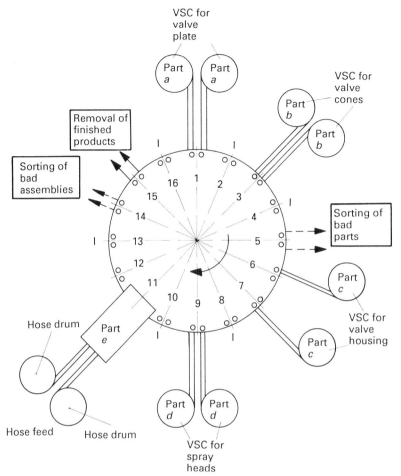

Figure 9.9 Layout of a circular cyclic assembly machine in duplex design for spray head assembly as in Figure 9.8 (Sortimat) (○ ○ = double workpiece holder, VSC = vibratory spiral conveyor, I = inspection station)

workpiece holder has a coding pin (see Figure 11.9) for recording the results of an inspection station. With individually recorded negative inspection results, displacement of the coding pin ensures the operations are not undertaken at subsequent stations, and therefore incorrect assembly is avoided. The assembly equipment is disabled by the occurrence of three successive negative inspection results and the fault point signalled.

The operational sequence is as follows:

Station 1:
The valve plates *a* are arranged by two vibratory spiral conveyors and fed to the double separating station via two discharge rails. A positioning unit (not shown) grasps both valve plates with a double gripping head and places them in the workpiece holders.
Station 2:
The inspection station checks if two valve plates have fed correctly.

Station 3:
The pre-assembled valve cones *b* are arranged by two further vibratory spiral conveyors and fed to a double separating station via discharge rails. Two valve cones are gripped and fitted into the valve plate by a positioning unit equipped with a double gripping head.

Station 4:
A double inspection station checks if the assembly operation on station 3 has been completed.

Station 5:
With a negative test result at station 4, the workpiece holder in which a part is absent is emptied, i.e. the other part is ejected and the coding pin set for disabling the respective further stations.

Station 6:
On account of the space requirement of these items of equipment, the arrangement, feed and assembly are split between stations 6 and 7. The valve housing *c* is arranged by a vibratory spiral conveyor and fed to the respective separating station via a discharge rail. A second positioning unit grasps the valve housing and assembles it over the valve cone into the valve plate which is located in one of the two workpiece holders.

After feed of the valve housing, the crimping operation is undertaken to produce a form-locking connection of the valve housing with the valve plate. The crimping unit is located beneath the indexing table and engages from beneath upwards into the valve plate for completion of the crimping operation.

Station 7:
The same operation as at station 6 is undertaken for assembly of the valve housing in the second workpiece holder.

Station 8:
A double inspection station checks if the operations at stations 6 and 7 have been performed correctly.

Station 9:
The spray head *d* is arranged by two vibratory spiral conveyors and fed to the respective separating stations beneath the indexing table via discharge rails. The spray heads are shot by compressed air out of the separating stations from beneath into the setting fixture and fitted by them on to the valve cones.

Station 10:
Completion of the operations at station 9 is checked by a double inspection station.

Station 11:
The houses *e* are fed in pairs from a drum, cut to length and then fitted on to the valve housing. The hose length is infinitely variable.

Station 12:
Vacant station.

Station 13:
A further double inspection station checks if the hose feed and fitting operations have been performed correctly.

Station 14:
By evaluation of the inspection results from station 13, unsatisfactorily assembled units are segregated by blowing them out of the workpiece carrier.

Station 15:
The correctly assembled spray nozzles–spray heads are blown by compressed air out of the workpiece carrier and removed via an ejection channel.

260 Practical examples

Station 16:
This double inspection station checks that no workpieces remain in the workpiece carriers. If this is not the case, the remaining workpieces are blown out of the workpiece carriers by compressed air.

Figure 9.10 shows a section of stations 6 and 7 for the separate feed of the valve housings into both workpiece carriers.

Figure 9.11 shows the whole machine. A double-position station 11 consisting of the hose feed with an infinitely variable feed and the fitting unit is shown on the left-hand side of the diagram.

The required output can only be achieved with a cycle time of 0.6 s and providing that the parts of pre-assembled units are free from contamination and foreign bodies and are of suitable supply quality.

The ongoing responsibility of a suitably qualified operative is a fundamental requirement for achieving the required production rate.

Figure 9.10 Stations 6 and 7 of an automatic assembly machine for spray heads as shown in Figure 9.8 (Sortimat)

9.1.4 Example 4: Terminal block

9.1.4.1 Objective
A terminal block as shown in Figure 9.12 is to be assembled fully automatically at a production rate of approximately 1400 units per hour. The function of the terminal block is to guide two wires into the clamp blocks and to secure them by a screw. The springs are used to lock the terminal on the terminal bar. A breakdown voltage test is

Assembly machines 261

Figure 9.11 Automatic assembly machine for spray heads as shown in Figure 9.8

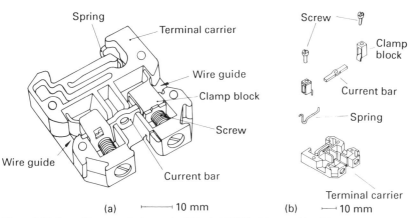

Figure 9.12 Assembly example 4: terminal block (OKU): (a) assembly; (b) parts

also necessary in addition to the assembly operations. At the usually assumed availability of 80%, when related to the required output a single cycle time of $3600 \div (1400 \div 0.8) \simeq 2$ s is required.

The total assembly procedure is subdivided into a pre-assembly and final assembly operation. The pre-assembly operation includes the fitting of the screws into the clamp blocks and the current bar to the clamp blocks. The final assembly operation comprises fitting the springs in the terminal carrier and fitting the pre-assembled unit consisting of the clamp blocks with screws and current bar in the terminal carrier.

In order to fit the current bar with both clamp blocks in the terminal carrier, it is necessary to screw the screws in up to the stop in the clamp blocks.

Following completed assembly of the assemblies in the terminal carrier, both screws are wound outwards so that the future user of the terminal block can insert their wires.

9.1.4.2 Operations to be performed
The operational sequence is as follows:

1. Place the current bar in a pre-assembly fixture.
2. Place two clamping blocks in the pre-assembly fixture and fit over the ends of the current bar.
3. Screw two screws in the clamping blocks. For later fitting of the assembly in the terminal carrier, the screws must be inserted to the thread end because a rigidly connected pre-assembly is formed.
4. Correctly position the terminal carrier in the workpiece carrier.
5. Fit the springs in the terminal carrier.
6. Place the pre-assembled unit consisting of the current bar and clamping blocks with screws in the terminal carrier.
7. Turn back the screws to a position so that the screw heads are inside the terminal carrier hole and the wire ingress is not obstructed by the screw shank.
8. Remove the incompletely assembled units.
9. Remove the fully assembled unit, place on a conveyor belt and transport to the electrical breakdown voltage test point and sort into good and bad parts.

Adequate inspection stations must be provided to safeguard assembly.

9.1.4.3 Criteria for method selection
The required cycle time of 2 s necessitates that the principal movements be cam-operated. The subdivision into a pre- and final assembly operation can be realized in that the pre-assembly operation is undertaken in one machine and the final assembly on another. In this case, however, the use of an auxiliary workpiece carrier would be necessary to ensure transport of the assembly from the pre-assembly to the final assembly machine. A second solution is to complete pre- and final assembly by a suitable workpiece carrier arrangement in one machine. Since the terminal block only consists of seven parts, this was implemented.

9.1.4.4 Description of equipment
Figure 9.13 shows the layout of the actual assembly machine as a circular cyclic automatic machine. The automatic assembly machine is constructed by the application of an assembly system as shown in Figures 6.29 and 6.30. All the main movements are mechanical via cams. The circular cyclic machine has 16 stations, and the workpiece carriers are designed so that they are suitable for holding the pre-assembled parts as well as completing the pre-assembly and also the final assembly. To increase the hopper capacity of the feed station, the vibratory spiral conveyors are equipped with additional hoppers. To increase the availability of the individual station and therefore the overall availability, the discharge rails between the vibratory spiral conveyors and separating station are designed so that buffers can form. The inspection stations do not function on the immediate shut-off principle.

If a negative inspection result is repeated three times, the machine is switched off and the fault point signalled.

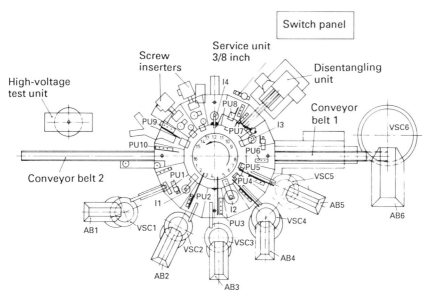

Figure 9.13 Layout of a circular cyclic automatic assembly machine for terminal blocks as shown in Figure 9.12 (OKU) (AB = additional bunker, VSC = vibratory spiral conveyor, I = inspection pit, PU = positioning unit)

The operational sequence of the assembly machine is as follows:

Station 1:
The current bar is arranged by vibratory spiral conveyor 1 and fed to the separating station via a discharge rail (not shown). The positioning unit 1 grasps the current bar and places it in the pre-assembly workpiece carrier.

Station 2:
Inspection station 1 checks the correct feed and also the position of the current bar.

Station 3:
The clamp blocks are arranged by vibratory spiral conveyor 2 and fed to the separating station via a discharge rail. Positioning unit 2 grasps the clamping block and places it in the slider of the pre-assembly workpiece carrier.

Station 4:
The same operation as at station 3 for feeding the second clamping block.

Station 5:
Inspection station 2 checks the correct feed of the clamping block. With a positive result, the two workpiece carrier sliders equipped with the clamping blocks are moved against the current bar so that the clamping blocks are fitted over it.

Station 6:
The screw is arranged by vibratory spiral conveyor 4 and fed to the separating station via a discharge rail. The positioning unit 4 grasps a screw and places it above the clamping block thread.

The screw is inserted by a pneumatic screwdriver.

Station 7:
The same operation as in station 6 for the feed and fitting of the second screw in the second clamping block.

Station 8:
The terminal carrier is arranged by vibratory spiral conveyor 6 and transported to the separating station by conveyor belt 1. Positioning unit 6 grasps the terminal carrier and places it in the final assembly workpiece carrier.

Station 9:
Inspection station 3 checks if the terminal carrier and both screws have been fitted.

Station 10:
The springs are separated from a pile by a disentangling unit, arranged by an electromagnetically operated discharge rail equipped with arrangement features and fed in an arranged manner to the separating station. Positioning unit 7 grasps a spring and places it in the terminal holder.

Station 11:
The pre-assembled unit is taken from the pre-assembly workpiece carrier by positioning unit 8 and placed in the terminal carrier.

Station 12:
Inspection station 4 checks if the spring and pre-assembled unit have been fitted correctly in the terminal carrier.

Station 13:
A pneumatically operated screwdriver turns the screw to the required inserted length, namely so far backwards that the screw head is located in the terminal carrier hole section and insertion of the wire is not obstructed by the screw shank.

Station 14:
As station 13 – winding back of the second screw.

Station 15:
Upon evaluation of the test results of the test stations, incorrectly assembled units are removed by positioning unit 9.

Station 16:
Positioning unit 10 removes the assembled unit from the workpiece carrier and places it in a specific position on conveyor belt 2. The terminal block is then transferred to the high-voltage testing station on conveyor belt 2. The parts are sorted into good and bad following high-voltage testing. (High-voltage testing itself is not a part of the assembly equipment.)

To achieve the required production rate, the output, complexity and level of difficulty of the assembly equipment necessitate the deployment of a qualified operator. A photograph of the circular cyclic automatic assembly machine is shown in Figure 9.14.

9.1.5 Example 5: High-pressure nozzle

9.1.5.1 Objective
The high-pressure nozzle shown in Figure 9.15 consists of six different single parts and is to be assembled fully automatically.

The required production rate is approximately 15 parts per minute.

The required level of cleanliness is an important criterion in the assembly of these nozzles. High demands are made on the housing and cone regarding surface quality, so these parts cannot be fed using automatic feed equipment. The required level of cleanliness necessitates that vacuuming operations be carried out in order to extract any foreign bodies. On account of this difficulty, the expected availability can only be estimated as 75%. Based on the required production rate and the achievable availability, the resulting cycle time is 3 s.

Assembly machines 265

Figure 9.14 Circular cyclic automatic assembly machine for terminal blocks as shown in Figure 9.12 (OKU)

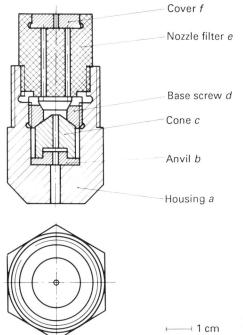

├──┤ 1 cm

Figure 9.15 Assembly example 5: high-pressure nozzle (Sortimat)

266 Practical examples

The assembly equipment must be integrated into an existing production operation. The high-pressure nozzle housing is fed directly via a conveyor belt from an upstream automatic production machine without intermediate storage. The finally assembled high-pressure nozzles must be fed directly to a palletting and packaging machine.

9.1.5.2 Operations to be performed
The operations must be performed in the following order:

1. Feed the housing a.
2. Vacuum-clean inside housing.
3. Feed the anvil b.
4. Fit the anvil into the housing.
5. Vacuum-clean inside of housing.
6. Oil inside of housing.
7. Feed cone c.
8. Feed base screw d.
9. Screw base screw into housing.
10. Vacuum-clean inside of housing.
11. Feed nozzle filter e and screw in.
12. Vacuum-clean nozzle filter.
13. Feed cover f and press into nozzle filter fully assembled.
14. Place fully assembled high-pressure nozzle on the conveyor belt in a specified position for transport to palletting and packing machine.

Adequate inspection stations must be integrated into the production operation.

9.1.5.3 Criteria for method selection
A longitudinal transfer arrangement is required for the integration of the assembly equipment in a line-type production line. At the same time, the supply of material is to be from one side so that an overhead- or underfloor-type longitudinal transfer unit is the most suitable for this application. The mechanical drive of all the main movements is made possible by the arrangement of a drive shaft parallel to the transfer equipment.

9.1.5.4 Description of equipment
Figure 9.16 shows the actual layout of the assembly machine as a longitudinal transfer arrangement. The basic unit is an overhead/underfloor longitudinal transfer unit which is driven via a cam-operated circular cyclic unit.

A synchronous drive shaft is arranged parallel to the transfer equipment. All the main movements are made via this shaft by cams and oscillatory lever systems. The space requirement of the individual stations necessitates that vacant stations be provided in the transfer system. Thirty-seven workpiece carriers are located in the working area of the longitudinal transfer unit; they are of simple design because the housing has a hexagonal outer contour and can therefore be reliably positioned in the workpiece carrier. The workpiece carriers are mounted on a steel band for cyclic transfer so that they can move by approximately 0.5 mm in the X-Y direction. A lift beam equipped with centring pins is arranged beneath the steel band guide over the entire length of the transfer machine. At the individual stations, the workpiece carriers are accurately aligned to the exact assembly position of the respective unit by the centring pin by raising the lift beam (Figure 9.17). The workpiece carriers have horizontally located coding pins for evaluation of the test results.

Assembly machines 267

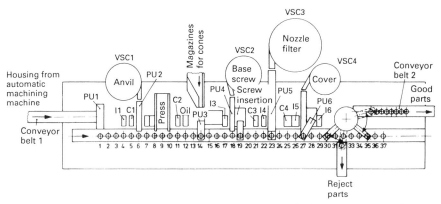

Figure 9.16 Layout of an automatic assembly machine with longitudinal transfer for high-pressure nozzles as shown in Figure 9.15 (PU = positioning unit, I = inspection pit, C = cleaning station, VSC = vibrating spiral conveyor)

The operational sequence is as follows:

Station 1:
The housing a is fed correctly positioned directly to the assembly machine from the upstream production machine via an intermediate cleaning unit on conveyor belt 1. Positioning unit 1 grasps a housing and places it in the workpiece holder.
Stations 2 and 3:
Vacant stations.
Station 4:
Inspection station 1 checks the availability of the housing.
Station 5:
The cleaning station 1 is placed over the housing by a vertical movement and the housing interior vacuum-cleaned.
Station 6:
The anvil b is arranged by the vibratory spiral conveyor 1 and fed to the separating station via a discharge rail (not shown). Positioning unit 2 grasps an anvil and fits it into a housing.
Station 7:
Inspection station 2 checks if an anvil has been fed and correctly positioned.
Station 8:
Vacant station.
Station 9:
The fed anvil is pressed into the housing by a pneumatic press. During this operation, to counteract the press forces, the workpiece carrier is supported by a slide-in support.
Station 10:
Vacant station.
Station 11:
Cleaning station 2 is lowered by a vertical movement on to the housing in order to remove any contaminants attributable to the pressing operation.
Station 12:
The metering unit is lowered into the inside of the housing to oil the internal thread.
Station 13:
Vacant station.

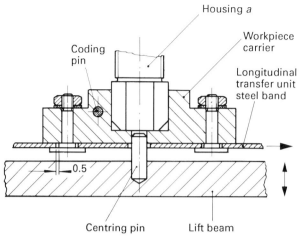

Figure 9.17 Workpiece carrier and centring device for automatic assembly machine for high-pressure nozzles as shown in Figure 9.15 (Sortimat)

Station 14:
The required high surface finish to the cone c does not permit arrangement and feed by a vibratory spiral conveyor. The cone is therefore already magazined on the production machine.

The cones are fed to the assembly machine in this magazine. The magazines are designed as duct magazines; a separating device pushes the lowest cone out of the magazine. Positioning unit 3 grasps the cone in this position and places it in the housing. To prevent shut-down of the assembly machine during a magazine change, the separating station is designed so that it forms an extension of the magazine change, and with the magazine empty can still hold ten workpieces. A replacement magazine can be fitted during the time that these ten workpieces are used up.

Stations 15 and 16:
Vacant stations.

Station 17:
Inspection station 3 checks the correct feed of the cone.

Station 18:
Base screw d is arranged by vibratory spiral conveyor 2, fed and separated via a discharge rail. Positioning unit 4 grasps a base screw and fits it into the housing.

Station 19:
The base screw is screwed into the housing at the required torque level by a pneumatically driven screwdriver.

Station 20:
Vacant station.

Station 21:
Cleaning section 3 vacuum-cleans the inside of the housing again.

Station 22:
Inspection station 4 checks if the base screw has been screwed in to the correct depth.

Station 23:
Nozzle filler e is arranged by vibratory spiral conveyor 3 and fed to the separating station via a discharge rail. After separating, positioning unit 5 grasps the nozzle filter,

Assembly machines 269

which is screwed into position in the housing by a separately arranged screw inserter using the positioning unit gripper tongs.
Station 24:
Vacant station.
Station 25:
Cleaning station 4 vacuum-cleans the inside of the nozzle filter.
Station 26:
Inspection station 5 checks the feed of the nozzle filter and also if it is inserted to the correct depth.
Station 27:
Cone f is arranged by vibratory spiral conveyor 4, fed by a discharge rail, separated and crimped into the nozzle filter by positioning unit 6. The crimping operation only requires a low level of force because the nozzle filter is made from an elastic plastic material.
Station 28:
Vacant station.
Station 29:
Inspection station 6 checks if the cover has been correctly fed and fitted.
Stations 30 to 35:
With a negative result at an inspection station, a double arm-positioning unit with two separately controlled gripper tongs and operated by a central movement grasps the assembly at station 30 and places it as an identified reject part in a scrap channel. With a positive inspection result, the second arm grasps the assembly identified as good at station 35 and places it on conveyor belt 2 in a defined position. The intermediate stations 31, 32, 33 and 34 remain free. Conveyor belt 2 transports the fully assembled high-pressure nozzles to a pallette-sizing and packaging machine.

In spite of the large number of operations to be undertaken and the permanent construction of the automatic assembly machine, the required output is achieved by easy accessibility at the individual stations and the deployment of an experienced operator.

9.1.6 Example 6: Audio cassettes

In the following example of audio cassette assembly, less emphasis is given to the detail of the individual automatic assembly machines and more to the combination of circular cyclic automatic assembly machines to form an assembly line.

9.1.6.1 Objective
The audio cassette as shown in exploded form in Figure 9.18 is required in much larger quantities, so that only fully automatic assembly is economic.

Depending upon their type, cassettes can contain up to 25 parts. Principally, a cassette consists of a lower section and a top section. To assist easy motion when winding the recording tape, foils are placed in the upper and lower sections. The guide rollers run on pins in the lower section; the lower section also carries the screening plate and spring clip. The spools with the leader tape rotate freely between the upper and lower sections. The leader tape is run out of the cassette so that when pulling in the actual magnetic tape, it can be cut for welding the magnetic tape without dismantling the cassette. The correct length of magnetic tape is then wound on to the cassette and welded on to the other end of the leader tape. Winding on of the magnetic tape is not a function of this assembly equipment.

270 Practical examples

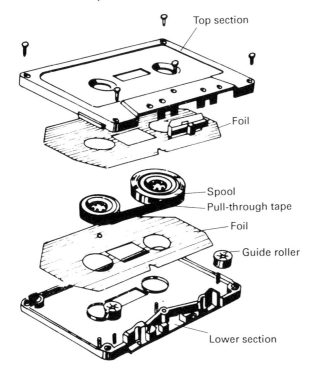

⊢——⊣ 1 cm
Figure 9.18 Assembly example 6: audio cassettes (OKU)

The requested production rate is approximately 2000 cassettes per hour. With an expected availability of 80% the theoretical cycle time is approximately 1.5 s.

9.1.6.2 Operations to be performed
The operations to be performed are shown schematically in Figure 9.19 and are therefore not explained in any further detail.

9.1.6.3 Criteria for method selection
The required cycle time of approximately 1.5 s can only be achieved by cam-operated automatic assembly machines. The product design enables distribution of the assembly process over three single machines. Parts which are difficult to handle, e.g. the foil, leader tape, etc. are produced by integrated production in the automatic assembly machines directly from the basic parts.

9.1.6.4 Description of equipment
Figure 9.20 shows the layout of the final assembly equipment. The equipment consists of the pre-assembly machine I for the assembly of the housing lower section, the pre-assembly machine II for the assembly of the housing upper section and the final assembly machine. The circular cyclic automatic machines are constructed from the basic modules as shown in Figures 6.29 and 6.30. All the main movements are cam-operated. Every assembly station is followed by an inspection station. The inspection

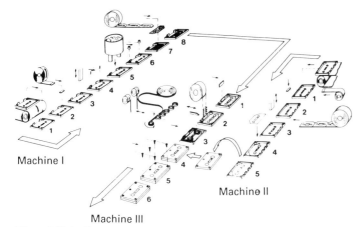

Machine I
Machine II
Machine III

Figure 9.19 Audio cassettes – assembly sequence (OKU)

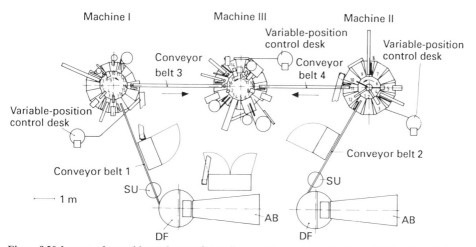

Figure 9.20 Layout of assembly equipment for audio cassettes as shown in Figure 9.18 and consisting of three circular cyclic machines as shown in Figure 9.21 (OKU) (DF = drum feed unit, SU = sorting unit, AB = additional bunker)

stations do not function on the immediate switch-off principle. The machine is switched off by pre-set counters following a pre-determined number of repeated negative inspection results. The fault points are indicated visually by centrally arranged signalling equipment. The lower section is pre-assembled on circular cyclic automatic machine I in the operational sequence as shown in Figure 9.19. The high production rate and the volume of the lower section make it necessary to provide an additional bunker (AB) for the storage of the parts. The lower section is arranged in a drum feed unit (DF) in terms of the upper and lower faces. Subsequent sorting (SU) for correct positioning occurs at a separate station. The correctly positioned lower section is fed to the automatic assembly machine via conveyor belt 1. The top section is pre-assembled on automatic assembly machine II in the sequence as shown in Figure 9.19. The top section is fed in the same manner as the lower section.

The geometrical form of the top and lower sections permits transport of the pre-assembled units from the automatic pre-assembly machines to the final automatic assembly machine III on conveyor belt 3 without using workpiece carriers because the upper and lower sections do not lose their arranged positions. The lower section is transported from automatic pre-assembly machine I to final automatic assembly machine III.

The pre-assembled top section is fed to the final automatic assembly machine on conveyor belt 4.

The linking of three circular cyclic automatic machines with conveyor belts is a loose arrangement and, on account of the length of the conveyor belts, gives a low buffer capacity between the individual automatic machines. With this buffer capacity level, short-duration faults on the individual machines do not have the same effect on the whole assembly line. Figure 9.21 shows a photograph of the assembly line, though without the feed equipment for the upper and lower sections.

Figure 9.21 Assembly equipment for audio cassettes (without feed equipment for the upper and lower sections) as shown in Figure 9.18 (OKU)

9.1.7 Example 7: Car fan motor

9.1.7.1 Objective

A car fan motor as shown by the exploded view drawing in Figure 9.22 is to be assembled at a production rate of approximately 380 parts per hour and also tested.

The low production rate permits manual work points to be included for difficult handling operations. Consequently, an availability of 85% can be expected, which corresponds to a cycle time of 8 s.

9.1.7.2 Operations to be performed

The following sequence of operations results from the product design:

1. Pre-assemble the bearing plates *a* with cup bearing *b* and retaining place *c* (see Figure 9.23).
2. Insert two magnets *d* and brushes *e* in a bearing plate.
3. Press in retaining spring *f* and magnetize housing.
4. Fit first bearing plate to housing *g*.
5. Pre-assemble armature *h* with thrust marker *i*, position second bearing plate and fit to housing.

Assembly machines 273

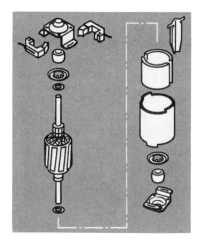

Figure 9.22 Assembly example 7: car fan motor (Siemens)

6. Caulk housing, measure armature axial play and press in as necessary.
7. Carry out test run, test insulation, separate good and bad units and sort at unloading station.

9.1.7.3 Criteria for method selection
As is evident from the operational sequence, pre-assembly operations are necessary, so the workpiece carriers must include the necessary location faces both for the pre-assembly operations and also the final assembly. The required cycle time of 8 s makes possible the application of a longitudinal transfer machine with a double belt system with frictional workpiece transfer. The double belt transfer system is more suitable for the integration of manual work points than is a cycled transfer system.

9.1.7.4 Description of equipment
The assembly equipment is shown in Figure 9.23. The equipment is of modular construction consisting of individual independent assembly machines which are flexibly connected together by a double belt transfer system. The workpiece carriers are equipped with coding equipment so that with non-completion or incorrect completion of an operation, the following operations are disabled by displacement of the coding pins and incorrect assembly is avoided. Remaining parts from incompletely assembled units are removed manually at the work point at station 1. With regard to part feed equipment, the following functions are covered by the individual modules of the assembly equipment: positioning, feed, separating, handling and assembly of the parts and also inspection for correct completion of the assembly operation.

The operational sequence is as follows:

Station 1:
This station is a manual work point for positioning both bearing plates a in the pre-assembly fixture of the workpiece carrier. If any parts of incompletely assembled units remain in the workpiece carrier, they are removed by the same operative.
Station 2:
This station is designed as a double station so that the workpiece carrier can be stopped and positioned at two different positions. The cup bearings b are arranged in pairs, fed and separated in the first position and fitted into the bearing shells by a handling unit

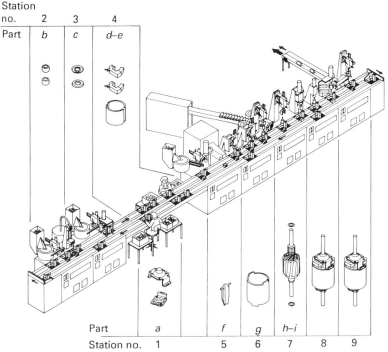

Figure 9.23 Assembly system for the assembly of car fan motors as shown in Figure 9.22

equipped with a double gripper. Both cup bearings are automatically lubricated with grease in this position.
Station 3:
Station 3 is also constructed as a double station. The workpiece carriers are also stopped and positioned. The retaining plates c are arranged by a vibratory spiral conveyor, fed to the separating station via a discharge rail, grasped in pairs by a positioning unit and placed in the bearing shells. The clamping plates are pressed into the bearing shells by a pneumatic press at the second station.
Station 4:
Station 4 is a manual workpoint for the fitting of the magnets d in pairs in the workpiece holder and placing brushes e in a bearing plate.
Station 5:
Station 5 is also designed as a double station so that the workpiece carriers can also be stopped and positioned at two points. The retaining spring f is arranged by a vibratory spiral conveyor at the first point, fed to the separating station via a discharge rail and placed in the workpiece holder by a handling unit. At the second position, the housing g is placed on the workpiece carrier via an inclined conveyor and magnetized.
Station 6:
The first pre-assembled bearing plate is transferred from the pre-assembly fixture into the final assembly fixture; the housing is then positioned on the bearing plate.
Station 7:
The armature h is removed from a column magazine by a webbing conveyor belt and fed to a stopping station. At this point, two thrust plates i are fed by a vibratory spiral

conveyor, transported to a separating station via discharge rails and fitted to the armature spindle ends. Then, pre-assembled with the thrust plates, the armature is fitted into the housing and the second bearing plate positioned.

Station 8:
Station 8 is also designed as a double station so that the workpiece carrier is also stopped and positioned twice at this point. The form-locking connection of the bearing plates with the housing is made at this first station by a pneumatically operated peening tool. The axial play of the armature in the housing is measured and set by pressing on to a bearing plate.

Station 9:
The fan motor is connected to the power supply. A test run is undertaken and also the insulation strength of the motor tested. The finally assembled fan motors are placed in a slide by a positioning unit. A deflector plate is fitted in the slide which, depending upon the test result, deflects reject motors. Good motors are transported via a belt system to the packing station. The empty workpiece carriers are transferred on to the return belt by a cross-pushing device for transport to station 1.

9.2 Flexible assembly systems

9.2.1 Example 1: Switch block

9.2.1.1 Objective
A switch block is to be assembled in four versions by programming. The product design is shown in Figure 9.24. The switch block is part of a control unit. Specific customer requirements can be fulfilled by a different fitting arrangement of the individual switch chambers. The switch chambers are operated by thermoelement-operated bimetal strips.

The variants relate to the different fitting of the switch chambers 1, 2 and 3 with current-conducting elements b to g. The following variants are possible:

- Fitting of switch chambers 1, 2 and 3
- Fitting of switch chambers 1 and 3
- Fitting of switch chambers 1 and 2
- Fitting of switch chambers 2 and 3.

Seven different single parts must be handled, and, with the maximum fitting arrangement, 22 parts are fitted together.

Depending upon the variant type, the production rate is 720 to 800 parts per working day. Two-shift working is planned. The assembly equipment planning is based on the variant with complete fitting of all switch chambers at a daily production rate of 720 units. With two-shift operation, the hourly production rate is 45.

Automatic handling of the base part a (ceramic) is not possible on account of its fracture sensitivity. The first assembly operation, fitting contact b in the base part a, is therefore undertaken manually. After fitting, the assembly operation is undertaken between parts a and b by forming. All other assembly operations are, however, undertaken manually.

9.2.1.2 Operations to be performed

1. Pre-assembly:
- Place contact b in the assembly fixture

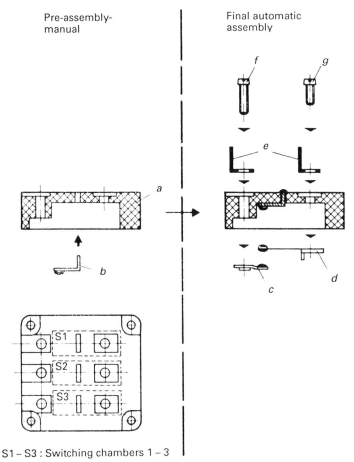

S1 – S3 : Switching chambers 1 – 3

Figure 9.24 Product design 1: switch block (a = base section, b = contact, c = mating contact, d = switch spring, e = connector, f = long screw, g = short screw)

- Place base a in the assembly fixture
- Assemble parts a and b by forming using a pneumatic press.
2. Final assembly:
- Place mating contact c in the assembly fixture
- Place switch springs d in the assembly fixture
- Place the pre-assembled assembly (base with contact) in the assembly fixture.
- Place connection parts e in the base
- Insert screw f
- Insert screw g.

9.2.1.3 Criteria for method selection

A semi-automatic flexible assembly cell is required for the performance of this assembly operation. This decision is based on the following considerations:

- On account of its fracture sensitivity, the base part a must be handled manually.
- Manually handling of the base part requires additional personnel. For economic considerations the personnel deployed must be fully occupied. Accordingly, the

pre-assembly operation – fit contact b to base part a – is undertaken manually. The integration of a press for this assembly operation in the flexible assembly cell is therefore avoided and its construction simplified.
- At the required production rate, the manual pre-assembly activity only occupies the operative up to approximately 45% so that the remaining time can be used for manually arrangings parts c, d and e and also for inserting the electromagnetic feed rails. The screws f and g are arranged by vibratory spiral conveyors and fed pneumatically through hoses.
- Since parts c up to and including g are handled and fitted by an assembly robot, the different fitting arrangements of the switch chambers can be programmed according to contractual requirements.

9.2.1.4 Description of equipment
Figure 9.25 shows the layout of the actual equipment. The SCARA-type industrial assembly robot R is mounted on an equipment table. All the production equipment for the parts feed, assembly equipment and tool provision for the robot and also for the manual work point are mounted on a common base plate which is itself mounted on the assembly cell so that, in the event of retooling, the entire equipment can be removed from base machine without time-consuming dismantling operations. For reasons of safety, the working area of the assembly robot R is screened from the manual working area by a guard. In order to avoid gripper changing and therefore also unnecessary secondary costs, the gripping tool is equipped with a triplicate system as shown in Figure 7.4.

All reaching and collecting movements in both the manual and robot activity which are associated with the individual parts or screw inserter are primary operations as defined by the primary–secondary fine-analysis principle. This was achieved by the arrangement of the gripping positions of the single parts and the tools with regard to the assembly equipment within a severely restricted area.

The sorting stations SS are equipped with sensors for checking the assembly sequence and so that the robot only makes a grasping action if a part is available. The assembly result is monitored in the assembly fixture by optical sensors. Incompletely assembled units are placed next to the conveyor belt.

The operational sequence is as follows.

1. Manual activity
Depending upon the variant type, the operative takes 1 to 3 contacts b from the parts container and places them in the pre-assembly fixture. The base part a is then taken from the parts container and placed over the contacts in the pre-assembly fixture. The assembly fixture together with the positioned parts is then placed in the press and the forming operation undertaken. After completion of this operation, the pre-assembled assembly is removed from the press together with the fixture, placed in its former position, the pre-assembled assembly removed and placed on conveyor belt 1. The work content of this pre-assembly is structured so that a number of base parts can be initially pre-assembled and then the single parts c, d and e placed in an arranged position in the electromagnetically operated discharge rail. The pre-assembled assembly is transported on conveyor belt 1 via a corner point to a stop point in the working area of the assembly robot R. Feed rails transport parts c, d and e to the separating stations SS in the working area of the assembly robot R. With regard to their buffer capacity, conveyor belt 1 and the feed rails are designed so that a reserve of

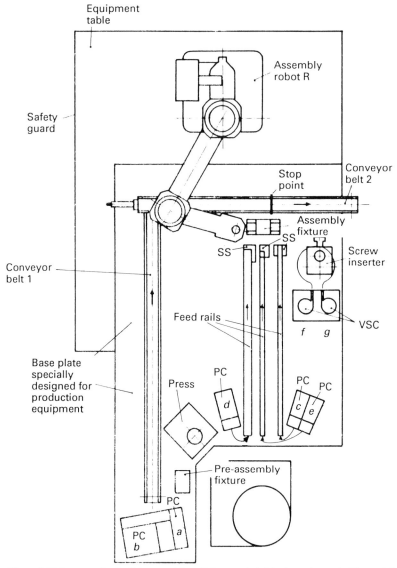

Figure 9.25 Layout of a flexible-assembly cell for a switch block as shown in Figure 9.24 (PC = parts container, VSC = vibratory spiral conveyor, SS = separating station, a = base part, b = contact, c = mating contact, d = switch spring, e = connector, f = long screw, g = short screw)

pre-assembled assemblies is available which is adequate for 20 min production of the assembly robot.

2. *Mode of operation of the assembly robot*
- Depending upon its programming, the assembly robot grasps mating contact c up to three times in succession at the separating station and places it in the assembly fixture.

- Grasps switch spring *d* up to three times (also in succession) and places it in the assembly fixture.
- Grasps a pre-assembled unit at the stop point and positions it over the mating contact and switch spring in the assembly fixture.
- Retention of the pre-assembled assemblies in the assembly fixture for clamping by pneumatically operated swing-in clamps.
- Grasps a connector *e* and places it on the base part *a* (up to six gripping actions, depending on the programming).
- Grasping of the screw inserter. Grasps screw *f* up to three times in succession, inserts it and then screw *g*; both screws are arranged by vibratory spiral conveyors and fed pneumatically via hoses into the screw-inserter head.
- Replacing the screw inserter in its static position following completion of the screw-insertion operations.
- Removal of the fully assembled switch block from the assembly fixture and placing on conveyor belt 2 for further transport.

Since the single parts are arranged manually, a 90% availability of the complete equipment is achieved. The assembly robot requires 39 s for fitting single parts *c*, *d* and *e* and also for the pre-assembled assembly. A time of 29 s is required for manipulating the screw inserter and fitting a maximum of six screws; 4 s are required for removing the completed component group.

The resulting work content of the assembly robot is a total requirement of 72 s per switch block. With an availability of 90% this corresponds to a production rate of 45 switch blocks per hour.

On account of the buffer formation between the manual work point and robot automated workpoint, with appropriate work structuring the assembly cell can function unmanned for approximately 15 min during work breaks or at the end of a shift. In this respect the automated work content is approximately 5% longer than the manual so that, with a suitable arrangement of breaks, there is always a lead in manual activity and therefore a buffer effect between the manual and automated activity. With this solution compensations can be made with respect, for example, to the reduction of the working week from 40 to 38.5 hours.

9.2.2 Example 2: Assembly of clips on car headlights

9.2.2.1 Objective

Vehicle headlights of different sizes and of various designs are assembled on the assembly line described in Section 4.6.3. The basic structure of the headlights consists of lower and upper sections. The lower and upper sections are form-locked together by clips. Depending upon the headlight size between four and six clips are necessary.

Figure 9.26 shows a clip schematically in the correct fitted position.

The fitting of these clips is to be automated. The workpiece carriers used for assembly of the headlights are fitted with coding equipment. The headlight type to be assembled is entered into the workpiece carrier coding equipment. For fitting the clips, the individual variants must be stored in the control of the automatic station, the details being recalled by the workpiece carrier coding equipment.

9.2.2.2 Operations to be performed
The following operations are necessary:

1. Arrange, feed and separate the clips.

280 Practical examples

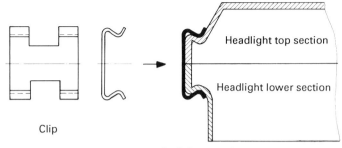

Clip

Figure 9.26 Assembly example 2: clip fitting (Bosch)

2. Pick up the clips using the fitting tool and fit.

Depending upon the headlight type, these operations are repeated up to six times.

9.2.2.3 Criteria for method selection
The differing number of clips to be fitted and the various fitting positions necessitate the use of a programmable unit and storable program controller. These requirements are fulfilled by the use of an assembly robot.

9.2.2.4 Description of equipment
Figure 9.27 shows the layout of the actual robot assembly station integrated into a manual assembly line. Figure 9.28 shows a photograph of this station. The assembly operation is preformed by a SCARA-type assembly robot. The clips are arranged in the correct position by a vibratory spiral conveyor and fed to gripping position G via a discharge rail. The pre-assembled headlights consisting of an upper and lower section

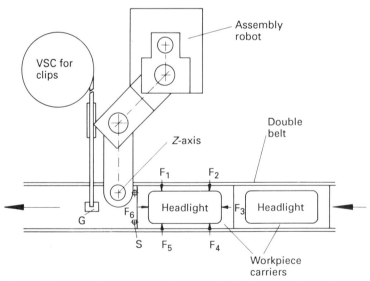

Figure 9.27 Layout of a robot assembly station for the fitting of clips as shown in Figure 9.26 (Bosch) (G = gripping position, S = stopper for workpiece carrier, F_1–F_6 = fitting stations for clips, VSC = vibratory spiral conveyor)

Flexible assembly systems 281

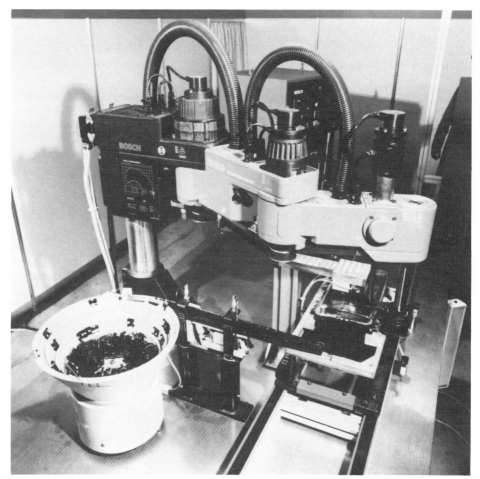

Figure 9.28 Robot station for the fitting of clips in a headlight assembly line (Bosch, Stuttgart)

are brought into the assembly position horizontally by a double belt system. A clip is grasped by the assembly robot gripper at the discharge rail gripping point. After gripping, the robot moves with its gripper in a direction corresponding to the fitting direction so that the open end of the clip is positioned in the fitting direction. The clip is forced over the headlight fitting faces of the upper and lower sections by a horizontal movement of the robot's arm. The assembly robot performs this operation between two and six times depending upon the headlight size. In order to achieve an advantageous start position for the reaching and bringing movements, the gripper position is arranged at the discharge point of the discharge rail in the centre of the double belt so that the fully assembled headlights can pass through the grasping position. Following the gripping operation and during movement from the gripping to the fitting positions, the Z-axis must make a vertical movement with the gripper downwards into the assembly position.

After the fitting, the Z-axis returns upwards again into the gripping position.

A cycle time of 10 s is achieved for a headlight size requiring the fitting of four clips.

9.2.3 Example 3: Domestic appliance drive

9.2.3.1 Objective
Two different speeds are required on a domestic appliance, namely a slow rotation and a fast rotation. Figure 9.29 shows such a drive and Figure 9.30 the design arrangement. The fan wheel is belt-driven via an electric motor.

Figure 9.29 Fully assembled domestic applicance drive (Messma–Kelch)

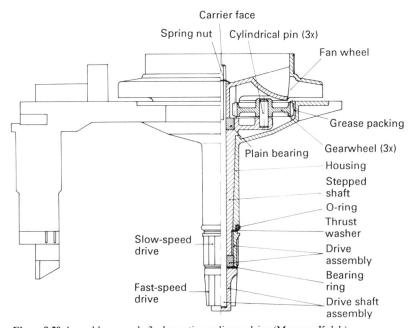

Figure 9.30 Assembly example 3: domestic appliance drive (Messma–Kelch)

The drive shaft and fan wheel assemblies are form-locked together so that the drive shaft assembly runs at the same speed as the fan wheel as a high-speed drive. This speed is reduced by the gearing, which transmits this speed via the stepped shaft to the drive assembly.

The slow-speed drive of the domestic appliance is achieved by this reduction.

A total of 140 000 units per annum are to be assembled with a cycle time of approximately 70 s on semi-automatic equipment operated with two-shift working and at an availability of 80%. The theoretical daily output is given by

$$[(16 \times 60 \times 60) \div 70] \times 0.8 = 658 \text{ parts per day}$$

This means that the required quantity can be assembled in 212 days per year (140 000 ÷ 658 = 212 days per year).

The domestic appliance drive consists of ten different single parts or pre-assembled units. Three gearwheels and three cylindrical pins are used in each assembly. The cylindrical pins, gearwheels and spring nuts are to be arranged and fed automatically. The bearing ring and thrust washer are provided by duct magazines. The pre-assembled units, drive shaft, drive and housing and also the fan wheel are arranged manually.

9.2.3.2 Operations to be performed

The design arrangement of the domestic applicance necessitates the following operational sequence:

1. Place pre-assembled 'drive shaft' unit in the assembly fixture.
2. Fit bearing ring over the drive shaft.
3. Fit 'drive' assembly over the drive shaft.
4. Place thrust washer on drive.
5. Place pre-assembled housing on workpiece holder and fit over the drive shaft.
6. Grease internal gear teeth in housing.
7. Place stepped shaft, pre-assembled with plain bearings, over the drive shaft and fit in the housing.
8. Fit three cylindrical pins into stepped shaft.
9. Fit three gearwheels on to cylindrical pins and, at the same time, engage the gearwheel teeth in the housing teeth.
10. Fit fan wheel to drive shaft.
11. Fit spring nut over drive shaft.
12. Remove fully assembled domestic appliance drive from assembly fixture and place to one side.

9.2.3.3 Criteria for method selection

The required production rate necessitates that all automatic handling operations be performed by an automatic handling machine. The application of an assembly robot is therefore necessary. The different sizes of the parts necessitates gripper changes. A multiple activity is necessary with every single operation in order to reduce the secondary activity requirement. The manual assembly activity must be decoupled from the automatic operations so that they can be performed outside the robot's working area. Decoupling in relation to the working cycle of the assembly robot must be achieved.

284 Practical examples

9.2.3.4 Description of equipment

Figure 9.31 shows the layout of the actual assembly equipment and Figure 9.32 a photograph of a section of this equipment.

A SCARA-design robot is arranged in the centre of the assembly equipment for the completion of the automatic operations. Gripping points G1 to G9 are positioned within the working range of this assembly robot. The gearwheels, cylindrical pins and spring nuts are arranged by vibratory spiral conveyors and fed to the gripping points via discharge rails. The bearing ring and thrust washer are withdrawn from chute magazines and then also fed to the gripping points. The drive shaft, drive, stepped shaft and fan wheel are placed and arranged manually on feed rails or double-belt systems and transported to the gripping points. In the case of the fan wheels, five wheels are stacked on top of each other.

A circular table with 18 workpiece carriers is positioned upstream of the assembly robot. This arrangement makes 18 similar operations possible so that the gripper change times are distributed over 18 similar operations. The operative has the task of removing the housing manually from a compartmentalized crate and placing it in the assembly fixture. In addition, adequate time remains to observe the automatic feed equipment and assembly operations and, if necessary, rectify any faults.

The different gripper systems required are placed in the immediate vicinity of the gripping point in order to achieve the shortest possible distances between gripper change actions and gripping.

The operational sequence is as follows:

1. The assembly 'drive shaft' is aligned manually to the carrier surface, placed on the feed rail and fed suspended to the gripping point. The assembly robot grasps a drive

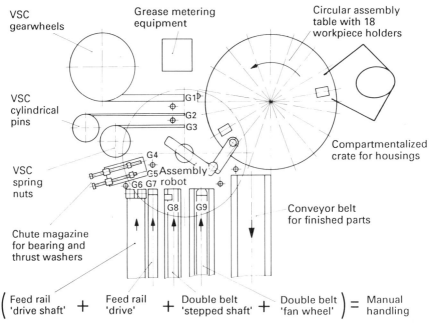

Figure 9.31 Automatic–manual assembly cell layout for the assembly of domestic appliance drives as shown in Figure 9.29 (VSC = vibratory spiral conveyor, G1...9 = gripping points, ⊕ = gripper) (Messma–Kelch)

shaft 18 times in succession from gripping point G6 and places them in the assembly fixture of the circular indexing table. Since the drive can only be transported suspended, the gripper must make a 180° slewing movement in order to move into the required fitting position.
2. A gripper change is made after completion of these 18 operations.
3. A bearing plate is gripped by a combination gripper at gripping point G4, then a drive at gripping point G7 followed by a thrust washer at gripping point G5; these three parts are fitted as a packet over the drive shaft. This operation is also repeated 18 times.
4. The operative takes a pre-assembled housing 18 times from the compartmentalized crate and places it manually over the drive shaft in the assembly fixture.
5. The next gripper change occurs parallel to the manual handling of the housing.
6. The robot grasps the grease metering unit and applies grease 18 times to the housing inner teeth.
7. The grease metering unit is replaced and a gripper change implemented. The new gripper is designed so that it is suitable for gripping and stepped shaft, cylindrical pins and the gearwheels.
8. A stepped shaft is grasped 18 times in sequence at gripping point 8 and fitted to the housing. For a form-locking connection between the drive and stepped shaft, the drive has inner teeth and the end of the stepped shaft outer teeth. To assist engagement the stepped shaft gripper makes an oscillatory rotational movement through a few degrees.

Figure 9.32 Section of the assembly cell for domestic appliance drives as shown in Figure 9.29

9. Three cylindrical pins are grasped in sequence at gripping point G2 and fitted into the stepped shaft. This procedure is repeated 18 times in sequence. The stage of assembly following this stage is shown in Figure 9.33.

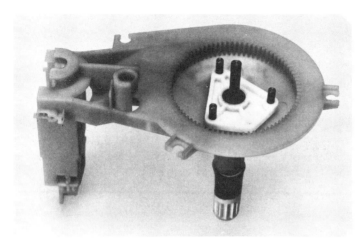

Figure 9.33 Assembly stage after fitting three cylindrical pins (Messma–Kelch)

10. Three gearwheels are grasped in succession at gripping point G1 and fitted on to the cylindrical pins whilst at the same time engaging the housing teeth. Here, also, the gripper makes an oscillatory movement to assist engagement. The procedure is repeated 18 times in sequence. Figure 9.34 shows the assembly stage following this operation.

Figure 9.34 Assembly stage after fitting three gearwheels (Messma–Kelch)

11. Gripper change.
12. A fan wheel is gripped 18 times at gripping point G9 and fitted to the drive shaft. For fitting the fan wheel on to the drive shaft the profiled inner bore of the fan wheel must be aligned to the drive shaft carrier face. Alignment is achieved by

positioning the fan wheel on to the drive shaft with a rotary movement. The rotary movement is continued until the fan wheel profile bore is aligned to the drive shaft.
13. Gripper change.
14. A spring nut is gripped 18 times at gripping point G3 and fitted to the drive shaft.
15. Gripper change.
16. The finally assembled unit is removed from the fixture and placed on a conveyor belt.
17. Gripper change (for drive shaft handling, first operation).

With continuous operation, this assembly cell combined two operatives, one for handling the housing and for observation of the automatic feed stations and the other for the manual arrangement of the drive shaft, drive, stepped shaft and fan wheel and placing on the feed systems, and also for removal of the fully assembled domestic appliance drives and placing in a compartmentalized crate. The required performance and availability is achieved by the distribution of the automatic and manual activity and the employment of two operatives to work together.

9.2.4 Example 4: Fitting out of printed circuit boards

9.2.4.1 Objective
In large-scale production, printed circuit boards are fed with electronic components by high-performance automatic feed machines. A programmable automatic feed machine must be provided for smaller quantities with a larger number of variants. The printed circuit boards are in the European format and are to be equipped with DIPs (dual in-line packages) and also components with axial or radial pins. The equipping time for one board must be 3 to 4 s.

9.2.4.2 Operations to be performed
The following operations must be performed:

1. Transport components from the magazine to the gripping points.
2. Grip the component and fit on the printed circuit boards.
3. With axial or radial components, bending the legs after positioning.

9.2.4.3 Criteria for method selection
The application of assembly robots for fitting out printed circuit boards is made necessary by the low production volume and their programmability.

In contrast to the DIPs, after positioning, the legs on the axial and radial components must be bent over. For this reason, it is necessary to divide the feed operation for these two component groups between two robot stations.

9.2.4.4 Description of equipment
Figure 9.35 shows a photograph of the actual printed circuit board feed equipment. Two SCARA-design assembly robots are used for the handling of these components. The European-format printed circuit boards are placed in multi-workpiece carriers (4 parts) which are equipped with coding equipment. The workpiece carriers are transported by a double-belt system.

The programmed coding of the workpiece carriers is read by reading heads and the work piece carriers fed to both robots accordingly. The components with axial or radial pins are handled on an assembly robot. The components are held in chute magazines, arranged in a magazine carousel and are fed to the gripping points in

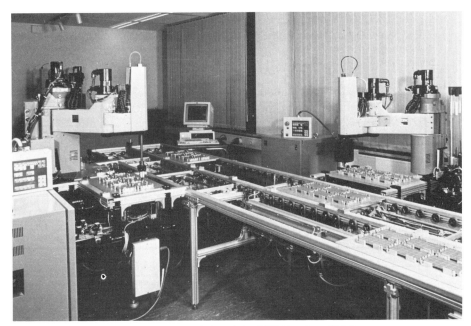

Figure 9.35 The fitting out of printed circuit boards by SCARA robots (IBM)

accordance with the program. The robot is equipped with a gripper changing system with two grippers. The components with radial pins are handled with one gripper and the ones with axial pins by the other. The bending station for the legs is beneath the fitting position; it is controlled by the robot control, is aligned to the fitting position of the respective component and performs the bending of the legs. The second robot handles the various DIPs. It is also equipped with a multi-purpose gripper which, controlled by the program, adjusts to the different dimensions of the various components.

9.2.5 Example 5: Auxiliary contact block

9.2.5.1 Objective
An auxiliary contact block (not shown) is to be assembled at a production rate of 80 parts per hour. The product consists of a housing, two closing contacts, two opening contacts, a contact guide, return spring and a cover. In addition to fitting these parts, the housing and cover must be joined together using ultrasonic welding equipment. A label is to be applied and the assembly tested on test equipment.

9.2.5.2 Operations to be performed
The product design necessitates the following sequence of operations:

1. Place the housing in the assembly fixture.
2. Fit two closing contacts into the housing.
3. Fit two opening contacts into the housing.
4. Fit one contact guide into the housing.
5. Fit the cover to the housing.

6. Test the pre-assembled unit on the test rig.
7. Join housing and cover by ultrasonic welding.
8. Attach label.
9. Produce return spring from spring wire, separate and fit into auxiliary contact block.
10. Test in test fixture to determine if contact function is fulfilled.

9.2.5.3 Criteria for method selection

The required production rate and the large number of different parts to be handled necessitates the application of an assembly robot for their automatic handling. The highest possible reusability of the modules integrated in the assembly equipment is a highly important criterion for method selection. To a certain degree, the selected method can be regarded as a pilot project for the application of new assembly techniques.

9.2.5.4 Description of equipment

The equipment design is shown in Figure 9.36. It principally consists of the following modules:

- Assembly robot (1), type ASEA, built on the Cardan principle (see Section 5.3.2.6.2, Figure 5.76) and fitted with a revolving-head multi-gripper system. The gripper change time is avoided by this system.
- Image processing system (2).
- Part feed systems (3 and 4) for housing and cover.
- Part feed systems (5 and 6) consisting of vibratory spiral conveyors, discharge rails and sorting stations for the feed of closing and opening contacts.
- Palette-type holder and magazine (7) for contact feed.
- Assembly fixture (8).
- Four-arm positioning unit (9) with a rotary main movement.
- Ultrasonic welding equipment (10).
- Labelling unit (11).
- Assembly fixture for positioning the return spring (12).
- Spring-coiling machine (13).
- Test device for function testing of assembly (14).
- Conveyor belt for removal of the fully assembled auxiliary contact blocks (15).
- Stored program controller (16).

The operational sequence is as follows:

1. The housings are fed to the robot grasping point at part feed station (3) from a hopper with a lift discharge, similar to a scoop design, lying flat with the housing aperture facing upwards via an electromagnetically operated discharge rail and an adjoining conveyor belt. At the same time, the housings are unarranged in relation to their internal contours. The image processing system (2) records the position of the housing and evaluates the image so the robot grasps the housing depending upon the evaluation and turns it into the correct position during transport into the assembly fixture (8) and then places it in the assembly fixture.
2. Two closing contacts are grasped at part feed station (5) and fitted in the housing.
3. Two opening contacts are grasped at part feed station (6) and fitted in the housing.
4. A contact guide is grasped from the palette type magazine (7) and fitted into the housing.

290 Practical examples

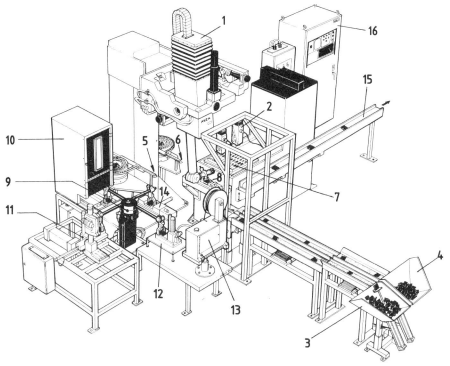

Figure 9.36 Flexible assembly cell for auxiliary contact block (ASEA) (1 = robot (ASEA), 2 = image processing system, 3–6 = part feed, 7 = palette magazine, 8 = assembly fixture, 9 = positioning unit, 10 = ultrasonic welding equipment, 11 = labelling unit, 12 = assembly fixture, return spring, 13 = spring-coiling machine, 14 = test fixture, 15 = conveyor belt, 16 = control)

5. The covers are also fed, arranged as with the housing, their position identified by the image processing system, grasped by the robot and fitted on to the housing.
6. The robot removes the pre-assembled unit from the assembly fixture and places it in the test fixture (14). A check is made at this point to determine if the assembly operations have been undertaken correctly.
7. The pre-assembled part is removed from the test fixture by a four-arm positioning unit and places it in the ultrasonic welding station (10).
8. The assembly previously welded in the welding station is removed by the positioning action of the four-arm positioning unit and is placed in the labelling unit (11) for labelling.
9. By the same positioning action of the positioning unit, another arm of the four-arm positioning unit removes the previously labelled assembly from the labelling unit and places it in the return spring assembly fixture (12). The return springs are produced individually in the spring-coiling machine (13) at the cycle rate of the assembly equipment and fed directly to the assembly fixture.
10. During a further operation, the four-arm positioning unit removes the assembled fixture from the return spring assembly fixture and replaces it in the test fixture (14) for final testing.
11. After testing, the assembly robot removes the finished assembly from the test fixture and places it on conveyor belt (15). Assemblies which have been identified as faulty by the test fixture are positioned separately adjacent to the conveyor belt.

The assembly of the next component group in the assembly fixture (8) is undertaken during the testing, ultrasonic welding and labelling operations and the fitting of the return spring.

With the complexity of this flexible assembly equipment, the continuous attention of an operative is necessary to achieve an availability of 80%.

Chapter 10

The integration of parts manufacturing processes into assembly equipment or of assembly operations into parts production equipment

10.1 Introduction

As already discussed with regard to assembly-oriented design, parts which are difficult to handle or cannot be handled automatically at all present large problems in the automation of assembly operations. Such parts are, for example:

- Highly flexible parts such as cables, soft wire parts, thin insulating paper parts, etc.
- Thin plates (less than 0.3 mm thickness) or punched metal parts similar to plates
- Interlocking parts such as hooks and clips
- Unstable formed punched parts, etc.

10.2 Integrated parts production

In accordance with the latest state-of-the-art automatic assembly machines with a production rate of 20 to 40 parts per minute are classified as high-performance machines. On the other hand, production rates of 150 to 600 parts per minute can be achieved producing punched or flexible wire parts on high-performance machines. These types of machines cannot be locally integrated into automatic assembly machines on account of these production rate differences because the single parts must be produced at the assembly machine cycle rate. On account of the short cycle times for this production process, it is considerably more advantageous to undertake production using very simple equipment.

Figure 10.1 shows schematically the design of a punching-bending unit for connection to an assembly machine. The operational sequence is as follows.

The raw-material strip e supplied in coil from a reel f is fed cyclically by a pneumatically operated feed d into a bell crank lever press c. Following operation of the tool for cutting and form-punching, the part remains attached to the strip by straps. During the same cycle, the pre-punched strip is fed to the cropping tool. The design of the whole tool must be such that, before cropping, the punched strip is freely accessible so that the positioning unit gripper b can grasp the part before it is cropped so that, after cropping, it can be placed in the assembly fixture a or in the base part already in the assembly fixture. At the same time as cropping the part from the strip, the strip is cut into several pieces and falls into a punching bin.

On account of the long cycle times for the punch-bend technique used, the application of bell crank level presses or pneumatic presses with hydraulic pressure convertors is possible for the generation of the required forces.

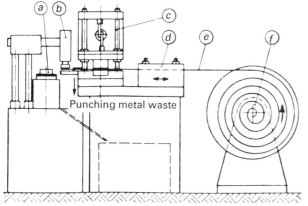

Figure 10.1 Punch-bending unit for an assembly machine (a = workpiece carrier, b = positioning unit, c = bell crank lever press, d = feed, e = raw-material strip, f = reel)

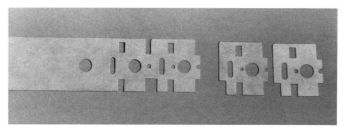

Figure 10.2 Insulating-plate punched strip

By way of example, Figure 10.2 shows the punched strips of an insulating plate made from 0.35 mm-thick insulating paper and produced in a machine as shown in Figure 10.1. Alternative solutions lie in the integration of the production of parts, e.g. helical springs, into the assembly process by the connection of spring-coiling equipment (see Section 5.2.4, Figure 5.44).

10.3 The production machining of parts in assembly equipment

If, on the one hand, the complete production of a part in one assembly system is uneconomic and, on the other, the mechanical handling of the finished part too difficult, the possibility exists of finishing parts which are in a semi-finished condition in an assembly system (providing, of course, that they can be handled easily).

Punched and bent parts which have contacts are a classic example in the electrical engineering industry. Primarily, these parts are economically produced on combination machines for punch-bending and contact welding. If the contact carrier material is extremely thin, unstable or very difficult to position on account of its form, then contact carriers can be pre-produced, i.e. the contours punched but still attached to the strip by bands. The contacts are then welded into position and the strip coiled. As shown in Figure 10.1, they are then fed through the assembly equipment by a feed unit where final forming and cropping from the strip occurs.

294 The integration of parts manufacturing processes

A further possibility for performing production machining operations in assembly equipment is illustrated by the next example. A connection tag as shown in Figure 10.3 is to be fed to three different positions and fitted in an assembly system by one feed unit. The output rate is 1200 parts per hour per position, so a total of 3600 parts per hour are required. The connection tag is made from a material with a thickness of 0.3 mm and is unstable. The part cannot be automatically handled by reason of its geometry and the required output rate. The following solution is possible:

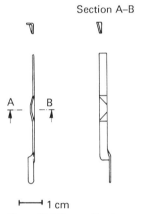

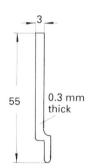

Figure 10.3 Connection tag

Figure 10.4 Plate for a part as shown in Figure 10.3

The connection tag as shown in Figure 10.3 is not finish-formed on an automatic punching machine but simply produced as a plate as shown in Figure 10.4. To avoid bulk material, a chute magazine, as shown in Figure 2.11(c) is brought from below the plate-punching tool and the plates are stored in stacks. The height of the chute magazine is 500 mm, so with a plate thickness of 0.3 mm they can hold approximately 1600 plates. The final forming of the plates thus produced to form connection tags as shown in Figure 10.3 is integrated in the automatic assembly machines. The automatic assembly machines must be equipped with an additional item of equipment for this purpose; this is shown schematically in Figure 10.5. This additional item of equipment consists of a circular indexing table which is connected in front of the circular cyclic machine.

The chute magazines with the plates are placed at station 1. The lowest plate is slid out of the chute magazine by the cyclic action of the additional circular cyclic table. The plate is formed into a connection tag by a pneumatic press at station 2. Stations 3 and 4 of the additional circular cyclic table remain vacant. The assembly machine positioning unit grasps the finish formed connection tag from the workpiece carrier of the additional circular cyclic table at station 5 and fits it to the assembly which is to be assembled on the circular cyclic assembly machine. An air-blast ejector is placed at station 6 to blow any ungripped connection tags out of the workpiece carrier so that it can receive a new plate at station 1.

10.4 Practical example: Assembly system with integrated parts production

The product shown in exploded view form in Figure 10.6 is an assembly for a relay switch. Assembly is to be fully automatic at a net rate of 20 parts per minute. The

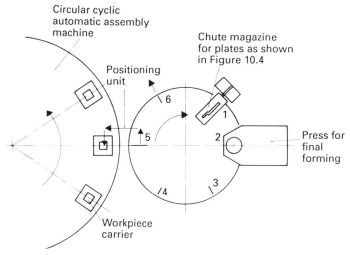

Figure 10.5 Additional circular cyclic table for the finish-forming of plates for a part as shown in Figure 10.3

contact rail parts *a* and *b* are very difficult to handle automatically, so production of these parts has been integrated into the assembly equipment. Part *a* is required once and part *b* twice per assembly.

Figure 10.7 shows the production equipment schematically. It principally consists of three units linked together:

- The production unit for parts *a* and *b* with simultaneous fitting of part *c*
- The washing unit
- The assembly centre with workpiece carriers and circulating in a rectangular pattern.

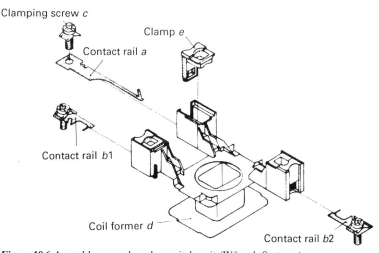

Figure 10.6 Assembly example: relay switch unit (Wünsch-Systeme)

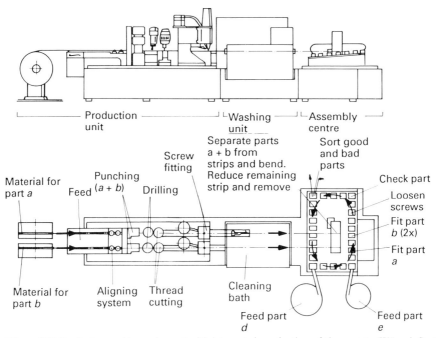

Figure 10.7 Product assembly equipment with integrated production of three parts (Wünsch-Systeme)

The operating sequence is as follows. The raw materials for parts a and b are provided in strip form on two coils and fed to the punching units via a feed and aligning system. The raw-material strip is 1.2 mm thick and can therefore be slid through the installation. Location holes are punched in the strips for accurate positioning.

Part a is punched in the punching unit so that it remains attached to the raw-material strip by narrow bands. Part b is required twice and is punched out twice per cycle and also remains attached to the raw-material strip by narrow bands. The strips are cyclically fed through the equipment by the feed unit. The core holes for the threads are drilled in another operational sequence and the threads cut at the next station so that the clamping screws c, pre-fitted in a clamping block, can be automatically arranged, fed and screwed into parts a and b. With the screws inserted, the strips pass through a washing unit where all traces of oil and metal swarf are removed.

The plastic part d is arranged by a vibratory spiral conveyor at the assembly centre, which is surrounded by workpiece carriers moving in a rectangular pattern and is then placed in the workpiece carrier by a positioning unit. The plastic part e is fed and positioned in the same manner. Parts a and b are pre-assembled at two further stations, separated with part c from the strip and positioned in part d. After separation, the remaining strip material is reduced by a cutting tool and the waste removed so that it does not disrupt the assembly process. The pre-fitted screws are loosened by a specified amount at a further station. The whole assembly operation is checked at the next station so that the assemblies can be sorted into good and bad and unloaded at a subsequent station.

A single availability of both stations for the production and feed of contact rails which is only dependent upon the coil size of the raw material is achieved by this equipment layout. Consequently, an availability of around 90% is achieved for the equipment as a whole.

10.5 The integration of assembly processes into parts production processes

A high level of efficiency can be achieved in large-scale series production if assembly processes can be largely coupled with parts production. The switch part as shown in Figure 10.8 is to be produced in this manner at a rate of 3500 parts per hour.

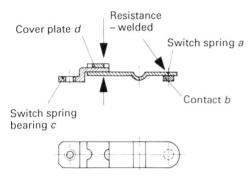

Figure 10.8 Assembly example: switch element

The following operational stages would be necessary with the usual method of production:

1. Parts production
- Cutting out and bending of switch spring a.
- Production of contact b by cutting from contact profile strip.
- Cutting out, bending and thread-cutting of switch spring bearing c.
- Cutting out cover plate d.
2. Assembly
- Fitting and welding of contact b to switch spring a.
- Fitting and welding of switch spring and contact to switch spring bearing c and cover plate d.

An analysis of the possibility of automation of assembly shows that, as a single part, the switch spring cannot be handled by an automated system because of its geometry and sensitivity; however, from the assembly point of view it is the base part of the assembly. The method of integration of production and assembly must, therefore, be oriented around this part. The other single parts can be handled automatically. The contact is very easy to produce by separation from the strip, and also the cover plate by cutting out. The requirements for integration therefore exist for both parts. On the other hand, the switch spring bearing c is a punched-bent part and also requires metal-cutting machining operation for the production of the internal thread. Its integration into the assembly equipment is not therefore economic because the punching forces required are several times higher than for parts a, b and d and also the single cycle time of thread-cutting is considerably longer than for the other operations. Based on this consideration, the main emphasis of switch-element production is to be based on the

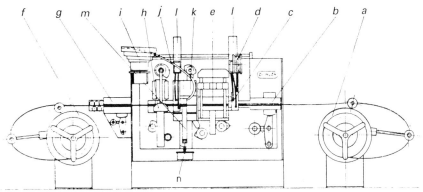

Figure 10.9 Automatic punching-bending machine with the integration of assembly for the production of a switch element as shown in Figure 10.8 (Bihler) (a = reel for switch spring strip, b = feed for switch spring strip, c = contact welding unit, d = roller for contact profile strip, e = punching press and tool for switch spring, f = reel for cover plate strip, g = feed for cover plate strip, h = punching press and tool for cover plate, i = vibratory spiral conveyor for switch spring bearing, j = feed rail for switch spring bearing, k = tool for fitting of parts with welding equipment, l = welding transformer, m = ejection channel for parts, n = hopper for finished assemblies)

production of parts and the necessary assembly operations (feed, assembly and welding) integrated.

This integrated production was achieved using a Bihler automatic punching-bending machine. The arrangement of the production equipment is shown in Figure 10.9.

The automatic punch-bending machine functions with an availability of 85% and a cycle time of 0.87 s per switch element. The 15% down time is required for tool grinding, changing the welding electrode and feeding in a new metal strip. To achieve synchronization of all movements, the principal movements of the equipment are cam-operated via a main drive.

Secondary and short stroke movements are made pneumatically and controlled by electronic interrogation elements for synchronization of the whole equipment. The cam-operated strip material feed has a repeat accuracy of 0.05 mm. With this high feed accuracy, it is possible to control the large number of switch spring machining stations so that the required quality of the individual part is achieved.

The material for the switch spring is fed incrementally at the machine cycle rate to the work points from a reel a. The contact profile strip is also fed incrementally from a reel d, and a contact is separated and welded on to the spring switch strip material. After welding of the contact, the switch springs are punched and formed in a punching-bending tool, but not, however, separated from the strip. The switch spring bearing is externally pre-finished, arranged by a vibratory spiral conveyor i and fed into the switch spring on the strip via the feed rail j. The cover plate is fed incrementally as a strip from reel f, cut but also not separated from the strip and then positioned over the switch spring and switch spring bearing. The parts are welded together and the cover plate separated from the strip. The finished switch element is then cut from the switch spring strip, automatically ejected and sorted into good and bad categories.

The different time requirements between the normal and integrated production methods as described is shown in Figure 10.10. With a production time reduction from 16.1 s to 0.24 s per switch element, a very high efficiency and therefore an extremely short amortization time is achieved [28].

Item no.	Operation	Production time in seconds/100 parts	
		Usual production	Production with integrated assembly
1.	Cut switch spring and form	27	—
2.	Cut out cover plate	9	—
3.	Cut out switch spring and bearing, form and cut threads	141	141
4.	Separate contact from profile strip	9	—
5.	Fit contact to switch spring and weld	420	—
6.	Fit switch spring with contact to switch spring bearing and cover plate and weld	1004	—
7.	Integrated production	—	100
	Total production time	1610	241

Figure 10.10 Comparison between usual production processes and integrated production of a switch element as shown in Figure 10.8 on production equipment as shown in Figure 10.9

10.6 The integration of parts production into assembly equipment within the concept of just-in-time production

The concept and purpose of just-in-time production is, based on assembly requirements, to regulate parts production so that the parts are produced at the time that they are required for assembly without intermediate storage. The possibilities for just-in-time production in large-scale series production in the fields of electrical and precision engineering are demonstrated in the following Section.

The product shown schematically in Figure 10.11, a contact-breaker, is to be produced at a production rate of 2 500 000 per annum. Assembly is fully automatic on a circular cyclic automatic machine with a unit cycle time of 4 s. The ultrasonic welding process in which part e is fitted to part a, largely determines the production time. With a possible availability of 80%, 3520 production hours are required annually in order to achieve the total production volume. With two-shift working, this gives a machine utilization of 220 days per annum.

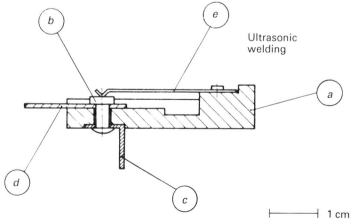

Figure 10.11 Assembly example: contact breaker

The results of the analysis of the single-part production process are shown in graphical form in Figure 10.12 in relation to the utilization of the automatic assembly machine.

1. Part a – plastic part
   Produced on a plastic injection-moulding press with a 24-chamber die
   60 shots per hour = 1440 parts/h.
   Machine occupation a:

   $$\frac{2\,500\,000 \text{ parts/annum}}{1440 \text{ parts/hour}} = \frac{1736 \text{ h/annum}}{16 \text{ h/day}} = 108.5 \text{ days/annum}$$

2. Part b – contact rivet
   This is a bought-in part.
3. Part c – punched part
   Produced on a high-performance automatic machine
   400 strokes/min = 24 000 parts per hour
   Machine occupation c:

   $$\frac{2\,500\,000 \text{ parts/annum}}{24\,000 \text{ parts/hour}} \div 16 \text{ h/day} = 6.5 \text{ days/annum}$$

4. Part d – punched part
   Produced as with part 3
   Machine occupation $d = 6.5$ days/annum
5. Part e – punched part
   Produced on a high-performance automatic machine
   200 strokes/min = 12 000 parts/hour
   Machine occupation e:

   $$\frac{2\,500\,000 \text{ parts/annum}}{12\,000 \text{ parts/hour}} \div 16 \text{ h/day} = 13 \text{ days/annum}$$

Based on these machine occupations for the production of the parts, consideration must be given to which most economic batch sizes approximate most closely to just-in-time production.

The production of 2 500 000 plastic parts a gives a machine occupation time of around 108 days per annum in comparison to 220 days' occupation of the automatic assembly machine. In this instance, optimization can be achieved between stock costs through part time and tooling costs since the occupation times are in a ratio of 1:2.

With the production of punched parts c, d and e, the position is basically reversed. When viewed in relation to machine occupation and tooling costs, the annual requirement of punched parts would have to be produced in one batch, which would result in high stores costs. High stores stocks do not only tie up capital, but also incur a risk that, with possible and necessary part alterations, large numbers of parts have to be scrapped. Market fluctuations can also significantly affect the storage time of the parts.

A subdivision of production batch sizes within the concept of just-in-time production, for example to a monthly requirement, would reduce the machine occupation time per part to 8.6 h but would, however, significantly increase the tooling costs. With an assumed tooling time of, on average, 3 h per part, with production in monthly batches the tooling time for the three punched parts is 108 h (3 parts × 12

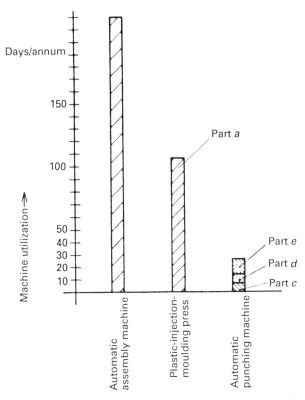

Figure 10.12 Comparison of machine occupation times for products as shown in Figure 10.11

monthly batch sizes of 3 h = 108 h) in comparison with a machine occupation of 416 h/annum (26 days × 16 hours/day). With capital-intensive equipment such as high-performance automatic machines, the tooling time costs are no longer in an economic relationship to the cost saving achieved by stock and throughput time reduction. Just-in-time production is therefore not only an organizational but also a technical production problem. With this type of product, just-in-time production is advantageous with the integration of the production of these parts by the implementation of simple equipment as, for example, shown in Figure 10.1. The provision of punched parts is then reduced to the stocking of the necessary raw material.

The operational sequence has the following effect on the design of the circular cyclic automatic machine:

1. Produce punched part c using a unit as shown in Figure 10.1 and place in the workpiece carrier.
2. Arrange plastic part a by a vibratory spiral conveyor, feed and place in the workpiece carrier above part c.
3. Produce punched part d and position as with punched part c.
4. Arrange contact rivet b by a vibratory spiral conveyor, feed and fit to parts c, a and d.
5. Complete riveting operation.
6. Produce punched part e, feed and position as punched parts c and d.

7. Complete ultrasonic welding operation for combining parts *a* and *e*.
8. Remove fully assembled unit.

With this solution, three of five components are produced just-in-time by the integration of part production processes in the circular cyclic automatic machine.

In comparison with 400 strokes per min, the production of punched parts in a 4 s cycle initially appears to be a production step backwards. However, with accurate economic appraisal it is evident that, not only is the correct part produced at the same time and in the required quantity, but also a considerable logistical expense is avoided. In addition, a by no means insignificant secondary effect of this production technique is the increase in the availability of the assembly equipment, since material feed in strip form is considerably more reliable than the feed of finished parts by feed units.

10.7 Limits for the integration of production processes

A condition for the combination of production processes for the production and assembly of parts in an installation is, for example, that no galvanic or heat-treatment processes are necessary between the production of the parts and their assembly. It such processes are necessary, then, up to the present, high-volume material must remain as loose material. The requirements for the integration of production can only be determined based on the product design and in which the material selection plays an important role. At this point, particular reference is again made to the details given in Chapter 2, 'Product design as a requirement for economic assembly'.

Chapter 11
Planning and efficiency of automated assembly systems

11.1 Introduction

Assembly is the melting pot for all the errors made in product design, planning and production. In comparison to the classical production technologies, e.g. cutting, forming and basic forming, the economic risk is higher with the automation of assembly processes. With automated assembly, whether in rigid, connected, flexible or other form, with the application of standardized modules, it must be assumed that the machines or equipment are special and are only suitable for a particular product or a relatively few variations of it. Considerable importance is therefore attached to careful planning.

Every product presents different requirements with regard to the realization of automatic assembly, so reference cannot be made to any standard planning method. The planning method as shown in Figure 11.1 has, however, proved itself as a general guideline on the basis of many years' experience; it can be modified depending upon the design, complexity and the production volume of a product. The result of planning stages 1 to 8 is combined in a specification for the finalization of the assembly equipment.

The planning procedure for automated assembly systems is described below based on this guideline.

11.2 Requirement list

The drafting of a requirement list for the planned assembly system, including specific data concerning the expected service life and production volume of the product to be assembled, is the basis of systematic planning. A product volume forecast must be drafted based on the above values and with particular reference to the introductory and run-down phases and expected product fluctuations (market conditions and season). The graph shown in Figure 11.2 illustrates the ideal service life of a product together with the expected production level, possible production fluctuations and the service life of the equipment. As a general rule, the start-up phase of production is undertaken manually along pilot lines. The period of use of the automated equipment is determined by the service life, i.e. envisaged sales life of a product excluding the start-up and, to some degree, run-down phases. The production volume determines the utilization factor by appropriate working time regulation (e.g. single- or double-shift

304 Planning and efficiency of automated assembly systems

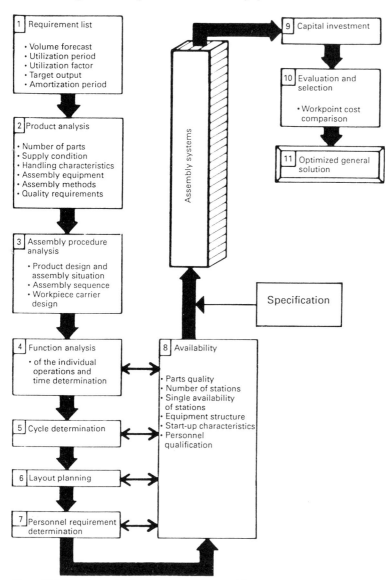

Figure 11.1 Planning guideline for automatic assembly system

operation). The required target output is determined by the production volume and utilization factor.

At this output rate and even with the lowest expected production volume the equipment must produce to target output with single-shift operation.

Depending upon the period of use, the level of investment is determined by the maximum amortization time. The time pattern of an investment project is represented linearly in Figure 11.3. The planning and construction times should run parallel to the introductory phase of the product so that the end of the procurement period coincides with the commencement of the maximum production volume. The procurement

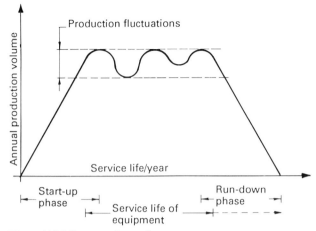

Figure 11.2 Life curve of a product

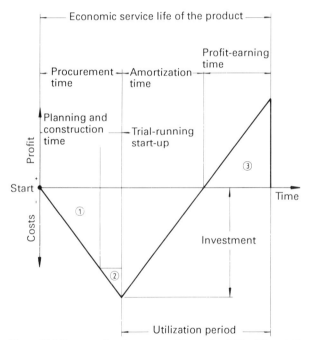

Figure 11.3 Pattern of an investment (Warnecke, Löhr, Kiener) (1 = procurement costs, 2 = trial-running and start-up costs, 1+2 = investment, 3 = profit)

period comprises the planning and construction time and the necessary requirement for trial running and commissioning. All the costs incurred during this period combine to form the investment capital. The utilization period must be arranged so that the invested capital is returned to the company by amortization and also that after reaching the amortization time a profit is obtained from the investment. A period of time is determined in the amortization calculation in which the capital expenditure is to be returned to the company via the proceeds of more rational production.

The amortization time is calculated as follows:

$$\text{Amortization time} = \frac{\text{Capital deployment}}{\text{Annual cost saving} + \text{Depreciation}} \qquad (11.1)$$

By comparison to manual assembly, the annual cost saving is principally achieved by personnel cost savings resulting from the automation of the assembly process. However, these savings are reduced by the capital costs and higher maintenance costs for maintenance of the automatic equipment.

11.3 Product analysis

The target output is given by the production capacity of the equipment to be planned in accordance with the details given in the requirement list.

The complexity of the equipment is determined by the product analysis. The production rate and number of parts in the product control factors for the size of the automated assembly equipment. The logical and economic degree of automation is determined by the supply condition, handling characteristics and quality of the parts, together with the required assembly direction and the necessary assembly processes.

The product analysis can be undertaken using the assembly-extended ABC analysis (see Section 2.2). The results of the assembly-extended ABC analysis should be arranged so that the following questions can be answered:

- How can or must the individual types of parts be provided for assembly, e.g. as bulk material, in magazines, packed and positioned correctly, etc?
- Depending upon their handling characteristics, required direction of assembly or assembly procedure, which types of parts are suitable for automatic assembly methods and which parts must be assembled manually?
- Which additional quality requirements must be placed on certain types of parts in order to facilitate automatic handling and assembly?

The costs which arise in the main field of assembly in order to implement automatic procedures must be determined by product analysis. These must be included in the overall appraisal. For example, the requirement for a different supply condition or even an additional quality requirement can not only incur additional parts costs but also an investment for new equipment for parts production.

11.4 Assembly sequence analysis

The product design and resulting assembly situation determine the assembly sequence and workpiece carrier design. With more complex products, the assembly sequence analysis determines whether or not a product can be assembled with one workpiece carrier or with several different ones. Depending upon the complexity of the product, it is shown whether or not the product can be assembled in one system or by several single installations loosely combined together.

11.4.1 Product design and assembly situation

The best overall synopsis of a product's design is given by a drawing. The drawing must be drawn so that the joining directions of the individual parts are clear. This is described below in further detail based on three examples.

11.4.1.1 Example 1

Figure 11.4 shows the product design and assembly situation of a high-pressure nozzle (see Section 9.1.5). The drawing shows that all the single parts to be fitted together can be fitted in sandwich form in a vertical line behind each other. Consequently, the complete assembly procedure can be undertaken in one workpiece holder; no pre-assembly operations are necessary.

This product design and the assembly situation do not necessitate distribution over several loosely combined assembly systems.

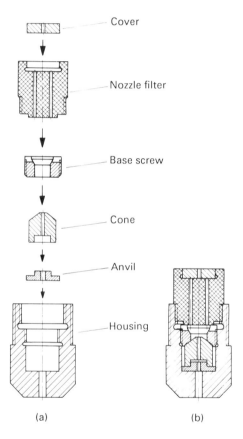

Figure 11.4 Product design and assembly situation of a high-pressure nozzle (Sortimat): (a) assembly order; (b) finished product

11.4.1.2 Example 2

Figure 11.5 shows the product design and the assembly directions of a switch consisting of eight single parts. The assembly direction for all parts is vertical and linear. The analysis shows further that the individual operations can be segregated into four separate part operations. The product design, assembly direction and operations such as rotating-mandrel riveting necessitate that the operations cannot be undertaken in one workpiece holder. For example, a form-locking connection between parts 2, 1 and 3 is formed by rotating-mandrel riveting. The riveting anvil for holding part 2 is exactly located in the fitting space of part 4. This means that the fitting of part 4 cannot be undertaken before part 2. The analysis further shows that the contact-gap setting operation requires rotation of the product through 180 degrees.

308 Planning and efficiency of automated assembly systems

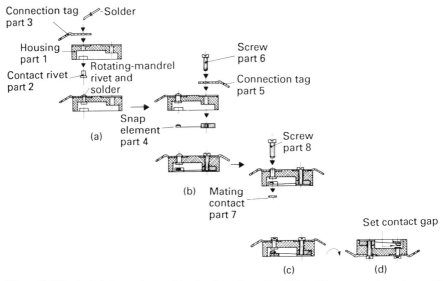

Figure 11.5 Product design and assembly situation of a switch ((a) to (d) are separate part operations; parts 7 and 8 are shown displaced)

11.4.1.3 Example 3
With more complex products, it is advisable to subdivide the product design drawing and the resulting assembly directions into component groups. Figure 11.6 shows a part section of a product design of a thermoswitch. Five further parts are to be fitted into the pre-assembled thermoswitch. The drawing shows that the assembly directions are not

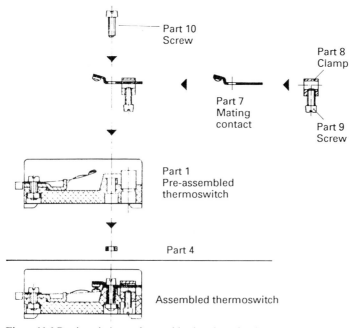

Figure 11.6 Product design and assembly situation of a thermoswitch

uniform and that pre-assembly processes are necessary. The screw (part 9) must be fitted in the clamp (part 8) before the clamp and screw can be fitted in a pre-assembled condition to the mating contact (part 7). The complete, pre-assembled unit consisting of parts 7, 8 and 9 can then be fitted in the pre-assembled thermoswitch.

11.4.2 Assembly sequence

An assembly sequence analysis is not necessary for parts which are constructed in sandwich form (see example in Figure 11.4). An assembly sequence analysis must, however, be undertaken for products with different assembly directions, which require several workpiece carriers and also for products of higher complexity. With an increasing complexity of a product and subdivision of the products into component groups (see Figure 2.3), the assembly sequence analysis shows the order in which the various assembly operations must be undertaken. The same also applies to pre-assembly operations and to the assembly of subassemblies as well as to the assembly of subassemblies to form an assembly of greater complexity. Figure 11.7 shows the assembly and work order sequence for the product shown in Figure 11.5.

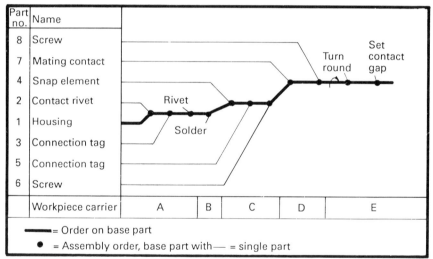

Figure 11.7 Assembly order and work order for a switch as shown in Figure 11.5 (part 1 is the base part)

The subdivision of the individual operations and their correlation to different workpiece carriers is based on the product design and the assembly and work order. Parts 1, 2 and 3 are assembled together in workpiece carrier A and rotating-mandrel workpiece carrier; however, on account of the danger of contamination it is advisable to use a different workpiece carrier B.

Parts 4, 5 and 6 are fitted to part 1 in workpiece carrier C; parts 7 and 8 are fitted to part 1 in workpiece carrier D. Workpiece carrier E then takes the completely assembled component group and swings it through 180°, for the contact-gap setting. Five different workpiece carriers are therefore necessary for this product.

Figure 11.8 shows the assembly sequence for the assembly of the component group shown in Figure 11.6. It is evident that the clamp (part 8) with the screw (part 9) must

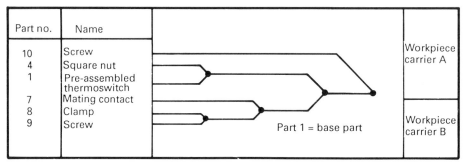

Figure 11.8 Assembly sequence of a thermoswitch as shown in Figure 11.6

be pre-assembled before the mating contact (part 7) can be fitted to this assembly. The square nut (part 4) must be fitted to the thermoswitch (part 1) before fitting this assembly into the base part (part 1). It is evident from the different positioning directions and the product structure, that the different assembly operations cannot be undertaken in one workpiece holder but that two workpiece carriers are necessary.

11.5 Workpiece carrier design

11.5.1 Introduction

The product design and assembly order determine the design form of the workpiece carrier, planned output rate, number of parts of a product to be assembled and the number of workpiece carriers in an assembly system. The availability of automatic assembly systems is, to a large degree, determined by the correct design of the workpiece carriers.

Workpiece carriers can principally be divided into two groups:

1. Workpiece carriers whose main function is in the arranged transport of the object to be assembled.

This is possible if the workpiece carriers are simply used for carrying the base part and if all the assembly, machining and inspection operations can be performed in or on the base part. A product example in this respect is shown in Figure 11.4.

2. Workpiece carriers which, in addition to performing a transport function for the part to be assembled, are also used as an assembly fixture.

Centring, retaining or clamping equipment is integrated into these workpiece carriers. They are necessary if, for example, before fitting of the base part, single parts must be positioned and held until these functions are taken over by the base part when it has been fitted.

Quite often, a large number of workpiece carriers are required to assemble complex products. For example, several hundred workpiece carriers can be necessary on longitudinal transfer lines, which represents a considerable capital investment.

Workpiece carriers should fulfil the following general requirements:

- The workpiece carriers should fulfil all the functions required of an assembly machine. If different workpiece holders or auxiliary equipment are required for pre-assembly operations in a machine, then as far as possible they should be integrated into one workpiece carrier.

- Movable centring equipment or clamps must be designed so that their operation can be by moving from station to station. The single cycle time of the station is not increased and no separate stations are necessary to operate these fixtures.
- The workpiece carriers must not only be suitable for holding the workpieces and assembly operations, but also facilitate a defined removal of the finished assemblies so that ordered placing can be achieved.
- The workpiece carriers must be designed so that, in the event of an incomplete assembly operation, residual parts can be removed without any technical difficulty before the restart of an assembly cycle.

For an optimum evaluation of the results of the inspection stations it is advisable to equip the workpiece carriers with coding equipment. As an example, Figure 11.9 shows the design arrangement of a vertical coding pin on a workpiece holder. The coding pins must be lockable in two positions, namely for a positive and negative result. With a negative result, for direct storing, e.g. by a compressed-air cylinder, the coding pin can be moved from the positive into a negative position. After unloading of the assembled unit, the coding pin must be returned to its initial position. To prevent incorrect assembly, sensors which disenable subsequent assembly stations are activated by the displacement of the coding pin.

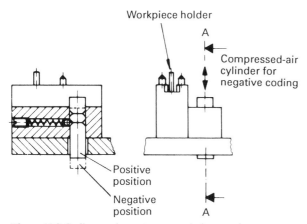

Figure 11.9 Coding equipment on a workpiece carrier

The coding pins can be arranged either horizontally or vertically. The horizontal arrangement is principally used on workpiece carriers without a permanent connection to the transfer equipment, for example on a double-belt circulatory system. Depending upon the design, vertical or horizontal coding pins can be used on workpiece carriers which are permanently attached to the transfer equipment.

11.5.2 Design examples of workpiece carriers

Design solutions are illustrated below on two examples of workpiece carriers which are not permanently attached to the transfer equipment.

11.5.2.1 Example 1
Figure 11.10 shows a workpiece carrier for the operations for assembling parts 1, 2 and 3 as shown in Figure 11.5. A contact rivet (part 2) is to be fitted on to part 1. A connection tag (part 3) is to be fitted over the rivet shank.

312 Planning and efficiency of automated assembly systems

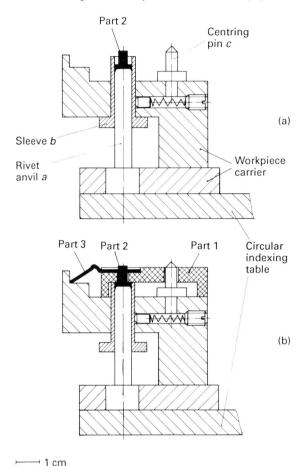

⊢⟶⊣ 1 cm

Figure 11.10 Workpiece carrier for the assembly of parts 1, 2 and 3 as shown in Figure 11.5

The parts are force-locked together by rotating-mandrel riveting. The product design and assembly sequence necessitate that the contact rivet (part 2) is placed first in the workpiece carrier. The contact rivet then stands with its head on the rivet anvil a in the workpiece carrier. The rivet shank points upward. The workpiece holder is fitted with a variable-height sleeve b over the rivet anvil a, which ensures that the positioned rivet does not lose its position during transport to the next station and is also safeguarded against being thrown out. Diagram (a) in Figure 11.10 shows the position of the sleeve b during positioning of the contact rivet. In this position, the sleeve and rivet anvil form a base for reliable holding of the contact rivet. When placing the base part (part 1) in the workpiece holder, the sleeve, which is only held by friction, is forced downwards so that the contact rivet is only surrounded in the region of its head. The contact rivet shank is then centred by the base part. In order to retain the base part in the correct position in the second plane, it is held in another bore by the centring pin c. The connection tag can then be placed over the contact rivet shank and located in the base part by its outer. Diagram (b) of Figure 11.10 shows the workpiece carrier situations following fittings of the three parts and before rotating-mandrel riveting. Since the base part is centred by the centring pin c and the contact rivet itself positioned by the sleeve b, support of the base part on its outer contours in the workpiece carrier is not necessary. The finished assembly is gripped on its outer contours for unloading.

11.5.2.2 Example 2

As shown by the product design and assembly sequence (Figure 11.8), two pre-assembly operations are necessary in order to complete the pre-assembled thermoswitch as shown in Figure 11.6 by the fitting of five further parts. The screw (part 9) is fitted in the clamp (part 8) outside the assembly equipment on a special-purpose assembly machine so that, in the design of the required workpiece carrier, this assembly can be regarded as an easy to handle single part.

To complete this operation, the workpiece carrier must be designed so that the pre-assembly of the square nut (part 4) with the pre-assembled thermoswitch (part 1), the pre-assembly of the clamp unit, consisting of the clamp and screw with the mating contact (part 7), and the assembly of the pre-assembled mating contact assembly with the clamp and screw in the thermoswitch, and also the fitting of the screw (part 10), are possible.

Figure 11.11 shows a plan view of the workpiece carrier developed for the thermoswitch. It is functionally divided into parts I and II. Part I is equipped for the pre-assembly of part 7 with the assembly consisting of parts 8 and 9. Part II is designed for the assembly of part 4 with part 1 and also for final assembly.

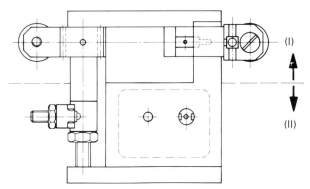

Figure 11.11 Workpiece carrier for the thermoswitch as shown in Figure 11.6 (plan view)

Details of the design and function of the workpiece carrier are given in Figure 11.12. The square nut (part 4) is guided by centring pin 1 (Figure 11.12, section A-A) and positioned in a holder. The holder has two areas for external centring of the square nut and holding the torque when tightening the screw. At the same time, this holder with its conical outer contour is used for holding the pre-assembled thermoswitch (part 1). The second fixing of the thermoswitch in the plane is by centring pin 2.

The mating contact (part 7) is held in a matching profile recess in the workpiece carrier by a stationary centring pin 3 (section B-B, Figure 11.12).

The workpiece carrier is equipped with a slider which moves in directions C and D. The pre-assembled clamp (part 8/9) is positioned in the respective slider location holder. With movement of the workpiece carrier the slider moves in a curve over the roller in the direction of arrow C so that the bores of the clamp (part 8) are pushed over the connection tag of the mating contact (part 7). The slider is a gear rack on the other

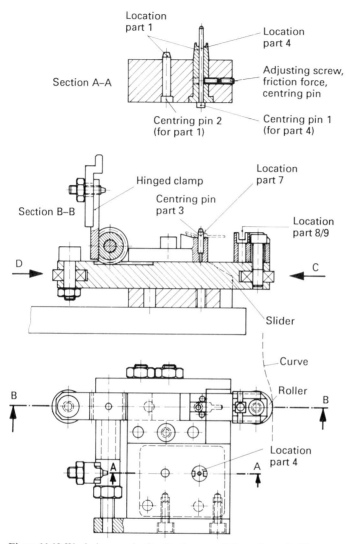

Figure 11.12 Workpiece carrier for the thermoswitch as shown in Figure 11.6

side and engages in a gearwheel. A hinged clamp is fitted on the gearwheel shaft. The clamp swings inwards by the slide motion and presses the micro-switch element down into the pre-assembled thermoswitch; otherwise the pre-assembled unit in the carrier (parts 7, 8 and 9) could not be fitted in the thermoswitch. When the component group, pre-assembled from parts 7, 8 and 9, is turned out into the thermoswitch, parts 8 and 9 are taken into the base part and become the mating contact which is aligned by the centring pin 1, which already has the square nut. The centring pin 1 is retracted with fitting of the screw (part 10) and the screw screwed into the square nut (part 4). Upon further transport to the next work station, the slider is returned to its original position in direction D by an opposed position cam bar. At the same time, the clamp swings upwards so that the fully assembled unit can be gripped and removed from the

Workpiece carrier design 315

workpiece carrier. Before recommencement of the cycle, the centring pin 1 is retracted into its start position for holding the square nut and for subsequent holding of the mating contact.

11.5.2.3 Example 3
Figure 11.13 shows a workpiece carrier for application in a longitudinal transfer system with the application of conveyor-belt systems, i.e. without permanent connection to the transfer equipment. The workpiece – the base part of the assembly to be assembled – is held and positioned on its outer contours. The workpiece carrier has no other function in this application since the parts to be assembled can be fitted directly into the base part. The workpiece carrier has two location holes for accurate positioning at the automatic stations. The workpiece carrier has a horizontal coding pin for storing the results of the inspection stations.

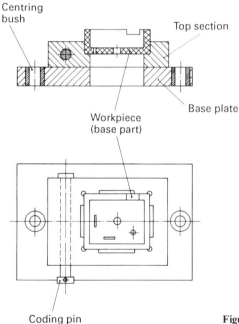

Figure 11.13 Workpiece carrier for double-belt systems

These types of workpiece carrier are preferably built in two sections. The base plate is generally available as a standard part from manufacturers of longitudinal transfer system equipment. If a large number of workpiece carriers are required, it is advisable to produce the workpiece carrier top section with the contours for holding the workpieces as a plastic injection-moulded part, and particularly so if the contours are complex.

If several hundred workpiece carriers are required in a circulatory system, then this is a highly recommendable possibility for the rational production of workpiece carriers.

11.6 Function analysis

The assembly procedure is determined by the positioning sequence. The object of the following function analysis is to divide the individual operations into their separate functional sequences and to determine their time requirements. A similar detailed analysis is not necessary for every part to be handled; similar operations can be combined into groups.

Figure 11.14 shows the pattern of a function analysis. Every assembly or machining operation is either tested separately or as a group. The procedure is as follows:

- The operation to be analysed must be clearly and completely described, for example:
  'arrange and feed parts and place in fixture.'
- Every operation must be resolved into part functions using the illustration of basic and part functions together with their symbols as described in Chapter 5, Figures 5.4 to 5.7.

At the same time it is also specified whether or not the part functions must be undertaken during the switching time t_s or the holding time t_h of the transfer equipment.

- The verbal symbolic representation is adequate for simple parts. A representation in functional sequence is necessary for parts which are difficult to handle (see Figure 5.7).
- If the integration of a manual operation is necessary on account of the product design and assembly sequence, the time requirement for this operation must be determined using pre-determined times (MTM/work factor, etc.) and also be included in the function analysis.

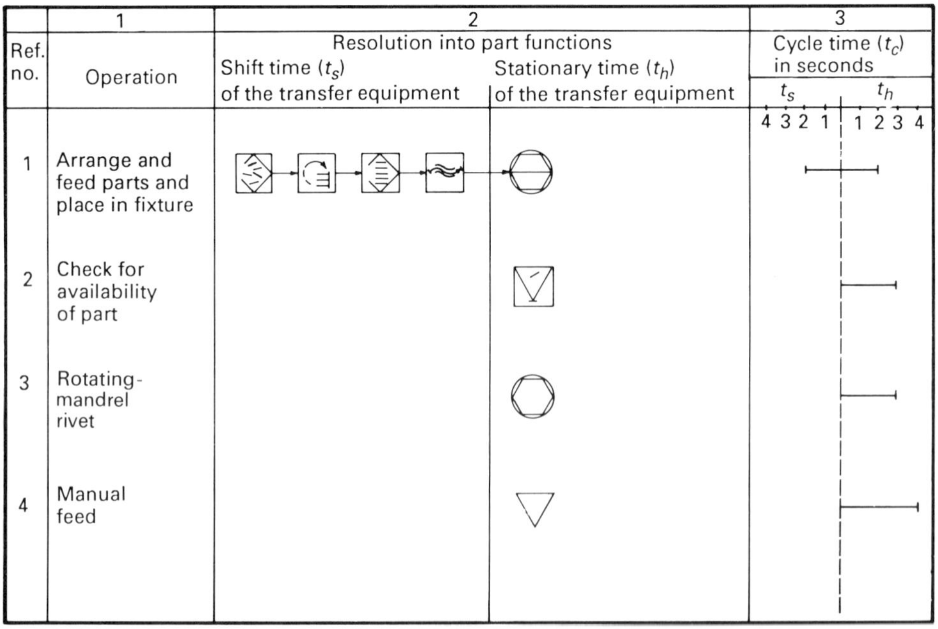

Ref. no.	1 Operation	2 Resolution into part functions		3 Cycle time (t_c) in seconds
		Shift time (t_s) of the transfer equipment	Stationary time (t_h) of the transfer equipment	t_s 4 3 2 1 | t_h 1 2 3 4
1	Arrange and feed parts and place in fixture	⟨symbols⟩	○	⊢—⊣
2	Check for availability of part		▽	⊢—
3	Rotating-mandrel rivet		⬡	⊢—
4	Manual feed		▽	⊢—

Figure 11.14 Function analysis

A distribution of the individual functions and their correlation to the shift and stationary times of the transfer equipment is obtained by this analysis and in which functions which must only be undertaken during the stationary period of the transfer equipment are highly important for the shortest possible cycle time. The longest single cycle time determined by this function analysis gives the technically shortest cycle time of a single station in an automated assembly system.

11.7 Determination of cycle time

The planned output is derived from the product volume structure and the service life of the equipment. The required cycle time t_c is calculated as follows from the planned output PO and the achievable static availability A_{STA}:

$$t_c = \frac{3600 \times A_{STA}[\text{seconds/hour}]}{\text{PO [parts/hour]}} \qquad (11.2)$$

So if PO = 500 parts/hour and $A_{STA} = 0.8$,

$$t_c = \frac{3600 \text{ s} \times 0.8}{500 \text{ parts/hr}} = 5.76 \text{ s/part}$$

The weak point of this calculation is in the assumption of a static availability. This must be obtained in detail and, on account of risks, not set too high (see Section 11.10). Typical values fall between 0.8 and 0.85.

If the shortest possible cycle time of an automated assembly system as determined by the function analysis is less than the calculated cycle time, the probability is very high that the actual output will be equal to the planned output. If, on the other hand, the cycle times determined by the function analysis are longer, a check must be made to determine if this is generally the case or only for certain operations. In the first case, this can mean that the planned performance cannot be achieved with the equipment. Then, the planned utilization period can be increased. However, if this is already exhausted, the possibility exists to design the equipment as a so-called duplex type. A duplex type means that all operations are undertaken twice per working cycle simultaneously. On transfer systems with workpiece carriers not permanently connected to the transfer equipment, two workpiece carriers can move into one station and the operation performed twice. If these possibilities are also exhausted, two assembly systems can then be necessary.

11.8 Layout planning

11.8.1 Principles of layout planning

The object of layout planning is to achieve an optimum arrangement of an assembly system with regard to workpiece flow and material provision in relation to the space available. At the same time, personnel requirements must be considered.

The experience gained from the assembly sequence, function analysis and the determination of cycle time including the determination of availability (see Section 11.10) are included in the layout planning as follows:

- Product design, assembly sequence and cycle time determine the use of the assembly system.

318 Planning and efficiency of automated assembly systems

- Product design, assembly situation and the resulting workpiece carrier design and also the availability expectation determine if the assembly object can be assembled in one single system or if the operations must be distributed over several machines.
- The required cycle time determines the kinematic structure of the system. With cycle times below 2.5 to 3 s the workpiece carriers should be permanently connected to the transfer equipment and not be carried by friction; otherwise the acceleration forces can give rise to problems.
- The assembly situation and function analysis determine the possible level of automation and therefore also the implementation of manual operation.

The overall arrangement must be planned so that the workpiece flow within the assembly system can be integrated into the overall sequence of production. The provision of material and associated gangways must also be considered.

In particular, as far as possible, material provision in large-volume containers should only be from one side. Such a requirement is generally more difficult to fulfil with the application of circular cyclic machines than for transfer arrangements. Furthermore, all factory rules and regulations must be obeyed.

11.8.2 Layout examples

The layout diagram can be subdivided into coarse and fine planning. Fine planning is only possible following finalization of the assembly systems and, as a general rule, is only possible for the individual station of an assembly plant. With coarse planning, the individual stations can be shown symbolically; it is, however, important that the overall dimensions be determined accurately.

11.8.2.1 Example 1
Figure 11.15 shows the rough layout of the assembly system for an electromechanical component group consisting of nine different parts. Six parts are arranged and fed

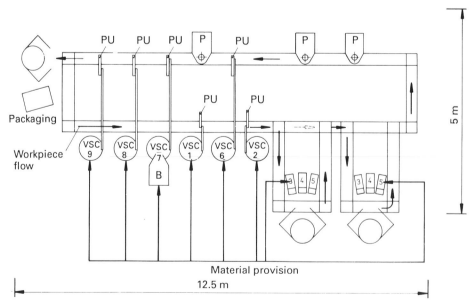

Figure 11.15 Layout of an assembly system in transfer line form with material provision from one side (PU = positioning unit, VSC = vibratory spiral conveyor, P = press, B = hopper, 1–9 = part numbers)

automatically by vibrating spiral conveyors, grasped and positioned by positioning units. Three other workpieces are positioned manually at two parallel work points. The system is constructed with the application of a double-belt system (Bosch type) with circulating workpiece carriers without a permanent connection to the transfer equipment. In order to achieve a limited decoupling for the manual assembly operations from the machine cycle, two parallel work points have been arranged on account of the work content of the manual work points.

The workpiece carriers are routed for these operations. All other functions are performed automatically.

The finally assembled products are removed and packaged manually. The single cycle time is approximately 7 s. The provision of material has been limited to one side by the selected arrangement of the vibratory spiral conveyers and manual work points on one side of the double-belt circulating system and the positioning of the automatic stations on the other. By reason of the rectangular arrangement of the double-belt system, the start and end of the assembly process are on the same narrow end of the arrangement.

Figure 11.16 shows an alternative arrangement schematically. In this solution the length of the arrangement is shortened but the width increased because the provision of material is distributed over two sides. The selection of both possibilities must be based on the prevailing availability of space or conditions for integration into a production

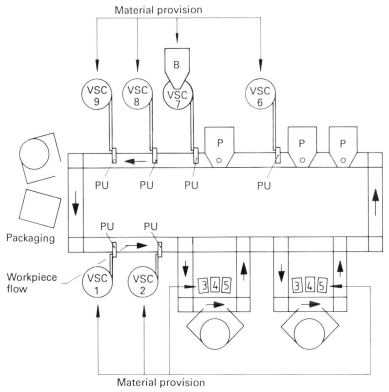

Figure 11.16 Layout of an assembly system in transfer line form with material provision from both sides (PU = positioning unit, VSC = vibratory spiral conveyor, P = press, B = hopper, 1–9 = part numbers)

sequence. The disadvantage in this case is that the part feed stations are arranged on both sides and consequently the distances for the maintenance person to rectify faults on the part feed stations are longer.

11.8.2.2 Example 2

Figure 11.17 shows the layout planning of an assembly line consisting of five loosely interconnected circular cyclic automatic assembly machines I to V for the assembly of the switch shown in Figure 11.5.

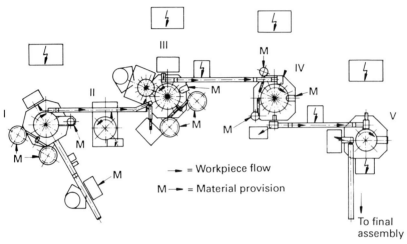

Figure 11.17 Layout of an assembly system for the assembly of switches as shown in Figure 11.5

The assembly sequence is shown in Figure 11.7. The assembly sequence analysis and the assembly sequence show that the part to be assembled must be assembled in five different workpiece carriers. The required cycle time of 3 s requires a permanent connection of the workpiece carriers to the transfer equipment and also a distribution of the working processes over five machines. The low work content per machine makes the application of circular cyclic units possible. The product assembly sequence and the function analysis necessitate the application of two manual work points for the arrangement and handling of two single parts which cannot be handled automatically. The decoupling of the individual machines is achieved by linking via conveyor belts.

On account of its geometry, the base part can be transported on conveyor belts without a workpiece carrier so that it does not lose its arranged position. At the same time, the conveyor belts between the individual machines form an intermediate buffer which is designed so that short-term disruptions on the machines do not result in shutdown of the whole system. The manual work points are equipped with reserve sections for decoupling from the cycle times of the machines (see Figure 6.40). With this arrangement, material provision is only possible from one side of the assembly system; however, with the small size of the individual parts, this is not an important factor. The workpiece flow is linear in one direction through the whole assembly system. The discharge of the fully assembled unit is for connection to the final assembly line, in this case at right-angles to the main material flow; however, under different conditions it could also be in the assembly system material flow line.

11.9 Determination of personnel requirement

A distinction must be drawn between the following in the determination of the personnel requirement on automated assembly equipment:

- Personnel for manual activities which are integrated in the assembly system
- Personnel for assembly system maintenance and supervision.

The requirement for integrated manual activities is derived from the number and duration of the operations to be undertaken and by application of the function analysis together with the application of pre-set time allowance methods (see Section 11.6).

The duties of the equipment maintenance personnel include the following activities:

- Maintenance of the functional safety of the equipment
- Action when faults occur in the feed systems, to avert shut-down of the system
- Action and fault elimination with shut-down of the system
- Maintenance of material availability in the feed systems.

The time required to complete the above activities is determined by the following factors:

- Space occupied by the system
- Accessibility to the stations
- Number of automatic stations
- Cycle time
- Complexity of the individual stations
- Degree of permanent interlinking
- Availability of the individual stations
- Handling quality of the parts to be assembled
- Optical or acoustic fault-warning equipment.

The uninterrupted operation of an automated assembly system is decisive for its economy; fault rectification ability is therefore one of the principal responsibilities of the equipment maintenance personnel. The fault occurrence time patterns on an assembly system and the maintenance personnel deployment are shown in Figure 11.18.

The reaction time is the time required by the maintenance person to observe the fault and identify the fault location. This time is dependent upon the optical or acoustic

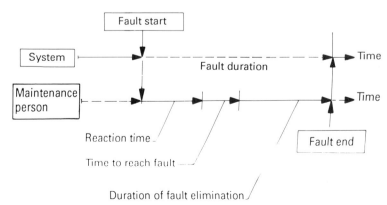

Figure 11.18 Time pattern of a fault (Herzlieb, IFA, Hanover Univeristy)

fault-warning system and the reaction time of the maintenance person. The time to reach the fault is determined by the distance between the maintenance person and the fault at the time of recognition. This implies that the work point of the maintenance person must be arranged as close as possible to the point at which, on average, the maximum number of faults occurs.

Figure 11.19 shows that the maintenance requirement increases with an increasing number of permanently interlinked stations. On the other hand, with a limited number of permanently interlinked stations and adequate buffer formation between the individual machines, and, in particular in the feed systems, with a favourable location of the maintenance personnel, many faults can be rectified before shut-down of the system is necessary. As is also shown in Figure 11.20, the maintenance requirement increases with shorter cycle times.

As an empirical guideline it can be assumed that with a single reliability of 99.5% per automatic station, good fault signalling and easy access to the fault points, with a 3 s cycle time a maintenance person can service 15 to 20 stations. The individual availability decreases with a decreasing cycle time, and by the same token the maintenance requirement increases.

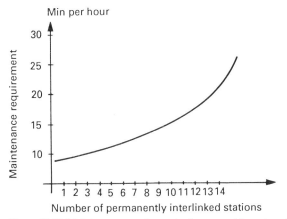

Figure 11.19 Maintenance requirement of an assembly system in relation to the number of rigidly interlinked stations (empirical values)

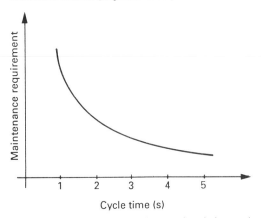

Figure 11.20 Maintenance requirement in relation to the cycle time (empirical values)

11.10 Determination of availability

A general problem with the operation of automated assembly systems is to maintain a high uniform availability over the longest possible period of time. This is absolutely vital with the high capital investment for such systems. With increasing complexity of such systems, the probability increases that a part of or the whole system can be stopped by the failure of one part. With automated assembly systems it is therefore important that the time-related characteristics be investigated during the planning stage and adequate measures implemented for high availability. Efforts to improve the availability of assembly systems should not be restricted to improving the reliability of the system components, but must also include the effect of the single parts to be machined and the workpieces for assembly.

The following factors adversely affect the availability characteristic and must therefore be included in the determination:

- Parts quality
- Number of stations
- Individual reliability of the stations
- System structure
- Start-up characteristic
- Personnel qualifications.

The experience gained from the availability determination during the planning stage is applied in the planning stage function analysis, cycle time determination, layout planning and personnel requirement determination [36].

11.10.1 Parts quality

Most of the causes of faults lie in the area of part feed and are the result of an unsatisfactory part quality or unreliability of the operating equipment. The highest reliability of the equipment is not itself sufficient if parts of non-uniform quality are to be arranged and fed automatically. In addition to the quality of the parts, cleanliness and the inclusion of foreign bodies play an important role (see Section 2.2.6).

The additional quality requirements for pre-production resulting from the automation must be written into a specification for the specification of the assembly systems. If these requirements result in additional investment costs, they should be included in the investment cost estimation and also in the cost comparison with alternative assembly systems.

11.10.2 Number of stations

The reliability of the assembly system is a product of the reliability factors of the single stations integrated into the assembly system. The reliability decreases with an increasing number of permanently connected stations (see Section 6.7.3.1).

In conjunction with the quality of the parts to be processed, the availability is determined to a very high degree by the individual availability of one such station and the overall structure of an assembly system.

11.10.3 Availability of individual stations

A high availability of the individual stations is a requirement for a high overall availability of the system. In inspection stations or assembly stations, pressing or riveting, for example, can be largely ignored in the availability determination since they have an availability of practically 100%. On the other hand, with resistance welding presses, for example, the electrode life is significant in the availability, since irregular stoppages for electrode changing or electrode cleaning occur. The part feed stations must also be critically examined since, in this case, a relatively large number of single functions occur in the availability calculation for one such station. The design of a parts feed station is described in Section 6.3.1, Figure 6.8. The overall reliability of one such station is given by the quality of the parts to be processed, the reliability of the feed equipment, the reliability of the transfer between the feed unit and discharge rail, the reliability of the discharge rail and the process of placing parts in magazines, and also the reliability of the mode of operation of the separating station and handling equipment. The availability determination of one such station can be made in accordance with the scheme as shown in Figure 6.9. The reliability coefficients to be correlated to the individual factors are dependent upon the degree of difficulty of the individual parts, the quality of these parts and also the selection of the equipment to be used. The actual level of expected overall reliability of a single part feed station is calculated on this basis.

11.10.4 System structure

The method of interlinking the stations is determined by the system structure. If workpiece carriers must be permanently connected to the transfer equipment on account of short cycle times or high positional accuracies, the method of interlinking must be rigid. A limited degree of decoupling is possible with equipment without a rigid connection of the workpiece carriers to the transfer equipment, depending upon the cycle time and the possibility for a buffer formation of the workpiece carriers between the individual stations. With similar systems of high complexity, linking of the individual machine systems together is critical for the overall structure of the system. The availability increases with the level of decoupling of individual system components or part systems. In this respect, the application of workpiece buffers is necessary. In an automatic production system, in the event of fault-related stoppages of certain system components, upstream or downstream connected stations are fed by the workpiece buffers for a limited period of time. The effect of these points of fault buffers is all the greater the higher the possibility of compensating for fault-related material flow interruptions on adjoining stations.

The location or capacity determination of fault buffers can only be made if the fault characteristics of the system are known or if reliable estimates can be made in this respect. In addition, the buffer capacities are also dependent upon the number and the location of the maintenance personnel [37]. All the same, fault buffers only have a limited effect and from the aspect of additional capital investment and space requirement should not be too large.

11.10.5 Initial operation characteristics

It can take several months or up to a year for automated assembly systems to reach their specified output and thus their planned output. The reasons for this are serious unforeseen or localized disruptions in the operations, and quite often also planning errors and inadequate consideration of the so-called short-duration faults of less than 2 min duration. The costs for the initial operating phase are added to the one off costs of the system and therefore increase the overall capital investment. The costs are generally attributable to the reduced system output and the increased personnel deployment during the initial operation phase. The end of the initial operation phase is the point at which the system reaches its planned output and from which no further economically justifiable production increases can be established.

The duration of the initial operation phase and therefore also its costs is dependent upon the complexity and structure of the system. Highly decoupled systems can be constructed in stages; this reduces the initial operating costs. On the other hand, complex systems and, in particular, those with permanent interlinking, are not so devisable in the course of their initial operation and therefore result in a higher lever of initial operating costs.

11.10.6 Personnel qualifications

To a very large degree, the availability of automated assembly systems is determined by the qualification and motivation of the maintenance personnel. This personnel unit must have specialist training and, in addition, be trained specifically for plant maintenance work. The focal point of this training is dependent upon the existing level of skill. For installations of a size which only require one maintenance person, the specialist training should be in the field of mechanics/machine-building. Additional training must then be given, in accordance with the type of plant, in pneumatics, electrics and electronics. The additional training must be such so that the maintenance person can identify and localize faults. He should be able to rectify faults up to a certain magnitude and, with the occurrence of more serious problems, be in a position to engage the correct expert by their ability to identify the cause of the fault. With larger systems which require several maintenance persons, a range of different skills is preferred. On a system which requires three maintenance persons, it is, for example, advantageous to deploy two skilled people in the field of mechanics/machine-building and one in the field of electrics/electronics.

The maintenance personnel must be trained so that they fully understand the product to be assembled and the assembly equipment. Involvement, for example, in the manual assembly of the product to be produced by automation at a later date or in the pilot production of a new product start-up should be included in the training programme.

The best practical experience of the system is gained by involvement of the future maintenance personnel in the final assembly and erection of the assembly system. Assembly system expertise is gained during initial operation of individual functions or stations [43].

11.11 Assembly systems

A specification based on the results of the individual planning stages of the planning project must be compiled for the selection between alternative solutions for the assembly systems. Quotations can then be obtained from manufacturers based on the specification.

With in-house construction, fine planning is undertaken based on the specification for the determination of the capital investment and procurement time.

Based on the results of the single analysis, the following factors must be included in the specification:

- Cycle time

- Integration of all necessary manual operations
- System structure
- Conditions relating to the periphery of automated assembly.

In addition, it is advisable to specify the individual stations as shown in the example in Figure 11.21.

Specification	Page 6 of 12 pages For project : Valve assembly	Single station no: 4 Feed of leaf-spring in accordance with drawing 507-01
Ref no.	Title	Requirement/type
1	Delivery condition of the parts	Loose bulk material in quantities of 10 000
2	Module for storage and arrangement	Vibratory spiral conveyor 300 mm dia. Discharge: left Inclination of discharge: 3° Noise damping cover
3	Discharge rail	Electromagnetic drive 3° inclination from vibratory spiral conveyor to separating station Semi side cover removable by snap connectors Length 400 mm = parts buffer for 2 minutes production time
4	Separation	Pneumatic transverse slide action
5	Sensors	Light barriers
6	Controller	Switch off vibratory spiral conveyor in the event of parts build-up in discharge rail.
7	Positioning unit	Pneumatic manufacture: NN 150 mm X - and 40 mm Z axis movement
Compiled by: Date: 12.3.86 Name: NN	Checked: Date: Name:	

Figure 11.21 Example: specification for a parts feed station

11.11.1 Cycle time

The result of the cycle time determination must be compared with the result of the availability determination and, if necessary, corrected accordingly. The finally determined cycle time affects the selection of the individual modules of an assembly system as follows:

- Cycled or uncycled transfer equipment
  With cycled transfer equipment, the method of drive (e.g. pneumatic or electric motor) must be specified.

- Selection of the handling equipment
  Programmable/not programmable; drive type (pneumatic, electric motor, cam-operated).
- Determination of the costs for monitoring and fault point signalling.

11.11.2 System structure – integration of necessary manual operations

The system structure is based on the product design, integrated number of parts, handling characteristics, assembly directions and the assembly processes, together with the assembly order. The degree to which assembly operations can be distributed over individual sections and also the possibility for buffer formation between the stations is determined by the assembly order and workpiece carrier design. The size of the workpiece buffer at the parts feed stations or between the individual assembly systems must be determined on the basis of the availability calculation and the function analysis. The integration of the manual assembly operations is determined by the product design and the assembly situation. The manual operations must be planned so that the person by whom they are performed is largely decoupled from the fixed cycle and the point of assembly constructed in accordance with ergonomic principles. The work content should be arranged to avoid monotony of work.

11.11.3 Conditions on the periphery of automatic assembly

The quality requirements relating to the delivered condition of the individual parts which are processed in the assembly system are based on the product analysis and the availability determination. At the same time, particular attention must be given to details which do not play any role in the function of the part but are, however, used as arrangement and sorting features.

11.11.4 Summary

A revised detailed layout planning is required based on the specification. At the same time, it is advisable to break down the overall layout of the assembly system into detailed layouts of the individual assembly machines. A short description and listing of the modules to be used should be compiled with the layout of the individual machines.

11.12 Investment calculations

It is a relatively simple matter to produce an investment calculation if the assembly system is offered completely by one specialized supplier. The installation costs, any expected costs in the periphery of assembly and the initial running costs only need to be added to the known system costs. To what extent a risk factor for unforeseen difficulties (e.g. longer initial running costs, late delivery, etc.) needs to be added to the calculated amount depends upon the complexity and degree of difficulty of such a system. With simple systems, a risk factor of 3% to 5% is common; with complex systems this can be up to 20%.

For the investment calculation with in-house construction of automatic assembly systems, a detailed procedure is necessary based on the specification and the system description in conjunction with the fine layout.

328 Planning and efficiency of automated assembly systems

Project: circular cyclic automatic assembly machine
Product: valve 172103
Operator:
Name:
Date:
Page: of
Pages

Item	Title	Bought-in part		Design (h)	Material (currency units)	In-house costs			
		Supplier	Currency units			Turning (h)	Milling (h)	Precision work (h)	Assembly (h)
1	Circular cyclic basic unit								
1.1	● Machine frame	NN	12 000	8					
1.2	● Circular cyclic unit 12-station type xxx	NN	7 500						
1.3	● Rotary table for 12 stations 700 dia.			10	150	10	15	14	20
1.4	● 12 workpiece holders			20	240	8		6	
2	Feed station 1: bracket								
2.1	● Vibratory spiral conveyor: 300 dia.	NN	2 600	10	20	3	6	1	5
2.2	● Discharge rail, separating station		300	15	30	6	8	2	15
2.3	● Positioning unit with gripper	NN	6 300	10	30				5
2.4	● Electrical installation		150						
3		x	x	x	x				
4		x	x	x	x				
5		x	x	x	x				
6	Total assembly								250

Intermediate sum Balance:

Figure 11.22 Investment cost determination for in-house construction of assembly systems

Figure 11.22 shows a practical proforma for this purpose. An assembly system – in this example, a circular cyclic automatic assembly machine – is broken down into its principal components and the costs determined in terms of bought-in parts and in-house requirements. To what extent the in-house requirements are detailed depends upon the organization of the in-house machine construction facilities. In the example, the distinction is made between design costs, material costs and production costs and in which the latter is subdivided into turning, milling, precision work and assembly. If only the unit costs for mechanical work are specified for mechanical machining, the detailed breakdown can be omitted. On account of the requirement for a detailed listing of the individual assemblies and their components and the correlation of the expected costs, the planner responsible is then obliged to estimate specifically the individual parameters so that, in comparison with the general cost estimation, a possible estimation error is considerably less.

The results of this detailed investment calculation are included in the general summary as shown in Figure 11.23 and the hours determined multiplied by the respective hourly rate or, with bought-in parts, the respective material procurement cost supplements added. The amount as determined in the intermediate total enables a comparison to be made with possible external tenders. The installation costs of the installation, the costs incurred for guaranteeing the function of an automated system and also the initial running cost must be added. A risk supplement must also be included for in-house construction and external supply.

Investment calculation for project:

Reference no.	Title	Currency units
1	Bought-in parts (CU) +total cost supplement	
2	Design requirement hours × hourly rate (CU)	
3	Costs for in-house construction:	
3(a)	Material costs + general cost supplement (%)	
3(b)	Turning (hours × CU-hourly rate)	
3(c)	Milling (hours × CU-hourly rate)	
3(d)	Precision machining (hours × CU-hourly rate)	
3(e)	Assembly (hours × CU-hourly rate)	
	Intermediate total:	
4	Installation costs	
5	Peripheral costs for automatic assembly	
6	Initial running costs	
7	Risk supplement for unforeseen difficulties	
8	Investment total:	

Figure 11.23 Calculation of investment costs for in-house construction of assembly systems

11.13 Evaluation and selection

In order to undertake an evaluation of the possible assembly systems, it is necessary to compile work-point cost calculations for the individual systems and by which the selection of the most favourable assembly system can be made on the basis of the achievable economy.

A space cost calculation includes all applications of the system; these costs are subdivided into the machine hourly rate and the personnel-related costs.

The cost calculation with the machine hourly rate can be undertaken in accordance with VDI guideline 3258 [47].

11.13.1 Machine hourly rate

Five cost classes which are explained in further detail in the following subsections are included in the machine hourly rate (C_{MH}).

11.13.1.1 Estimated depreciation (C_D)

Within the context of a space cost calculation, the estimated depreciation is in linear form. The replacement cost must initially be determined for the calculation of the depreciation. It is calculated from the procurement cost, installation and possible initial running costs multiplied by a correction factor for the expected inflation rate. The service life of an installation in years is dependent upon the sales life (the so-called economic service life) of the produced product. Since, on account of their structure, assembly systems fall within the field of special machines, it is advisable to determine the service life differently for reusable components of the system and for the product-specific components. For example, the depreciation for product-specific components can be three years, and eight years for standardized components.

11.13.1.2 Estimated interest (C_I)

Interest costs are incurred in respect of the capital invested in the assembly system, and which must be included in the cost calculation as estimated interest charges. They arise on account of the linear depreciation for 50% of the replacement cost of the equipment.

11.13.1.3 Space costs (C_S)

The space costs are dependent upon the space occupied and the costs per unit area. The space component is determined by the area of the system, the required operating area and the standing area required for the provision of the workpieces. The space which is required for repair purposes and servicing of the system is also a part of the operating area.

11.13.1.4 Energy costs (C_E)

In this case, all types of energy such as current, compressed air, water and, if applicable, gas costs must be determined. Their consumption must be calculated over the period of use.

11.13.1.5 Maintenance costs (C_M)

All measures which are used for maintaining the planned operating ability during the period of use are a part of maintenance. The maintenance service must also be included. On account of the different design types, guidelines for maintenance costs on automatic assembly systems have not been published. Commencing with semi-automatic, uncomplicated systems, it is, however, advisable to assume a value of 5% of the replacement cost per annum for maintenance costs and, with fully automatic systems, a value of 10% per annum.

11.13.16 Calculation of the machine hourly rate

The utilization time (T_U) per annum must be determined for the calculation of the machine hourly rate. It is calculated based on the utilization period per working day and the number of working days per annum.

The machine hourly rate of an automated assembly system is calculated by the following formula:

$$C_{MH} = \frac{C_D + C_I + C_S + C_E + C_M}{T_U} \text{ (currency units/hour)} \tag{11.3}$$

11.13.2 Personnel-related costs

With personnel-related costs, a distinction must be made between personnel associated with assembly and indirect personnel. Under the latter the proportion of costs relating to supervisors or chief operators needs to be considered.

It is also important to differentiate between the different activities of the employed workforce. The costs to be determined are based on the number of persons and the respective wage rate groups (= wage + all necessary additional costs). With indirect related personnel-dependent costs, the respective percentages must be included. The personnel costs of salaried employees must be converted to hourly costs. If a system is operated on a shift basis, the shift allowances C_{SA} are calculated by the following formula:

$$C_{SA} = \frac{\text{shift allowance/day}}{16 \text{ h}} \tag{11.4}$$

These costs must be included in the personnel-related costs.

11.13.3 Work-point cost calculation

The work-point costs are derived by the addition of the machine hourly rate and the respective personnel-related costs.

The assembly costs per assembled part C_{AP} are calculated by the following formula:

$$C_{AP} = \frac{\text{work-point costs (CU/h)}}{\text{net output of system (parts/hour)}} \text{ (CU/part)} \tag{11.5}$$

Figure 11.24 shows the formula for the determination of the work-point costs for assembly systems.

A similar work-point cost calculation must be undertaken specifically for each alternative solution so that the most favourable solution can be selected on the basis of the result, e.g. by calculation of the amortization times.

Figure 11.25 shows one such example. The rationalization of the manual assembly for a product which is produced at a rate of 800 000 parts per annum resulting in manual assembly costs of 1.20 CU per part was selected. Two alternative solutions were investigated: solution no. 1 as semi-automatic and solution no. 2 as a fully automatic system. As shown by the amortization time calculations, the semi-automatic solution generates a return on deployed capital by production savings in 1.04 years, and the fully automatic solution in 1.17 years. This result clearly indicates that, in this case, the semi-automatic solution is more economic than the fully automatic solution.

332 Planning and efficiency of automated assembly systems

	Work-point cost calculation	
	Project:	Solution no.
	Replacement cost of system (C_R) _____ CU	
	Service period n _____ years	
	Service period n _____ h/day	
	Net output _____ parts/h	
Machine hourly rate	$C_D = \dfrac{C_R}{n} =$ _____	CU/annum _____
	$C_I = \dfrac{C_R}{2} \times p =$ _____	CU/annum _____
	$C_S =$ _____ m² × CU/m² × month	CU/annum _____
	C_E: Current costs _____ kW × CU/kWh _____ = CU/h _____	
	Compressed air _____ m³/h × CU/m³ _____ = CU/h _____	
	Water _____ m³/h × CU/m³ _____ = CU/h _____	
	_____ CU _____ = CU/h _____	
	h/annum × CU/h _____	= CU/annum _____
	C_M = replacement cost × _____ %	= CU/annum _____
	Total SYSTEM COSTS/ANNUM	= CU/annum _____
	$C_{MH} = \dfrac{\text{system costs/annum}}{\text{h/annum}}$	= CU/h _____
Personnel-related costs	_____ Persons × wage CU/h _____ + _____ % wages additional costs	CU/h _____
	_____ Persons × wage CU/h _____ + _____ % wages additional costs	CU/h _____
	Persons × shift premium = $\dfrac{\text{CU shift premium}}{\text{h/day}}$ =	CU/h _____
	Total personnel costs/h =	CU/h _____
Assembly costs	Machine hourly rate C_{MH}	CU/h _____
	+ Personnel costs/h	CU/h _____
	= Assembly costs/h	CU/h _____
	$\dfrac{\text{Assembly costs (CU/h)}}{\text{Net output (parts/h)}}$ = Assembly costs/part	CU/part _____

Figure 11.24 Assembly cost determination (work-point cost calculation)

Project:

Production per annum: 800 000 parts   Comparative costs (currency units)

	Manual assembly	Solution 1: semi-automatic assembly	Solution 2: automatic assembly
Capital deployment (CU) (replacement cost)	150 000	450 000	730 000
Depreciation over 4 years (CU)	37 500	112 500	182 500
Assembly costs (CU/part)	1.2	0.80	0.65
Savings (CU/part)		0.40	0.55
Savings (CU/annum)		320 000	440 000
Amortization time $= \dfrac{\text{capital deployment}}{\text{savings/annum} + C_D}$		1.04 years	1.17 years

Figure 11.25 Comparative calculation for different solutions

11.14 Optimized overall solution

As a result of the evaluation of the different assembly systems, either a direct solution can be made or a revision and optimization of a solution including part solutions of various assembly systems. With such a revision, reference is made to the detailed results of the individual planning stages of the different assembly systems in order to compile a new overall concept and layout. The means of achieving the most economic optimized overall solution is the recompilation of a work-point cost calculation for comparison with the work-point cost calculation of the initial basis.

11.15 Computer-aided planning of automated assembly systems

Computer-controlled assembly systems are used today in the field of assembly technology. The application of computer-aided methods in the planning and the layout of assembly systems gives increasing momentum to this trend.

The object lies in the interlinking of part functions to a specialized concept of integrated data processing [39]. The work areas involved are shown in Figure 11.26. The basis for this procedure is a central data bank. The reference point for the application of these integrated systems is the analysis of the assembly operation. The assembly planning is based on the analysis. The requirements for modules are defined and determine the boundary conditions for the CAD specification of the part systems. Any possible collision problems can be identified during the planning phase by the graphical simulation of an automatic assembly station. Assembly cells can be designed by application of the data of the individual modules which are, in fact, modules of more complex assembly lines. These methods and systems are at present in the early development stage but are gradually gaining in importance for practical application.

11.15.1 CAD layout planning

By application of the CAD systems currently available on the market it is also possible to produce layout plans and the arrangement of assembly systems. An example in this

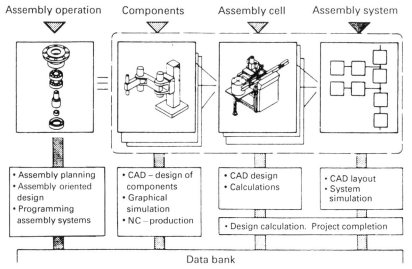

Figure 11.26 Computer-aided methods in the planning of assembly systems (Feldmann, Erlangen University)

respect is the CADLAS system developed at IPA Stuttgart; this is a computer-aided method for the layout arrangement of automatic assembly systems (CADLAS = Computer Aided Design of Layouts of Automated Assembly Systems) [45].

The system is aimed at these objectives:

- Simplification of the optimization process in the production of layouts and therefore qualitative and quantitative improved planning results
- Improvement of the visual presentation
- Possibility of three-dimensional presentation
- Increase in planning effectiveness
- Time saving in layout production
- Simple and rapid alterations.

The CADLAS program package for computer-aided layout production is subdivided into the following three programs:

- Data and command input in menu technology
- ROMULUS for the realization of all graphics commands
- Switching program ROMPIPE for the transfer of input commands of ROMULUS.

The hardware and software structures for the program package are shown in Figure 11.27. A VAX 11/780 computer is used as the control computer for ROMULUS, for the program for the data and command input and for switching between programs. All graphics functions are generated by a graphics processor PS 300 and a high-resolution graphics screen. Operator control (menu technique) supplemented by text input and output is undertaken on a terminal. A disc memory is used for the recording of all data, and the completed layout can be printed using a simple plotter.

The program for data and command input is subdivided into a definition and a manipulating mode. In the definer mode the user selects all the available system

Computer-aided planning of automated assembly systems   335

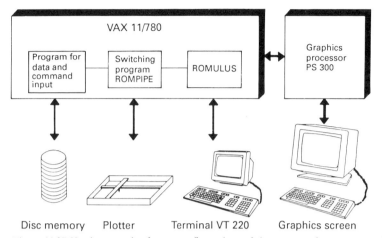

Figure 11.27 Hardware and software configurations of the program for computer-aided design of layouts of automatic assembly systems (CADLAS) (IPA Stuttgart)

elements from a list, for example conveyor-belt sections, circular indexing tables, rotational units, vibratory spiral conveyors, box magazines, etc. and also the element required for the construction of the stations. The respective geometrical details are stored in a library, and these can be added to at any time. Part systems of the individual stations are defined in a further work section. Total systems can be formed from the part systems.

Following input of all data, all graphically shown elements of the first station are made available to the user on the screen in the manipulation mode.

These can be arbitrarily manipulated and combined to form one station. Following the construction of the individual stations, the part systems and overall system are constructed by the same procedure.

Figure 11.28 shows a section from a manually constructed layout and an automatically constructed assembly. The planned modules are shown to scale in a plan view. In the computer-aided layout, any components which were not available in the

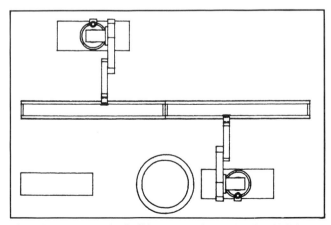

Figure 11.28 Layout of a flexible automated assembly line (IPA Stuttgart)

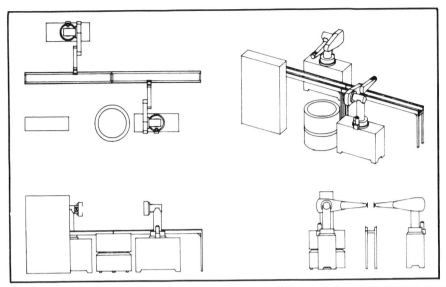

Figure 11.29 Computer-aided layout generation of an assembly system using CADLAS (IPA Stuttgart)

element library were generated and then stored there. Following definition of the required elements, they were assembled to form the same assembly station. The result is shown in Figure 11.29 [45].

11.15.2 Simulation technique

Simulation offers the possibility of reproducing known systems and modules and also of investigating the effect of different arrangement strategies and structural changes. The operational characteristics of systems which are still in the project stage can be analysed in terms of trend by their application and therefore weak points identified at an early stage and also their quantitative effect on the total system estimated. The availability factor of the system can be increased during this planning stage by system structure alterations, buffer dimensioning or by placing a higher requirement on the individual availabilities.

Initially, the abstract of an actual system is defined by a model produced in a computer-compatible form. This model shows the system structure and also the functional interrelationship of the individual modules. The stations, buffers, transport equipment, etc. must be characterized by parameters such as cycle time, buffer size, conveyor-belt speed, single part quality, etc. and also the resulting faults.

The procedure for a CAD simulation of assembly units is shown in Figure 11.30. Various programs have been developed for the simulation of assembly systems and which are generally structured for specific applications. An example in this respect is the general simulation language 'GPSS-Fortran' for the investigation of branched assembly systems. The program is of modular construction and permits the inclusion of problem-related subprograms of a higher-level language, Fortran [39].

An important principle for the interpretation of simulation results must always be observed. The result of a simulation study is only as good as the accuracy of the data on

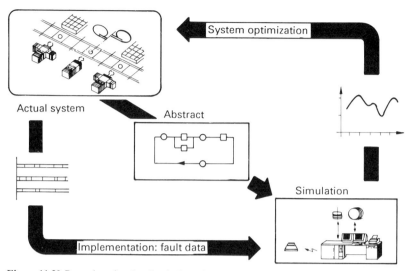

Figure 11.30 Procedure for the simulation of assembly systems (Feldmann, Erlangen University)

which it is based. In addition to the quite often expensive and complex construction of systems analysis models, this is the actual problem of the simulation technique.

With regard to simulation as a project aid, it can be difficult to determine fault parameters for part components. Values can, however, be determined based on the application of similar modules at other points [30, 37, 41].

The correct estimation or determination of possible causes of fault and the resulting fault times is an important factor for the correct assessment of the expected availability of an automated assembly system. This cannot be replaced by a simulation technique, but confirmation of the data material is achieved [39]. Similarly to the application of CAD, simulation technology is still in the laboratory stage; it is, however, expected that it will become widely used in practice in the course of time.

Chapter 12
Practical example: planning and realization of an automated assembly system

12.1 Introduction

The planning and realization of an automated assembly system for the assembly of the thermoswitch as shown in Figure 12.1 is described below, based on the application of the planning guidelines as outlined in Chapter 11, Figure 11.1.

The switch consists of 37 single parts (21 different parts and the application of a locking paint). In addition to the handling and assembly of the single parts, operations such as measuring the contact gap and adjustment, etc. must be undertaken. Assembly is manual during the introductory phase of the product. Automated assembly is, however, necessary on account of the expected production volume.

12.2 Planning procedure

12.2.1 Requirement list

The product sales volume forecast indicates that, following an introductory phase of approximately one to two years, the expected annual production is 3.5 million switches. The economic service life of the product is estimated as six to seven years, so that, after deduction of the introductory phase, the expected service life of the assembly system is four years. On account of the high number of switches required, the system is to be designed for two-shift operation so that, in the event of market fluctuations, single-shift operation can be introduced as necessary.

At an annual production rate of 3.5 million switches over 16-hour working days, a planned production rate P_{pl} is calculated as follows:

$$P_{pl} = \frac{3\,500\,000 \text{ parts}}{220 \text{ days} \times 16 \text{ h/day}} = 994 \text{ parts/hour}$$

This production rate requires a high level of automation, so a target automation time of one year is specified.

12.2.2 Product analysis

The product analysis is undertaken in accordance with the assembly-extended ABC analysis as described in Section 2.2. The results of the analysis in relation to the supply

Planning procedure 339

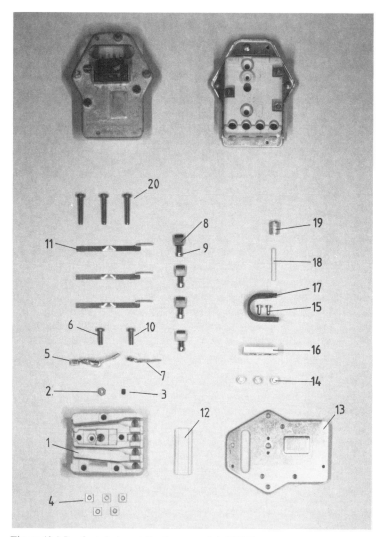

Figure 12.1 Product design of the thermoswitch (EGO)

condition, handling characteristics and quality requirements of the individual parts are summarized in Figure 12.2.

Based on the analysis, the following alterations on the product are necessary in order to create the conditions for automated assembly:

Socket (part 1)
This part is the base part of the assembled part. It cannot be automatically arranged on account of the ceramic material. Simple manual handling is, however, possible on account of the positionally arranged supply by the subcontractors. Tolerance reductions between the workpiece holder and the assembly positions are necessary in order to facilitate the assembly operations.

Part no.	Name	Number per product	Delivery condition	Handling characteristics	Quality requirement	Remarks
1	Socket	1	packed positioned	●	◐	initial manual handling
2	Square nuts	1	loose material	○	◐	
3	Grub screw	1	loose material	○	◐	
4	Hexagon nut	5	loose material	○	◐	manual handling
5	Snap element	1	loose material	●	○	manual positioning
6	Screw	1	loose material	○	◐	
7	Mating contact	1	loose material	○	○	
8	Clamp	4	loose material	○	○	
9	Screw	4	loose material	○	◐	
10	Screw	1	loose material	○	◐	
11	Connection tag	3	loose material	●	●	
12	Key	1	loose material	○	○	
13	Cover	1	loose material	○	○	
14	Rivet	3	loose material	○	○	cleanliness
15	Rivet	2	loose material	○	○	cleanliness foreign bodies
16	Cap	1	loose material	○	○	
17	Bimetal strip	1	loose material	◐	○	resorting
18	Pin	1	loose material	○	○	
19	Adjusting screw	1	loose material	○	○	
20	Screw	3	loose material	○	◐	
21	Locking paint	1	tube	○	○	

○ suitable for automation
◐ conditionally suitable for automation
● not suitable for automatic handling

Figure 12.2 Results of an assembly-extended ABC analysis for the thermoswitch as shown in Figure 12.1

Hexagon nut (part 2) and square nut (part 4)
With regard to their handling characteristics, these standard parts are suitable for automation, although the normal tolerances do not comply with the requirements of the assembly operation. On both parts, the tolerances of the permissible eccentricity of the thread bore in relation to the outer contour are too large, so, as a departure from the DIN tolerance, appropriately reduced tolerances must be agreed with the suppliers.

Snap element (part 5)
The snap element is a pre-assembled unit, since, on account of its basic design and also the pre-tensioning of the snap element spring, it cannot be arranged automatically. Manual positioning in the feed section is therefore necessary.

Screws (parts 6, 9, 10 and 20)
With regard to their dimensions, the screws required for the product are suitable for automated handling. Square nuts are used as mating threads for all screws. Since the

Planning procedure 341

nuts do not have a lead-in chamfer to increase the reliability of assembly, the screws must be provided with an assembly aid in the form of a suitably formed screw shank tip. With regard to the slot depth and screw head height, a tolerance reduction is necessary in comparison with the DIN tolerances so that the screw-in depth can be controlled using the automatic screw-insertion unit on the assembly equipment.

Grub screw (part 3)
The grub screw must be arranged on the slot end. The screw slot must be burr-free so that this arrangement feature can be used without difficulty. An appropriate agreement on quality must be made with the initial supplier.

Connection tag (part 11)
The connection tag, 0.3 mm-thick sheet-metal part, cannot be arranged automatically. With bulk material storage, the instability of the part would result in deformation of the parts, so the sensitivity of this single part, which is important for the quality of the end product, does not permit automatic handling. In this case, an investigation should be undertaken to establish if the connection tag can be provided punched as a plate and magazined with the object of integrating the forming operation into the assembly machine (see Chapter 10).

Rivets (parts 14 and 15)
Both rivets are produced in a forming process. Since, in this case, the danger exists of the presence of foreign bodies (falling, swarf, etc.), agreements must be concluded with the suppliers regarding the removal of foreign bodies and also a guarantee of cleanliness.

Bimetal strip (part 17)
On account of its design, the bimetal strip cannot be arranged in its final position by normal feed equipment. Resorting is therefore necessary.

The assembly processes required by the design are reforming and assembly by pressing on and in. These processes must be retained on account of the specified product characteristics and environmental conditions. The assembly equipment is investigated in the course of the assembly sequence analysis.

12.2.3 Assembly sequence analysis

12.2.3.1 Product design and assembly situation
The overall design of the product can be subdivided into the assemblies of socket and cover and also in terms of the complete assembly. Figure 12.3 shows the product design and the assembly situation for the socket assembly. The analysis shows that a number of pre-assembly operations must be undertaken. For example, the screw (part 9) must be fitted into the clamp (part 8). Both parts must be assembled with part 7 so that they can be fitted as an assembly into the base (part 1). The assembly direction indicates that horizontal assembly operations are necessary.

The cover assembly is shown in Figure 12.4. All the assembly operations for this unit can be performed linearly and vertically. However, for completion of the final assembly, the fully assembled unit must be turned through 180°.

Figure 12.5 finally shows the assembly situation and direction of final assembly. All the assembly operations can be undertaken vertically from above downwards.

342 Practical example: planning and realization of an automated assembly system

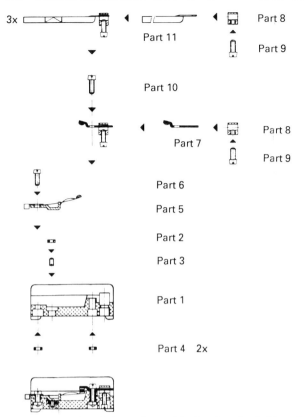

Figure 12.3 Product design and assembly situation of unit (a): socket

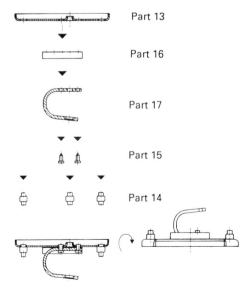

Figure 12.4 Product design and assembly situation of unit (b): cover

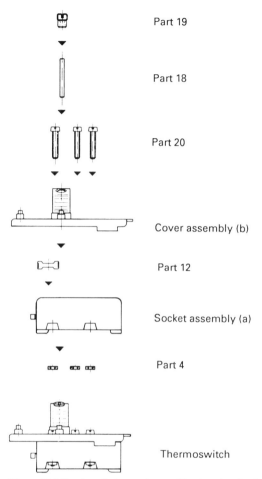

Figure 12.5 Product design and assembly situation for final assembly of the thermoswitch

12.2.3.2 Assembly sequence
The assembly sequence of the thermoswitch is shown in Figure 12.6. As shown here, six assembly processes are necessary for the assembly of the socket unit.

Since part 1, the socket, is not only the base part for the socket assembly but, as a complete assembly is also the base part for final assembly, the assembly of the socket and the final assembly could be undertaken on one machine. All necessary pre-assembly operations and the assembly of the cover must, however, be undertaken on subassembly assembly machines. This would mean that, in accordance with the assembly sequence, 18 single operations would have to be integrated into one such machine. A low availability is inevitably incurred with the interlinking of so many operations in one machine. Subdivision is therefore necessary, so that, following the last assembly operation, the condition of the subassemblies facilitates further transport of the assembled part.

The geometry of both basic parts, socket and cover, permits transport from machine to machine without using workpiece carriers and also without losing the arranged

344 Practical example: planning and realization of an automated assembly system

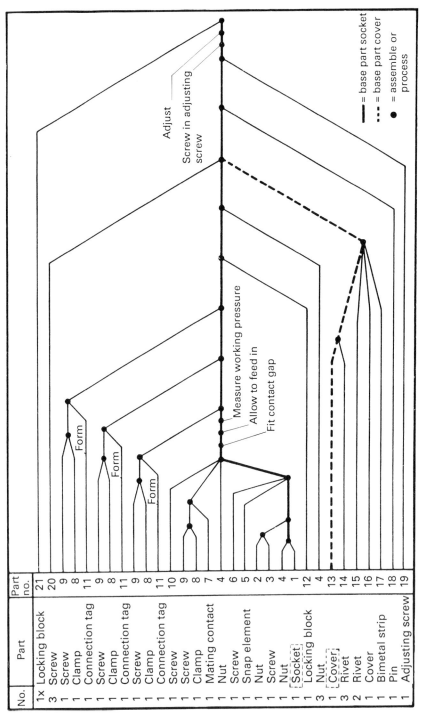

Figure 12.6 Assembly sequence of the thermoswitch as shown in Figure 12.1

position. Separate assembly sequence diagrams must be produced for individual sections (see Sections 11.4.1.3 and 11.4.2, Figure 11.8).

12.2.3.3 Workpiece carrier design
Workpiece carriers must be designed suitably for individual assembly stages. In this respect, consideration must be given to the aspect as to whether or not work is possible with uniform or different workpiece carriers. The product design of the thermoswitch requires different workpiece carriers. An example in this respect was given in Section 11.5.2.2 in diagrams 11.11 and 11.12. Different workpiece carriers necessitate distribution of the assembly operation for the final assembly of this product among several machines.

12.2.4 Function analysis

The time requirement for the individual operations for the assembly of the thermoswitch is determined by the function analysis as given in Section 11.6.

- The single parts 2, 3, 4, 6, 7, 8, 9, 10, 12, 13, 14, 15, 16, 17, 18, 19 and 20 can each be arranged and handled in 2.8 s.
- Part 11 can only be handled in 2.8 s under the assumption that it is pre-punched as a plate and fed in a magazine and also that the re-forming operation is undertaken in parallel with the assembly operation.
- In accordance with an MTM study, parts 1 and 5 can also be handled manually in 2.8 s.
- The operation 'set contact gap' requires 3.5 s. This operation can only be undertaken during the dwell time of the transfer equipment.
- The operations 'feed in snap element' and 'measure working pressure' require 2.5 s and can only be undertaken during the dwell time of the transfer equipment.
- The operation 'screw in adjusting screw' can be undertaken in 2 s and must be performed during the dwell time of the transfer equipment.
- The operation 'adjust' requires 3.5 s and can only be undertaken during the dwell time.

12.2.5 Determination of cycle time

The determination of the planned output in accordance with 12.2.1 indicated 994 thermoswitches per hour, so when rounded off to 1000 parts per hour, the cycle time can be calculated as:

$$t_c = \frac{3600 \text{ s/hour} \times 0.8}{1000 \text{ parts/hour}} = 2.88 \text{ s}$$

12.2.6 Layout planning

Layout planning is based on the data obtained from previous planning stages, in particular relating to the assembly sequence, function analysis, cycle time determination and the space available. The planned space in this case measures 15×20 m.

The complexity and assembly require subdivision of the assembly procedure into individual sections. The cycle time determination indicated a planned cycle time of

2.88 s. The function analysis indicates that, with the exception of the operations 'set contact gap', 'feed in snap element' and also 'measure working pressure and adjust', all handling and assembly operations relating to the single parts can be undertaken within this cycle time. As shown by the assembly sequence, the operations 'set contact gap', 'positioning of snap element' and 'measure working pressure' are undertaken immediately after each other, and the operation 'set' is the second last assembly operation.

The following subdivision can be implemented as shown by the assembly sequence in Figure 12.6.

Section I
Assembly of parts 1, 2, 3, 5, 6 and once part 4; parts 2 and 3 must be pre-assembled.

Section II
Assembly of parts 7, 8, 9, 10 and once part 4 in part 1.

Section III
The following operations can be undertaken: set contact gap, feed in snap element ten times in sequence and measure the working pressure.

Since these operations can only be undertaken during the dwell time of the transfer equipment, and, as shown by the function analysis, they require cycle times longer than 2.88 s, this section must be built in duplex form, i.e. the assembly operations must be undertaken in pairs.

Section IV
Parts 8, 9 and 11 must be pre-assembled twice and fitted into the base part 1 as a pre-assembled unit. At the same time, the re-forming operation to form part 11 from the plate into its final condition must be integrated into the assembly operation of part 11.

Section V
As section IV: the same operations, but only performed once.

Section VI
Handling and assembly of part 13 with part 14 three times.

Section VII
Assembly of parts 15, 16 and 17 with the pre-assembled unit from section VI.

Section VIII
Handling of the socket assembly, locking block part 12 and cover and also assembling three times with parts 14 and 20.

Section IX
This section includes the adjustment operation which requires a longer cycle time than the planned cycle time so that this assembly section must also be constructed in duplex form. Parts 18 and 19 are to be fitted into the subassembly assembled in section VIII. The adjusting operation must then be undertaken and the adjustment locked by the application of locking paint. Automatic unloading of the fully assembled thermoswitch for packing then takes place.

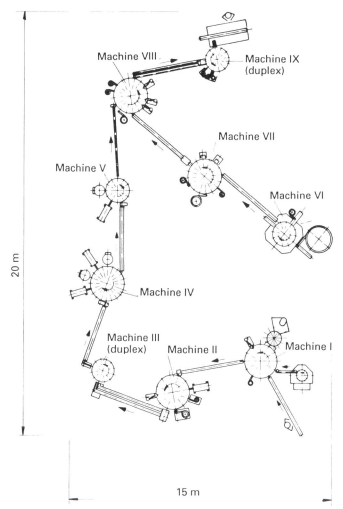

Figure 12.7 Layout of the assembly system for the thermoswitch as shown in Figure 12.1

Figure 12.7 shows the overall layout of this assembly system as described by the subdivision of the operational sequences. The machines are interconnected by simple webbing belts which, at the same time, act as fault-elimination buffers.

12.2.7 Determination of personnel requirement

On account of the assembly system size and the level of manual activity, the personnel requirement determination is as follows:

1. 3 persons are required for the manual operations
- One person for handling part 1
- One person for handling part 5
- One person for packing the fully assembled thermoswitches.

2. 5 persons are required for equipment supervision and maintenance
- One wholly responsible supervisor
- Two system maintenance persons (mechanic/electrician)
- Two auxiliaries for system maintenance whose functions are the provision of parts in the vibratory spiral conveyors and the elimination of simple faults, for example, the removal of foreign bodies or deformed parts in the feed areas.

12.2.8 Determination of availability

An availability of 80% was assumed in the determination of the cycle time. To achieve this value, the factors as listed in Section 11.10 must be critically analysed and the resulting measures clearly defined. This arrangement can then be used as a specification for the selection of the correct assembly systems and also for the structuring of detailed planning.

12.2.8.1 Parts quality
As shown by the product analysis, using the assembly-extended ABC analysis in accordance with Figure 12.2, a tolerance reduction in comparison with the standard DIN quality is necessary for a number of individual parts or the quality level as given by the AQL must be increased considerably in comparison with manual assembly. In some cases, secondary dimensions which relate to the function of the parts become principal dimensions for handling. All the respective points must be detailed by the system planner in the pre-production stage in conjunction with product development and must be agreed with suppliers.

12.2.8.2 Number of stations
The subdivision of the assembly operations into nine sections ensures that, with the exception of section VIII, no section is allocated more than seven assembly operations. (Nine assembly operations are planned in section VIII; these, however, cannot be further subdivided on account of the product design.) The distribution of the assembly parts 8, 9 and 11, each three times over sections IV and V, is advantageous, so not more than ten assembly operations are allocated to one section.

12.2.8.3 Individual availability
Efforts must be made to achieve an individual availability of 99.4% to 99.5% for the lead stations. In order to achieve this target, the design of the feed stations incorporates adequate reserve sections as shown in Figure 6.9, Section 6.3.1. They should hold a buffer capacity for approximately 2 min production time, which corresponds to approximately 40 single parts. To eliminate faults, the covers of vibratory spiral conveyors and cover strips of feed rails, etc. must be rapidly removable by snap-on connectors. The vibratory spiral conveyors must be dimensioned so that a workpiece buffer capacity of at least 30 min production time is available. Stepped-type vibratory spiral conveyors must be provided in order to prevent jamming of the parts between the screw-form plates.

12.2.8.4 System structure
The application of circular cyclic units is made possible by the distribution of the assembly operations over nine assembly machines and also the resultant limited number of operations per assembly machine. The required cycle time of 2.88 s necessitates an electric-motor-driven Maltese cross or cam drive of the circular cyclic

units. Under consideration of the vacant stations, the number of stations on the circular cyclic unit should be as large as possible so that the angular accelerations which occur during indexing are as low as possible. For mechanical recording of the inspection results at the inspection stations, the workpiece carriers should be equipped with coding pins. Inspection stations must be provided at every work point, irrespective of whether it is a parts feed or assembly station for direct inspection of the previously completed operation. The assemblies which are found to be faulty must be separated after leaving every assembly machine so that only good assemblies are fed to the next machine. The inspection stations do not function on the immediate switch-off principle. An assembly machine should preferably only be switched off following three successive unsatisfactory results with simultaneous signalling of the fault location. The individual assembly machines should be non-permanently connected together by conveyor belts. The belts function as intermediate buffers between the individual machines, and should be dimensioned so that a minimum of 2 min production time can be buffered.

If such an intermediate buffer is full on the next machine on account of faults, the previous machine must be stopped by a suitable control device. The machine must be automatically switched on when the workpiece flow is resumed. Reserve items of equipment as described in Section 6.5.1, Figure 6.40, must be provided for the manual handling operations which are to be integrated so that decoupling from the single cycle is achieved.

12.2.8.5 Initial operation characteristics
The distribution of the assembly operations over nine different assembly machines permits construction of the system in stages so that the planning of completion of the individual assembly machines must be in the same order as the assembly sequence. Upon delivery of the first machine, this section is transferred from manual pre-assembly on to the machine so that the machine can be set and the operators complete their training schedule before installation of the next machine. A start-up phase of two to three weeks can be expected per machine. With the spaced start-up of new assembly machines, it should be possible to contain the start-up phase within a period of 18 weeks minimum to a maximum of 25 weeks.

12.2.8.6 Personnel qualifications
A specialized professional qualification is absolutely necessary for the supervisor with general responsibility and the two system attendants. The following training plan is applicable to this group:

- Six to eight weeks' involvement in manual assembly for familiarization with the product-related requirements and assembly conditions.
- Involvement in the trial and installation of the separate assembly machines with the object of being present from the first test run and during any faults which occur.
- Depending upon the professional qualification of the operatives, additional training should be given, e.g. in pneumatics or electronics.

12.3 Detailed planning of assembly system

12.3.1 Introduction

Based on the results of the planning stages, circular cyclic systems with indirect Maltese-cross drives with 12 and 16 stations are to be used for the individual assembly

350 Practical example: planning and realization of an automated assembly system

machines. The pre-assembly operations necessary are undertaken on upstream satellite assembly machines. In this case, circular cyclic units with a direct Maltese drive are used (see Section 5.4.1.1.2, Figures 5.81 and 5.82).

Pneumatically operated equipment is used on account of the short Z- and X-movements of the positioning equipment. The gripper designs are such that the holding force is generated by springs and, on the other hand, the release force is pneumatic. This ensures that, in the event of failure of the compressed air supply, the grippers do not release the parts.

The vibratory spiral conveyors are a stepped design. The feed sections are designed for capacity equivalent to 2 min' production. Irrespective of their individual functions, the inspection stations are rigid or equipped with a lift movement.

Every individual machine has its own control and forms a completely independent unit.

12.3.2 Machine I

Figure 12.8 shows a detailed layout of assembly machine I. The working sequence is as follows:

- A square nut (part 4) – parts and part numbers are shown in Figure 12.3 – is correctly arranged by a vibratory spiral conveyor, fed to the separating station via

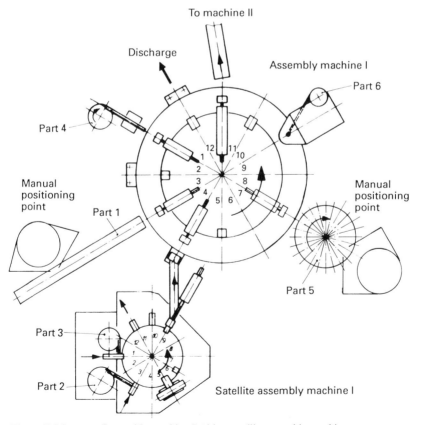

Figure 12.8 Layout of assembly machine I with a satellite assembly machine

Detailed planning of assembly system 351

a discharge rail and then fed into the workpiece carrier at station 1 by a handling unit.
- An inspection station at station 2 checks the availability and the correct fitting of the square nut.
- The base part socket (part 1) is taken from the packaging carton by the operative at station 3 and likewise placed on the conveyor belt in the correct position. The conveyor belt is designed so that 40 parts can be buffered end-to-end. The operative is therefore decoupled from the assembly machine cycle. The part is separated at the end of the conveyor belt by a transverse movement and placed in the workpiece carrier over the square nut (part 4) by a handling unit.
- At station 4, a subassembly which is assembled on an upstream satellite machine and consists of a hexagon nut (part 2) and a grub screw (part 3), is placed in the socket (part 1) by a positioning unit.

The mode of operation of the satellite assembly machine for pre-assembly is as follows.

The M3 × 4 mm grub screw (part 3) is arranged in a vibratory spiral conveyor at station 1 and fed to the separating device by a discharge rail. The grub screw slides in the discharge rail in its slot and its position thus corresponds to its final assembly position in the assembly. Upon separating the grub screw, it is fitted into a bore in the workpiece carrier with its slot pointing downwards. A check is made at station 4 to determine if the hexagon nut is available. The grub screw is screwed up into the hexagon nut from beneath at station 6. Stations 7 and 8 remain vacant. The pre-assembled subassembly is removed from the workpiece carrier at station 9 by a positioning unit and placed on a small circular-cord double conveyor belt for transport to the separating station at assembly machine I. The circular-cord double arrangement of the conveyor belt is necessary so that the grub screw does not contact the conveyor belt with its slot. With transport by an electromagnetically driven discharge rail, the danger exists that the grub screw can wind free on account of the vibrations. At station 10 a check is made to determine if the workpiece carrier is free. Any unfitted residual parts are automatically ejected at station 11. Station 12 remains vacant. This subassembly is shown in detail in Figure 12.9 and also the assembly situation in which reference is made to the screw-in dimension (3 mm with a tolerance of ±0.1 mm).

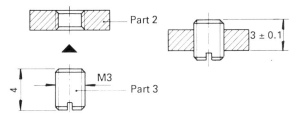

Figure 12.9 Pre-assembly of a grub screw into a hexagon nut

- A check is made at station 5 of assembly machine I to determine if the assembly has been fed and correctly positioned at station 4.
- Station 6 remains unoccupied.
- A 24-pitch indexing table is positioned in front of station 7, in which the snap elements (part 5) are placed manually by an operative. A positioning unit grips a snap element, removes it from the indexing table and fits it into the socket (part 1).

352 Practical example: planning and realization of an automated assembly system

- Station 8 remains unoccupied.
- The screw (part 6) is arranged in the correct position at station 9, magazined, separated and fitted (see Figure 5.94). The automatic screw inserter is fitted with a torque and screw-in depth control.
- A check is made at station 10 to determine if the screw has been fitted and screwed in to the correct depth.
- The positioning unit removes the fully assembled unit from the workpiece carrier at station 11 and places it in an arranged position on the conveyor belt for transport to machine II.
- In order to ensure that a sequence of operations at station 1 can recommence without difficulty, any unfitted single parts are automatically rejected at station 12.

12.3.3 Machine II

Figure 12.10 shows a detailed layout of assembly machine II. The operations and pre-assembly operations to be undertaken on the machine have already been described in detail in Section 11.5.2.2 with the workpiece carrier design as shown in Figures 11.11

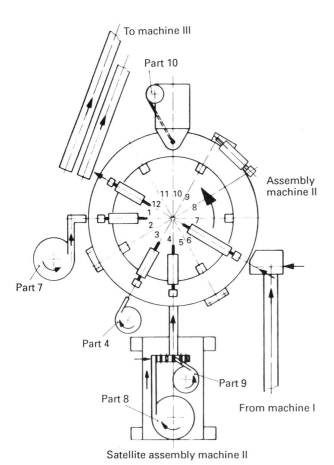

Figure 12.10 Layout of assembly machine II together with the satellite assembly machine

Detailed planning of assembly system   353

and 11.12 (additional reference to Section 11.4.1.3, Figures 11.6 and 11.8), so in the following, details will only be given relating to the overall operational sequence and the prefitting of the clamp (part 8) with the screw (part 9). The operational sequence is as follows:

- The mating contact (part 7) is arranged by a vibratory spiral conveyor at station 1 and fed by a discharge rail to the separating station. The part is placed in a workpiece holder by a positioning unit (see Figures 11.11 and 11.12).
- A check is made at station 2 to determine if the mating contact has been fed and positioned.
- A square nut (part 4) is arranged by a vibratory spiral conveyor at station 3, fed to the separating station by a discharge rail and placed in the workpiece holder (part II) by a positioning unit.
- At station 4, the subassembly, pre-assembled in an upstream satellite assembly machine and consisting of the clamp (part 8) and screw (part 9), is placed in the workpiece holder (part I) by a positioning unit. The mode of operation of the satellite assembly machine is shown schematically in Figure 12.11. The clamp (part 8) is arranged in a vertical 4-station indexing plate by a vibratory spiral conveyor, fed to the separating station via a discharge rail and then pushed directly from the separating station into the indexing plate workpiece holder. An automatic screw inserter for arranging, feeding and fitting of the screw (part 9) is arranged at station 2. The inserter only screws the screw in to such a depth that the clamp bore still remains unobstructed. A check is made at station 3 to determine if the screw was fed. The pre-assembled unit is removed from the workpiece holder rotary plate at station 4 and pushed into a discharge rail. At this point, the unit is in the correct position for gripping and placing in the assembly machine, i.e. the screw head downwards. The electromagnetically operated discharge rail forms an intermediate buffer between the satellite machine for pre-assembly of this unit on assembly machine II.
- Station 5 is an inspection station to check if fitting and positioning has been correctly undertaken at stations 3 and 4.

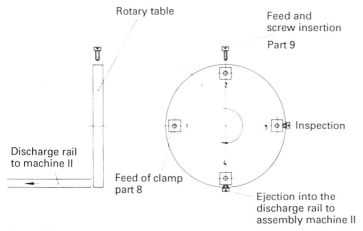

Figure 12.11 Mode of operation of the satellite assembly machine on assembly machine II for assembly of a screw with a clamp

354 Practical example: planning and realization of an automated assembly system

- The units as pre-assembled on machine I arrive at station 6 on a conveyor belt and are separated at this point. A unit is placed in a workpiece holder by a positioning unit (see Figures 11.11 and 11.12).
- The screw (part 9) is fitted to the clamp (part 8) over the mating contact (part 7) by a stationary cam profile during rotation of the indexing table arrangement from station 6 to station 7.
- A positioning unit arranged tangentially to the indexing table at station 8 removes the pre-assembled unit from the workpiece carrier and places it in the socket positioned in the workpiece holder.
- A check is made at station 9 to determine if the unit has been correctly assembled.
- At station 10, an automatic screw inserter is arranged for the positioning, magazining, feed and screwing in of screws (part 10). The screw inserter is equipped with a torque and screw-in depth control.
- A check is made at station 11 to determine if the screw has been fitted.
- The positioning unit removes the fully assembled unit from the workpiece holder at station 12 and places it on a conveyor belt. The width of the conveyor belt is designed so that it can be divided into two. The sorting slider pushes every second positioned unit on to the second section of the conveyor belt so that, by the conveyor-belt design, the pre-assembled units are fed in pairs to machine III. In the course of an evaluation of the inspection results, incompletely assembled units are segregated on the conveyor belt by a gate.

12.3.4 Machine III

The operations 'set contact gap', 'feed in snap element' and 'measure working pressure' are undertaken on this machine. As shown by the function analysis, the cycle times are required at this point which exceed the specified cycle time so that these operations must be undertaken twice per station with double workpiece carriers. The detailed layout is shown in Figure 12.12.

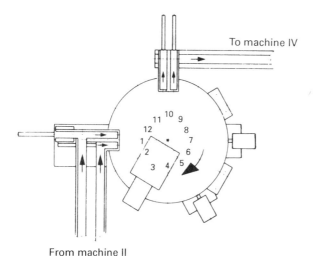

From machine II

Figure 12.12 Layout of assembly machine III

Detailed planning of assembly system 355

The operational sequence is as follows:

- The pre-assembled units arrive from machine II in pairs via the double-conveyor-belt design and workpiece carriers at station 1.
- Station 2 remains unoccupied.
- The contact gap is set at station 3. A second gauge is placed between the contact and the mating contact pressed against the setting gauge by a press. The setting gauge is retracted following completion of this operation.
- Station 4 remains unoccupied.
- The working pressure of the switch element is measured at stations 5 and 7; a workpiece at station 5 is measured in the left-hand and that at station 7 in the right-hand workpiece holder. Station 6 remains unoccupied during this time.
- Stations 8 and 9 remain unoccupied.
- The subassemblies found to be correct are ejected at station 10 in pairs and are then fed to machine IV behind each other as a unit.
- Subassemblies found to be incorrect are ejected at station 11.
- Station 12 remains unoccupied.

Sufficient space must be provided at station 11 so that, with the spasmodic occurrence of a high scrap level, adequate space is available so that the manual resetting procedure by an operative can be implemented for a limited period of time.

12.3.5 Machine IV

The clamp (part 8) is pre-assembled with the screw (part 9) by machine IV and then fitted with the connection tag (part 11) into the socket (part 1). As shown by the product analysis, in its final form the connection tag (part 11) cannot be handled automatically and is therefore in the assembly machine as a pre-punched plate. Two stations, consisting of a six-pitch circular cyclic arrangement, a holder for the magazine of the plates and form presses, are provided in the assembly machine for separating and finish forming.

This mode of operation is described in Section 10.3 by Figures 10.3, 10.4 and 10.5. The same satellite assembly machines are used for the pre-assembly of the clamp with the screw as were used on assembly machine II.

Further details relating to the description of the individual pre-assembly or forming operations are discussed below. The workpiece carrier is similar to that as fitted to machine II. The slider design for pre-assembly of the screwed clamp with the connection tag is designed as a double workpiece. Figure 12.13 shows a detailed layout of machine IV.

The operational sequence is as follows:

- A check is made at station 1 to determine if the workpiece holder is free for reuse.
- Station 2 remains unoccupied.
- At station 3, a positioning unit grips the formed connection tag (part 11) on an upstream machine B and places it in the workpiece holder.
- A check is made at station 4 to determine if part 11 has been fitted.
- At station 5, the positioning unit grasps the subassembly consisting of the clamp (part 8) and screw (part 9) which have been pre-assembled on the satellite assembly machine, and places it in the workpiece holder.
- Station 6 checks if the operation at station 5 has been completed.
- Station 7: operation of station 3.

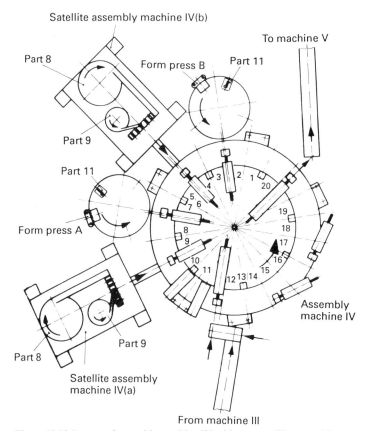

Figure 12.13 Layout of assembly machine IV with two satellite assembly machines

- Station 8: operation of station 4.
- Station 9: operation of station 5.
- With indexing of the rotary table from station 10 to station 11, the workpiece holder slider for assembling the two clamps fitted with screws is moved over the connection tags by a stationary cam profile (see also Figure 11.12).
- The pre-assembled units approaching from machine III are separated at station 12 and placed in the workpiece holder by a handling unit.
- A check is made at station 13 to determine if the pre-assembled unit has been positioned.
- Station 14 remains unoccupied.
- A positioning unit is arranged tangentially to the rotary table at station 15 to place the pre-assembled unit, consisting of a clamp with screw and connection tag, in the socket (part 1). This is achieved by rotation inside the workpiece carrier (see Figure 11.12).
- Station 16 checks if the assembly operation has been completed.
- Station 17: operation of station 15.
- Station 18: operation of station 16.
- Station 19 remains unoccupied.
- A positioning unit removes the pre-assembled unit from the workpiece holder at

station 20 and places it in an arranged position via a gate on a conveyor belt for further transport to machine V. Subsequent to an evaluation of the inspection results, any incompletely assembled units are separated out by movement of the gate.

12.3.6 Machine V

If two connection tags (part 11) are assembled with the screw-inserted clamps on machine IV, the third connection tag is then pre-assembled with the clamp and screw on machine V. A detailed layout of this machine is shown in Figure 12.14. The design and mode of operation is the same as for machine IV, but only for one connection tag.

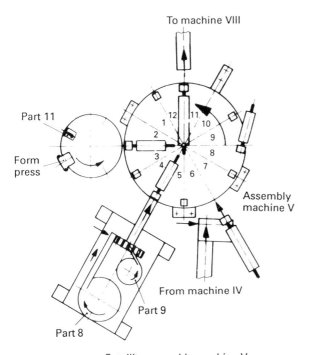

Figure 12.14 Layout of assembly machine V with a satellite assembly machine

12.3.7 Machine VI

The three rivets (part 14) are fitted into the cover (part 13) by this machine. Figure 12.15 shows the detailed layout of this machine.

The operational sequence is as follows:

- The rivets are positionally arranged in a vibratory spiral conveyor with three discharge channels at station 1 and then fed to a separating station via a triple discharge rail. The separating station is designed so that the position of the separated rivets corresponds to the assembly position of the workpiece holder and the cover (part 13). A workpiece holder with a triple gripper grasps the rivets in the sorting position and places them in the workpiece holder.

358 Practical example: planning and realization of an automated assembly system

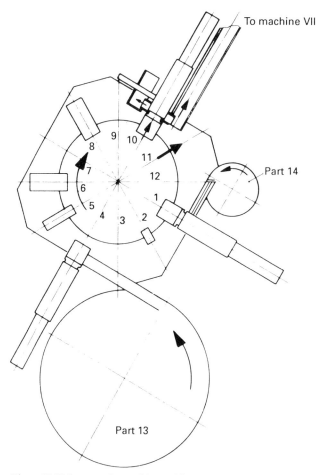

Figure 12.15 Layout of assembly machine VI

- A check is made at station 2 to determine if the assembly operation at station 1 has been completed.
- Station 3 remains unoccupied.
- The cover (part 13) is positionally arranged at station 4 by a vibratory spiral conveyor, fed to a separating device via a discharge rail, grasped by a positioning unit and then fitted over the rivets which were fed to station 1.
- A check is made at station 5 to determine if the assembly operation at station 4 has been completed.
- Assembly of the three rivets to the cover is by re-forming in a rotating-mandrel riveting process. Since the rivet pattern does not permit the use of the triple rotating-mandrel riveting unit, the operation must be distributed over two stations. The riveting operation is undertaken at station 6 by a double rotating-mandrel riveting unit and a single rotating-mandrel riveting unit at station 8.
- Station 7 remains unoccupied.
- A check is made at station 9 to determine if the rotating-mandrel riveting operations have been undertaken at stations 6 and 8.

- A positioning unit grasps the fully assembled unit at station 10 and places it on a transfer slider.
  Depending upon the inspection result at station 9, with a positive result the assembly is pushed on to the conveyor belt for transport to machine VII, whereas with a negative result it is pushed to a reject gate.
- Non-assembled parts are ejected at station 11.
- Station 12 remains unoccupied.

12.3.8 Machine VII

Parts 15, 16 and 17 are assembled with the pre-assembled unit consisting of parts 13 and 14 on this machine. The assembly operation is by re-forming of the rivets (part 15) by a rotating-mandrel riveting procedure. Figure 12.16 shows a detailed layout of the machine.

The difficulty of the assembly operation is that as a first operation both rivets (part 15) must be positioned on their rivet heads in a workpiece carrier and must not lose their position during indexing of the circular cyclic unit. Figure 12.17 shows schematically the design and mode of operation of the required workpiece carrier. Both rivets (part 15) are placed with the rivet head in two suitably sized blind holes in the workpiece carrier. The depth of the blind holes is half the height of the rivet head. A

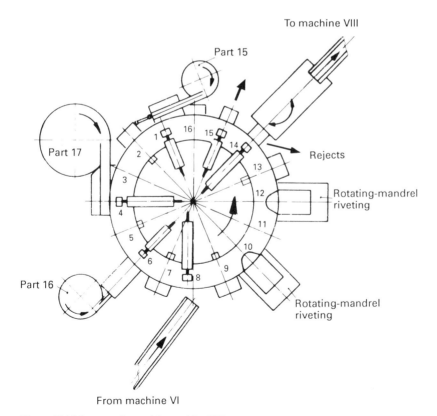

Figure 12.16 Layout of assembly machine VII

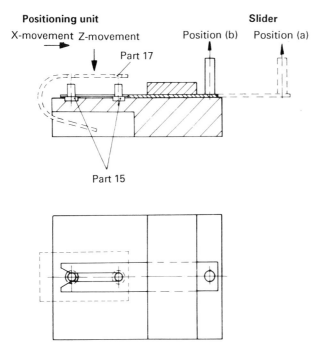

Figure 12.17 Schematic arrangement of the workpiece carrier for assembly machine VII as shown in Figure 12.16

forked slider is in a forward position (a) during the positioning operation. After positioning the rivets but before indexing of the circular cyclic unit, the forked slider is moved along the shaft of the rivets into position (b). Both rivets are then securely held in their blind holes. If the bimetal strip (part 17) is fitted, the positioning unit correctly positions the part with its X-movement above both rivet shanks. The bimetal strip is fitted over the shanks by the Z-movement of the positioning unit. Following completion of this assembly operation and with indexing of the circular cyclic unit, the forked slider is returned to its start position (a) by a stationary cam profile. A bimetal strip drops by the height of the slider on to the lower face of the rivet head and then positions both rivets for the following assembly operations.

The operational sequence as shown in Figure 12.16 of this machine is as follows:

- The rivets (part 15) are correctly arranged at station 1 by a vibratory spiral conveyor and fed to a separating station spaced at the assembly spacing by two discharge channels with a double discharge rail. The positioning unit with a double gripper grips two rivets and places them in the holder of the workpiece carrier. Before opening of the gripper, the forked slider for holding the rivets is moved into the workpiece holder.
- A check is made at station 2 to determine if the operation at station 1 has been completed.
- Station 3 remains unoccupied.
- The bimetal strip (part 17) is pre-arranged at station 4 by a vibratory spiral conveyor and fed to the separating station via a feed rail. The separating station is designed as a swing-action station for bringing the bimetal strip into the correct

assembly position. The bimetal strip is held by a positioning unit and fitted over the shank of both rivets. The forked slider for holding the rivets is returned to its start position by the indexing of the circular cyclic unit from station 4 to station 5.
- A check is made at station 5 to determine if the operation at station 4 has been completed.
- The cover (part 16) is correctly positioned at station 6 and fed to a separating station via a discharge rail and then gripped by a positioning unit and fitted over the shank of the rivets.
- A check is made at station 7 to determine if the operation at station 6 has been completed.
- The pre-assembled cover assembly (part 13) complete with rivets (part 14) is fed from machine VI at station 8 to a separating station and placed in the workpiece holder over the shanks of the rivets by a positioning unit.
- Station 9 checks if the operation at station 8 has been undertaken.
- The forming operation is undertaken at station 10 and station 12 by rotating-mandrel riveting units.
- In the meantime station 11 remains unoccupied.
- A check is made at station 13 to determine if the operations at stations 10 and 12 have been completed.
- The fully assembled unit is removed from the workpiece holder at station 14 by a positioning unit and placed at an intermediate storage point. With a negative test result at station 14, the assembly is rejected from the intermediate storage point. With a positive result, an overhead swing-action positioning unit grips the assembly and places it on a conveyor belt to machine VIII. During this overhead swing action, the assembly is turned through 180° to bring it into the correct assembly position on the following machine.
- A positioning unit is installed at station 15 to remove any unassembled single parts.
- Station 16 remains unoccupied.

12.3.9 Machine VIII

Figure 12.18 shows a detailed layout of this machine. As shown in Figure 12.5 and the assembly sequence, the socket assembly and cover assembly and also the locking block (part 12) are assembled together with the application of auxiliary parts such as three square nuts (part 4) and three screws (part 20).

The operational sequence is as follows:

- A square nut (part 4) is arranged in its correct position by a vibratory spiral conveyor at station 1, fed to a separating station via a discharge rail and placed in the workpiece holder by a positioning unit.
- The same operation is undertaken at station 2, but at this point the square nut is placed in pairs in the workpiece holder.
- A check is made at station 3 to determine if the operations at stations 1 and 2 have been completed.
- At station 4, the socket, as pre-assembled on machine VII, is fed to a separating station via a conveyor belt and then placed in the workpiece holder over the pre-positioned square nut by a positioning unit.
- The locking block (part 12) is arranged in its correct position at station 5 by a vibratory spiral conveyor, fed to a separating station via a discharge rail and fitted into the socket assembly by a positioning unit.

362 Practical example: planning and realization of an automated assembly system

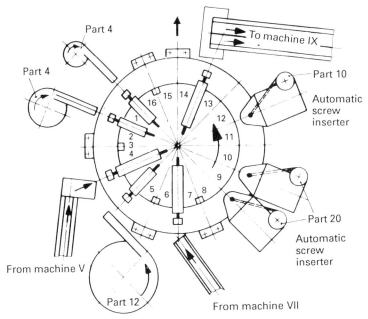

Figure 12.18 Layout of assembly machine VIII

- A check is made at station 6 to determine if the operations at stations 4 and 5 have been completed.
- At station 7, the pre-assembled cover assembly is fed by the conveyor of machine VII, separated and placed in the workpiece holder over the socket assembly by a positioning unit.
- A check is made at station 8 to determine if the operation at station 7 has been completed.
- The screws (part 20) are arranged, magazined, fed and inserted by automatic screw inserters at stations 9, 10 and 12. Automatic screw inserters as shown in Figures 5.94 and 5.95 are used. The controls for the screw-insertion operation and required torque are integrated in the automatic screw inserter so that additional inspection stations are not necessary.
- Stations 11 and 13 remain unoccupied.
- At station 14, the fully assembled units are removed from the workpiece holder by a positioning unit and transferred to a slide station. The slide station distributes the fully assembled units alternately on a dual-track double conveyor belt so that the assemblies can be placed in pairs in machine IX.
- Any single parts not fitted are ejected at station 15.
- A check is made at station 16 to determine if the workpiece carrier is free from parts for further use.

12.3.10 Machine IX

Parts 18 and 19 are fitted on this machine; in addition, the adjusting operation is also undertaken.

The adjustment is locked by the application of a locking paint. As shown by the function analysis (see Section 12.2.4), the adjustment operation requires 3.5 s and can

Detailed planning of assembly system 363

only be undertaken during the dwell time of the transfer equipment. In order to achieve this, the machine must be designed so that all operations can be undertaken in pairs; this necessitates doubling of the single-cycle time. Figure 12.19 shows a detailed layout of this machine.

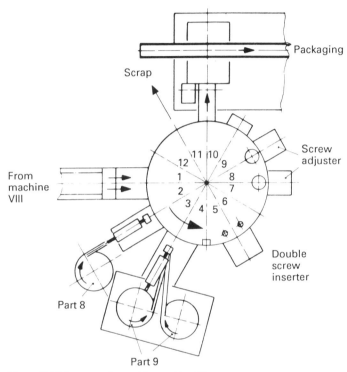

Figure 12.19 Layout of assembly machine IX

The operation sequence is as follows:

- At station 1, the pre-assembled units approaching from machine VIII are inserted in pairs into the double workpiece holder of machine IX. The threaded hole of the bimetal strip for the adjusting screw (part 19) has a 10° inclined position. In order to facilitate the fitting of the adjusting screw vertically and in a straight line, the location faces of the workpiece carriers must be inclined so that the female thread in the bimetal strip is brought into a horizontal position.
- The pins (part 18) are arranged at station 2 by a vibratory spiral conveyor and fed to the separating station via a double discharge rail. The separating station rotates the pins from a horizontal into a vertical position. The positioning unit which has a double gripper tong grips a pair of pins and passes them through the threaded hole in the bimetal strip into the thermoswitch cover.
- The adjusting screws (part 19) are arranged in pairs at station 3 by two vibratory spiral conveyors and then fed to the discharge rails which are equipped with separating devices.
  The adjusting nut has a shank for pre-assembly into the bimetal strip thread. The positioning unit with a double tong gripper grasps the separated adjusting screws

in pairs and places them with the conical internal bore of the adjusting screw over the pins (18) and with the shank of the adjusting screw placed in the bimetal strip thread bore.
- A check is made at station 4 to determine if the operations at stations 2 and 3 have been completed.
- The adjusting screw is screwed into the bimetal strip to a depth of two threads at station 5 by a double screw inserter.
- Station 6 remains unoccupied.
- The adjusting operation is undertaken at stations 7 and 8. On account of the dimensions of the adjusting screw, the screw spindles for adjustment must not be arranged too closely together, so that both thermoswitches can be adjusted at one single station. The screw adjusters are therefore placed at two stations: at one station the thermoswitch in the left-hand workpiece carrier half is adjusted and at the other station the thermoswitch in the right-hand workpiece carrier half is adjusted. The screw adjuster guides the blade into the adjusting screw slot and turns the adjusting screw to a depth until the pin (18) contacts the snap element (part 5). This operation is repeated twice for the precise determination of the switching point of the snap element. In accordance with the specified setting of the thermoswitch, the screw adjuster winds the screw outwards from this point by a number of angular degrees as determined by the computer so that the specified setting is obtained (e.g. 460 angular degrees). The screw-adjusted drive is by driven incremental motors for an exact definition of the thermoswitch position during the adjusting operation; the screw adjusters are equipped with advanced acting spring-loaded centring devices so that the thermoswtiches are held in a precise position during the adjustment operation. The adjustment is immediately evaluated by the screw-adjuster control as a Yes/No result and the workpiece carriers coded accordingly.
- Station 9 remains unoccupied.
- The satisfactorily adjusted thermoswitches are automatically sealed by drops of paint at station 10 by a sealing paint metering unit. The finally assembled thermoswitches are ejected from the workpiece carrier by an ejector and slide into good and bad areas via a controllable gate to the manual packing table.
- The unadjusted or unsatisfactorily adjusted thermoswitches are ejected at station 11.
- Station 12 remains unoccupied.

12.4 Investment calculations

The planning, design and also erection of the assembly system as described in the preceding Section should be an in-house activity with internally produced parts together with the purchase of commercially available modules. An investment calculation was undertaken for the designs as detailed in Section 11.12. The cost determined for this system was around 2.9 million currency units. This includes the costs for planning, design, erection, trial running and start-up and other costs occurring in the main field of activity of assembly.

12.5 Evaluation and selection/work-point cost comparison

The planning of an optimized manual assembly with partial mechanization of the assembly processes and the material flow was undertaken parallel to the planning of the previously described automated assembly system. The individual work points are constructed in accordance with MTM and optimized by the primary and secondary assembly operation analysis (see Section 4). The investment for manual assembly is calculated at 580 000 currency units. The time allowance per average part is 3.9 min. Under the assumption of an output rate of 135%, this gives an effective working time of 2.88 min per part. At a required rate of 1000 parts per hour, the total of 2880 min working time must be available per hour; this means $2880 \div 60 = 48$ manual work points.

In accordance with the requirement list (Section 12.2.1) a utilization period of 4 years is assumed so that the work-point cost calculation is based upon a four year depreciation. Under the assumption of an annual price increase rate of 5%, with an initial procurement value of V_P, an interest rate p and a depreciation period n, the reprocurement value V_R is calculated as follows:

$$V_R = V_P \times (1 + p)^n \tag{12.1}$$

Accordingly, the reprocurement value of the manual assembly system is given by:

$$V_{RM} = 580\,000 \times (1.05)^4 = 580\,000 \times 1.215 = 704\,700 \text{ CU}$$

On the other hand, the reprocurement value of the automated assembly system is given by:

$$V_{RA} = 2\,900\,000 \times 1.215 = 3\,523\,500 \text{ CU}$$

The work-point cost calculation for both alternative assembly systems is calculated in accordance with the details given in Section 11.13. The results for solution I, 'manual assembly', with assembly costs of 1.38 CU per thermoswitch, are given in Figure 12.20. The results for solution II, 'automated assembly' (as described in Section 12.3), with assembly costs of 0.65 CU per thermoswitch, are given in Figure 12.21. The comparison of the relevant factors of both solutions is given in Figure 12.22. With automated assembly, in comparison with manual assembly the assembly costs can be reduced from 1.38 CU to 0.65 CU, which is equivalent to a saving of 0.73 CU per part. Based on the capital deployment, annual saving and theoretical depreciation, for automatic assembly the amortization time is 0.97 years; this therefore fulfils the requirements as based on the requirement list in accordance with Section 12.2.1.

12.6 Investment risks

Reference has already been made to the economic risks of such investment in the introduction to Chapter 11, 'Planning and efficiency of automated assembly systems'. The risk factors comprise a series of assumptions; they are principally the following:

- On account of the fact that the equipment is a special machine, the service period is of particular importance. The time-related pattern of procurement, commissioning and the start-up characteristic is therefore a risk factor.
- The planned availability cannot always be achieved.
- In comparison to the planned personnel training, the training required to achieve the specified production output is often higher.

If the procurement time alters, for example by inadequate planning or the occurrence of difficulties during construction or commissioning, the effective time decreases with a constant product service life. The investment costs also increase with a lengthening of

366 Practical example: planning and realization of an automated assembly system

WORK-POINT COST CALCULATION

Project: Thermoswitch Manual assembly   Solution no. I

Reprocurement value of system C_R/CU = 704 700
Service period n: 4 years at 220 days
Service period n: 16 h/day
Net output: 1000 parts/h

| Machine hourly rate | $C_D = \dfrac{C_R}{n} = \dfrac{\text{CU } 704\,700}{4 \text{ years}}$ | | | | CU/annum 176 175 |

$$C_I = \frac{C_R}{2} \times p = 10\% \qquad \qquad \text{CU/annum} \quad 35\,235$$

$$C_S = 280 \text{ m}^2 \times 10.00 \text{ CU/m}^2 \times \text{month} \qquad \text{CU/annum} \quad 33\,600$$

	C_E Electricity costs	30 kW × CU/kWh 0.20 = CU/h	6.00		
	Compressed air	60 m³/h × CU/m³ 0.08 = CU/h	4.80		
	Water	__ m³/h × CU/m³ __ CU/h	__		
		CU	CU/h		
	220 days × 16 hours/day = 3520 h/annum × CU/h		10.80	= CU/annum	38 016
	C_M Reprocurement value × 10%			= CU/annum	70 470
	Total system costs/annum			= CU/annum	353 496

$$C_{MH} \quad \frac{\text{System costs/annum}}{\text{h/annum}} = \frac{353\,496}{3520} \qquad = \text{CU/h} \qquad 100.42$$

Personnel-related costs	48 persons × wage:	CU/h 11.42 + 110% additional wage costs	CU/h 1151.14
	2 persons × wage:	CU/h 13.90 + 110% additional wage costs	CU/h 58.38
	1 foreman:	CU/h + % additional wage costs	CU/h 43.86
	50 persons × shift premium = $\dfrac{\text{CU shift premium}}{\text{h/day}}$ =		CU/h 26.74
	Total personnel costs/h =		CU/h 1280.12

Assembly costs	Machine hourly rate C_{MH}	= CU/h 100.42
	Personnel costs/h	= CU/h 1280.12
	Total assembly costs/h	= CU/h 1380.54

$$\frac{\text{Assembly costs CU/h}}{\text{Net output (parts)/h}} = \text{CU} \; \frac{1380.54}{1000} = \text{Assembly costs/part}: \qquad \text{CU/part} \quad 1.38$$

Figure 12.20 Work-point cost calculation, solution I

the procurement time. The amortization time increases with a decrease in the utilization time and an increase in the investment costs.

The assembly system as shown in its fine planning detail (Section 12.3) consists of the following units:

- 18 circular cyclic units
- 44 positioning units
- 29 vibratory spiral conveyors

WORK-POINT COST CALCULATION

Project: Thermoswitch Automatic assembly   Solution no. II

Reprocurement value of system C_R/CU = 3 523 500
Service period n: 4 years at 220 days
Service period n: 16 h/day
Net output: 1000 parts/h

Machine hourly rate

$$C_D = \frac{C_R}{n} = CU\ \frac{3\ 523\ 500}{4\ \text{years}}$$ CU/annum 880 875

$$C_I = \frac{C_R}{2} \times p = 10\%$$ CU/annum 176 175

$$C_S = 160\ m^2 \times 10.00\ CU/m^2 \times \text{month}$$ CU/annum 19 200

Machine hourly rate

C_E Electricity costs   40 kW × CU/kWh 0.20 = CU/h 8.00
     Compressed air   130 m³/h × CU/m³ 0.08 = CU/h 10.40
     Water   ___ m³/h × CU/m³ ___   = CU/h ___
                       ___ CU   = CU/h ___

3520 h/annum × CU/h   = CU/annum   64 768
C_M Reprocurement value × 10%   18.40   = CU/annum   352 350
Total system costs/annum   = CU/annum 1 493 368

$$C_{MH}\ \frac{\text{System costs/annum}}{\text{h/annum}} = \frac{1\ 493\ 368}{3520}$$ = CU/h 424.25

Personnel-related costs

5 persons × wage: CU/h 11.42 + 110% additional wage costs   CU/h 119.91
2 persons × wage: CU/h 15.20 + 110% additional wage costs   CU/h 63.84
1 supervisor: CU/h 18.00 + 110% additional wage costs   CU/h 37.80

$$\_\_ \text{persons} \times \text{shift premium} = \frac{\text{CU shift premium}}{\text{h/day}} =$$ CU/h 4.07

Total personnel costs/h = CU/h 225.62

Assembly costs

Machine hourly rate C_{MH} = CU/h 424.25
Personnel costs/h   = CU/h 225.62
Total assembly costs/h   = CU/h 649.87

$$\frac{\text{Assembly costs CU/h}}{\text{Net output (parts)/h}} = \frac{\text{CU 649.87}}{1000\ \text{parts}} = \text{Assembly costs/part:}$$ CU/part 0.65

Figure 12.21 Work-point cost calculation, solution II

- 52 inspection stations
- 9 automatic screw inserters
- 5 rotating-mandrel riveting units
- 4 calibrating units
- 4 adjusting units
- 1 locking paint metering device.

With such a complex assembly system, it is possible that, in comparison to the expected

Project: Thermoswitch as shown in Figure 11.1

Production per annum: 3 500 000 parts	Currency units	
	Solution I: Manual assembly	Solution II: Automatic assembly
Capital deployment CU (Reprocurement value)	704 700	3 323 500
$C_D = 4$ years = CU	176 175	880 875
Assembly cost CU/part	1.38	0.65
Saving CU/part over solution I		0.73
Saving CU/annum		2 555 000
Amortization time = $\dfrac{\text{Capital deployment}}{\text{Saving/annum} + C_D}$ = years		0.97

Figure 12.22 Comparison of cost calculations for manual and automatic assembly

procurement time, the actual time can be extended by one year, which would result in a reduction in the period of utilization from four to three years. On account of the complexity of the system and the handling of 21 different parts, in this case, the expected availability of 80% cannot be achieved, so that an output reduction of 100 parts down to 900 parts per hour occurs. As shown by this type of modified work point cost calculation, in accordance with Figure 12.23 the assembly costs per thermoswitch are 0.91 CU in comparison with 0.65 CU as shown by the 'planned' work point cost calculation in Figure 12.21. If all the pessimistic assumptions were to occur simultaneously, the saving in comparison to manual assembly would decrease from 0.73 CU to 0.47 CU per part. On account of the higher capital deployment, a shorter utilization period and lower saving in comparison to manual assembly, the amortization time would increase from 0.97 years to 1.4 years.

These figures clearly illustrate the nature of the considerable variation which can occur from the planned assumptions with inadequate basic planning. If the utilization period decreases from 4 years to 3 years and consequently the amortization period increases from 0.97 years to 1.4 years, this would be of serious economic importance, and particularly so if, in addition, by reason of external effects, for example, the onset of a recession or the necessity for technological revision of the product, the utilization period of the system of 3 years could no longer be realized. In the planning and selection of assembly equipment, it is therefore highly important that the largest possible proportion of the system be constructed from commercially available units because they can be reused in the construction of new assembly systems. The risk of non-utilization of this type of equipment over the full depreciation period by the inclusion of a proportion of the investment for new applications and projects is thereby reduced. With new systems attempts should be made to achieve reusability factors of up to 80%.

Investment risks 369

WORK-POINT COST CALCULATION

Project: Thermoswitch Automatic assembly	Solution no. Risk estimation

Reprocurement value of system $C_R/\text{CU} = \underline{4\,150\,000}$
Service period n: $\underline{3}$ years at $\underline{220}$ days
Service period n: $\underline{16}$ h/day
Net output: $\underline{900}$ parts/h

Machine hourly rate	$C_D = \dfrac{C_R}{n} = \text{CU} \dfrac{4\,150\,000}{3 \text{ years}}$		CU/annum $\underline{1\,383\,333}$
	$C_I = \dfrac{C_R}{2} \times p = \underline{10\%}$		CU/annum $\underline{207\,499}$
	$C_S = \underline{160} \text{ m}^2 \times 10.00 \text{ CU/m}^2 \times \text{month}$		CU/annum $\underline{19\,200}$

	C_E Electricity costs	$\underline{40}$ kW × CU/kWh	$\underline{0.20}$	= CU/h	$\underline{8.00}$
	Compressed air	$\underline{130}$ m³/h × CU/m³	$\underline{0.08}$	= CU/h	$\underline{10.40}$
	Water	___ m³/		= CU/h	___
			CU	= CU/h	___

 3520 h/annum × CU/h $\underline{18.40}$ = CU/annum $\underline{64\,768}$
C_M Reprocurement value × $\underline{10\%}$ = CU/annum $\underline{415\,000}$
Total system costs/annum = CU/annum $\underline{2\,089\,800}$

$$C_{MH} = \dfrac{\text{System costs/annum}}{\text{h/annum}} \quad \dfrac{2\,089\,800}{3520} \qquad = \text{CU/h} \quad \underline{593.69}$$

Personnel- related costs	$\underline{5}$ persons × wage CU/h $\underline{11.42}$ + $\underline{110}$% additional wage costs	CU/h $\underline{119.91}$
	$\underline{2}$ persons × wage CU/h $\underline{15.20}$ + $\underline{110}$% additional wage costs	CU/h $\underline{63.84}$
	$\underline{1}$ supersivor CU/h $\underline{18.00}$ + $\underline{110}$% additional wage costs	CU/h $\underline{37.80}$
	Persons × shift premium = $\dfrac{\text{CU shift premium}}{\text{h/day}}$ =	CU h $\underline{4.07}$
	Total personnel costs/h =	CU/h $\underline{225.62}$

Assembly costs	Machine hourly rate C_{MH}	= CU/h $\underline{593.69}$
	Personnel costs/h	= CU/h $\underline{225.62}$
	Total assembly costs/h	= CU/h $\underline{819.31}$
	$\dfrac{\text{Assembly costs CU/h}}{\text{Net output (parts)/h}} = \text{CU} \dfrac{819.31}{900} = \text{Assembly costs/part:}$	CU/part $\underline{0.91}$

Figure 12.23 Work-point cost calculation for risk estimation

Chapter 13

The operation of automated assembly systems

With automated assembly systems and even with the application of standardized units, the assembly machines are special machines; it cannot therefore be expected that during initial use trouble-free operation will be experienced and the planned output achieved immediately. Further to the details given in Chapters 11 and 12, strategies relating to the initial start-up and operation of these systems will be discussed below.

13.1 Prerequisites for initial running

The synchronous functional running of an assembly system can definitely be expected by the operator during initial operation. It cannot, however, be expected that all the faults which can occur during continuous operation of such a system can be identified and rectified during a relatively short test run by the system manufacturer. Furthermore a prerequisite is that the quality of the parts to be handled complies with the quality characteristics as specified in the planning phase. The effect of the quality level of the parts to be processed on the functional reliability of the assembly system should be examined during the initial running period by a systematic procedure for investigation of weak points [38].

13.1.1 Quality of the parts

The effect of the quality level of the parts to be processed was discussed in Sections 2.2.6.1, 2.6.6.2, 11.3 and 12.2.2. As a general rule, the specified quality level is determined by AQL (acceptable quality level) [10]. The specified AQL level is only a limit value; exceeding the limit can, however, result in a return of a supplied batch of parts to the supplier. The AQL does hot, however, give any indication relating to the actual quantity of effective parts in a batch. A lower AQL value gives a higher guarantee for a lower proportion of defective parts than a higher AQL value. Data relating to the occurring fault types and fault quantities must be determined during the initial running period. The effect on the quality level of deliveries can only be determined on the basis of this data.

This data determination can be made in a simple manner in that at every feed station for supplied parts, the parts which result in a shut-down are collected and inspected and the results evaluated. One can therefore determine very quickly the magnitude of the proportion of foreign bodies and also the nature of the foreign bodies (e.g. debris or influx of other parts), whether the deformation of single parts is the principal cause of the stoppages or quite simply an unsatisfactory maintenance of tolerances. The inspection plans for the individual parts, inspection processes or the manufacturing processes and handling procedures should be examined on the basis of these results and if necessary revised.

13.1.2 Functional reliability of the system

The availability factor of an automated assembly system is the product of all the individual reliabilities of the functions integrated into the system within the production process. During the initial operation of such systems, the investigation of the fault characteristic should be undertaken in two stages by a rough and fine analysis. Approximately 60% to 70% of the principal causes of faults should be determined by the rough analysis during initial running, the remainder being determined by the fine analysis.

13.1.2.1 Rough analysis of the fault operation characteristic
In the rough analysis for the investigation of the fault operation characteristic, the assembly system is subdivided into the observation segments and the analysis undertaken in the same order as the assembly operation.

An observer initially records the frequency of the occurring faults in the observed system section by simply compiling bar chart lists. In the rough analysis, only the fault point and the number of faults are recorded but not the fault time. For example, on a feed station an observer subdivides the station into the feed unit (e.g. vibratory spiral conveyor), discharge rail, separating stations and the handling characteristics of the positioning unit. The procedure for determining the quality of the parts in accordance with Section 13.1.1 also falls within the field of activity of the observer and, in which respect, in determining the fault characteristic of a vibratory spiral conveyor, a distinction must be made as to whether the parts quality or the design of the vibratory conveyor is responsible. The separating station must be critically observed since, in this case, any possible tilting of the head parts and incorrect positioning of the sensor can result in a multiplicity of faults.

By way of example, Figure 13.1 shows a separating station. A single part is fed to the separating station by a conveyor belt. A sensor arranged in position as shown in diagram (a) signals an approaching part; however, it cannot be separated by a

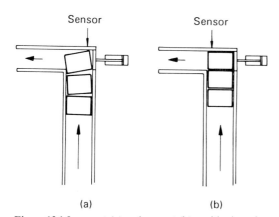

Figure 13.1 Incorrect (a) and correct (b) positioning of a sensor on a separating device

transverse movement on account of its inclined position because the position of the part would result in jamming. By repositioning the sensor as shown in diagram (b) the sensor initially records the correct availability of the part for transverse movement at a point when the part is positioned so that tilting can no longer occur.

Following evaluation of the individual stations, the following points relating to the overall system must be investigated:

- Workpiece carriers
- Transfer unit
- Fault signalling.

Considerable importance is attached to the workpiece carrier with regard to the functional reliability of the system. An important point in evaluation is the danger of contamination of the workpiece carriers. Contamination can, for example, occur by the accumulation of contaminants from the individual part, by deposits or by wear particles. Burrs can break off when handling plastic parts and result in contamination of the workpiece carriers. This can result in inaccuracies during positioning of the parts. If such a danger of contamination is identified, it is recommended at the end of the assembly operations that the workpiece carriers be cleaned by suitable equipment, e.g. a vacuum-cleaning unit. The danger of contamination of automated assembly systems can possibly have a detrimental effect on the optical sensors. If errors occur in the positioning accuracies, the transfer equipment should be checked for its pitch indexing accuracy.

If errors occur in the positioning accuracy, the transfer equipment must be checked in respect of its pitch indexing accuracy. The fault signalling equipment must be thoroughly investigated to determine whether or not it is clearly visible by the plant attendants from any point in the assembly system.

To summarize a graph analysis, the application of the RASED principle (record, arrange, supplement, evaluate, decide) is recommended in order to formulate necessary decisions from the results obtained.

As shown in Figure 13.2, the start-up phase of an automated assembly system is characterized by a low output and high personnel deployment. The low output is attributable to start-up difficulties, e.g. parts quality and the functional reliability of the system. The high personnel deployment results from the number of faults which occur and the necessary implementation of a rough analysis of the fault running characteristics. The measures derived from the rough analysis and their implementation must be monitored by a schedule. The increase in output and production in the personnel deployment can be verified on the basis of this plan. If the identified objectives are not achieved, a detailed weak point analysis must be implemented.

13.1.2.2 Detailed analysis of the fault operation characteristic
Experience shows that following elimination of the start-up difficulties by a rough analysis, the achievable availability can be 5% to 10% below the planned availability. Details of the causes must be determined by detailed analyses of the fault operation characteristic.

The following methods can be applied in the course of a detailed analysis:

- The installation of fault operation recorders
- Chronological survey
- MANALYS fault time recording systems.

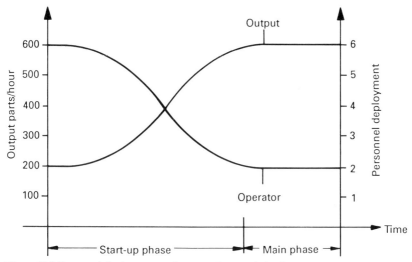

Figure 13.2 Personnel deployment and output characteristic of automated assembly systems in the start-up phase

The installation of fault time recorders gives a recording of the operational shutdowns in relation to the fault locations, but no indication relating to the fault causes. Its installation is by agreement and in accordance with the factory rules and regulations.

With distributed time-lapse recordings in accordance with REFA all events occurring in the assembly system are recorded and their duration measured [40] within a specified period of time and also in a number of randomly selected time intervals.

The MANALYS fault time determination system (assembly system analysis) was developed under the direction of Professor Wiendahl at the Institute for Factory Plant and Equipment at Hanover University [30]. It has been shown that, for an accurate analysis of the time-related operational characteristic, observation by a person is indispensable, since many faults are not recordable by any other method. The fault characteristic of assembly systems with a short cycle time is principally determined by a large number of faults of likewise short duration. Short-duration remedial action in the feed process with removal of foreign bodies or jammed workpieces is typical in this respect. Based on the data available from an operational investigation (500–1000 items of fault data can be expected for a 16 h investigation period), a considerable period of time is required for data evaluation. This fact makes the application of microprocessors absolutely necessary. The microprocessor can be used directly on the assembly line. The items of data are entered manually by the observer via a keyboard without intermediate data storage.

The evaluation takes place in dialogue, i.e. access can be made during the evaluation phase and the evaluation results displayed graphically immediately. The equipment required is a microcomputer with a small memory, disc drives for mass storage and a dot matrix printer (see Figure 13.3).

The investigation report of a fault time investigation is shown schematically in Figure 13.4 by MANALYS. This report is useful to the operator of automatic assembly systems as a basis for decision for the elimination of identified bottle-necks. The

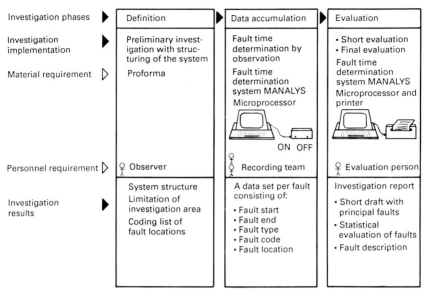

Figure 13.3 Assembly system analysis MANALYS (Wiendahl, Ziersch; IFA, Hanover University)

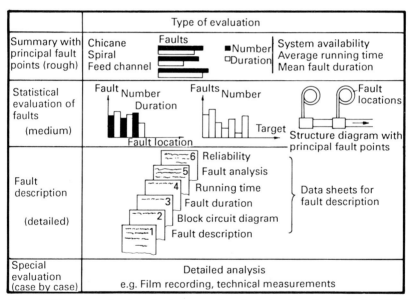

Figure 13.4 Structure of an investigation report for a fault time investigation by MANALYS (Wiendahl, Ziersch; IFA, Hanover University)

principal aspects of the report are the principal fault points in terms of number and duration, the statistical evaluation of faults and the fault description. The statistical evaluation contains the duration of the faults relative to the total fault time and also the time-related distribution of every fault. For purposes of clarity, the values are shown in tabular and histogram form. The detailed fault description enables the technician to determine and eliminate the weak points on site. If these descriptions prove to be inadequate, additional measures such as film recordings or technical measurement can be implemented [41].

13.1.3 Practical example of fault time determination by MANALYS on an automatic assembly system

An assembly consisting of eight single parts and five integrated processing operations is assembled automatically at a cycle time of 3 s on an assembly system consisting of five circular cyclic type automatic assembly machines (see Figure 13.5). Two manual assembly workpoints are included in the assembly system for the manual handling of two single parts which cannot be arranged automatically and fed on account of their material and sensitivity.

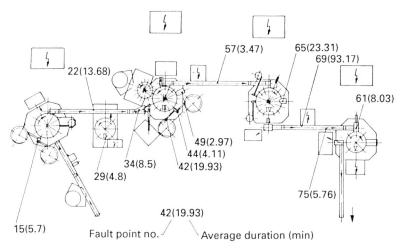

Figure 13.5 Layout of an assembly system with fault details and correlation of the fault-free running time by MANALYS (IFA, Hanover University)

The overall arrangement of this system includes six circular cyclic drives, 16 handling units, nine vibratory spiral conveyors, 17 inspection stations, two automatic screw-insertion stations, one soldering unit, one rotating-mandrel riveting unit and various calibrating and adjustment units [42].

In accordance with the schedule planning, the required production rate of 833 parts per hour at an assumed targeted availability of 80% should be achieved following initial provision of the assembly system and a start-up period of three months. The system is operated for 12 hours per working day, since a daily production of 10 000 parts is necessary. A rough analysis is undertaken during the start-up phase for identification of the principal weak points. The result necessitates theories of changes within the system and also measures for quality improvement of the individual parts.

Following implementation of these measures, the average daily output was 7865 assembled units with considerable fluctuations from day to day. The output corresponds to an availability of 63% and, based on the results of the rough analysis, could not be permanently improved any further.

A fine analysis of the fault operational characteristic was undertaken by the fault time determination system MANALYS. The investigation period covered a period of two weeks during which 80 separate points of the assembly system were investigated with regard to their fault characteristic. A mean fault-free running period of 0.81 min was determined for the overall system by the analysis. The individual stations of the assembly system exhibited mean fault-free running periods between 2.97 and 576 min.

The identified fault parts are shown in the layout of this assembly system in Figure 13.5 and are shown in relation to the times determined for the average fault-free running period.

A fault description was undertaken for every fault point based on the statistical overall evaluation. By way of example, Figure 13.6 shows the fault description for fault point 15 as shown in Figure 13.5.

A program of 26 individual measures of a technical or organizational nature was drafted and implemented for the elimination of the principal causes of faults based on the overall analysis of 80 separate positions investigated. The output characteristic of the system investigated before a MANALYS analysis and following implementation of the resulting measures is shown in Figure 13.7 and demonstrates that the daily output of, on average, 7865 parts was seen to be increased to 10 200. This corresponds to an improvement in the availability from 63% to 82%. This system is now operated with an availability which is 2% higher than that planned.

13.2 Payment

With regard to personnel deployment and qualifications, reference is made to Sections 11.9 and 11.10.6.

Payment for the personnel integrated into an assembly system for the manual activity can be on a group piece-work basis related to the number of parts, in which system-related faults must not be taken into account. Payment of the system attendants must not be on a hourly rate, but in relation to the performance of the system. The availability of an automated assembly system which is to be achieved is, to a large extent, also determined by the qualifications of the system attendants.

The time required for the rectification of sources of faults which result in stoppages of the system if not immediately rectified is dependent upon the rapidity of action and qualification of the system attendants. It is therefore advizable that these persons be paid in relation to the system availability. Such methods of payment are not, however, to be found within the current wages agreements, so appropriate conditions of work must be formulated and agreed. At the same time, a differential must be made between occurrences which can or cannot be influenced by the system attendants.

Retooling times with a part-type change, the verifiable delivery of substandard quality parts or the breakdown of individual stations by defects which were not identifiable by preventive maintenance are, for example, non-affectable occurrences. The basic wage should be based on a technical availability of the systems to be maintained of, for example, 80%. With a higher availability, a premium should be added to the basic wage depending upon the availability achieved. The amount of the premium should be determined on the basis of a premium curve which exhibits a flatter tendency after a certain result. Figure 13.8 shows a possible wage premium curve in relation to the technical availability of the assembly system. This procedure avoids the high technical availability being achieved at the expense of product quality or overloading of the production equipment.

Location:	No. 15: Contact tag separator, Machine I.
Description:	The separator does not move forwards.
Cause:	The light barrier is contaminated or set incorrectly.
	The contact tags do not interlock with the separator because, in some cases, they slide over each other.
	The contact tags are bent.
	The light barrier signals contact tags which are not completely engaged in the separator.
Note:	See also no. 14. In order to eliminate the fault, the control must be switched off and the separator operated manually in order to remove the jammed contact tags. Occasionally this procedure is not successful. The separator must then be forced forward using a screwdriver.
Reference:	Number of faults: 165; Cycle time: 3 s.
	Fault time range: 754–107 174 s.
	Recording period: 34.7 h.
	$A_{T+A} = 93.41\%$, $A_T = 93.46\%$, $A_A = 99.95\%$
	Average fault-free running time: 6.43 min.
	(A_T = technical availability, A_A = action availability)

Figure 13.6 Fault description (IFA, Hanover University)

13.3 Maintenance

Quite naturally, the effects of wear and tear occur during the operation of automated assembly systems. Depending upon the complexity of a system, a particular ratio exists between the capital deployed and the costs required for maintenance. Practical experience indicates that the maintenance cost per annum is between 5% and 10% of

378 The operation of automated assembly systems

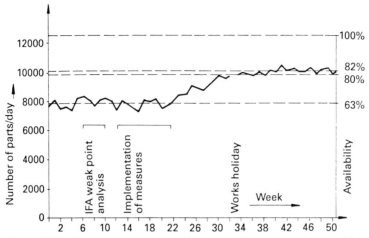

Figure 13.7 Performance characteristics of the assembly system as shown in Figure 13.5

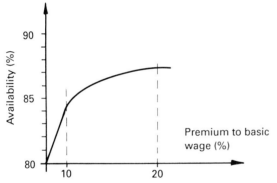

Figure 13.8 Availability-related wage premium for system attendants

the investment costs. Increasing complexity, shorter cycle times and longer periods of utilization per day result in higher maintenance costs. Flexible or pneumatically operated systems with cycle times below 4 s to 5 s have a higher rate of wear than mechanically constructed systems.

With maintenance costs, approximately 60%–70% are attributable to personnel costs, and the rest to spare parts. Spare parts which are not available or cannot be immediately procured result in additional costs by loss of production. To avoid such additional costs, planning for a logical spare parts and supply stock is necessary.

For spare parts stock planning, details of the parts of which failure would result in shut-down of the system should be obtained from the manufacturer.

Two alternative methods for spare parts stocking are possible:

- The system supplier undertakes to provide spare parts delivery within hours.
- The operator retains his own spare parts stock.

On account of the different costs for spare parts, practical distribution of spare parts stocking is recommended between supplier and operator. For expensive parts, the supplier agrees to supply immediately as and when required. On the other hand, the lower-cost, typically high-wear parts such as cylinders, valves, switches, sensors and printed circuit boards for control, etc. can be held in stock by the operator as spare parts.

The spare parts and maintenance requirement patterns of an automatic system fall into the categories of early failures, a steady period with few failures and the wear phase due to natural use. The spare parts and maintenance requirements over the service life of a system are shown schematically in Figure 13.9.

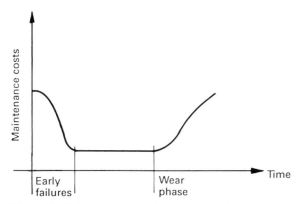

Figure 13.9 Pattern of repair costs for automated assembly systems

With long service life, the steady period and the wear phase can be repeated. As a general rule, the early life failures can be identified during the initial running period by a rough analysis of the fault characteristics and also result in the compilation of spare parts lists.

All the daily maintenance operations undertaken by the system attendants must be detailed in a maintenance plan:

- Lubrication of bearing points, etc.
- Cleaning plan, e.g. of optical sensors, vibratory spiral conveyors, hoppers, etc.
- Draining of pneumatic systems.
- Sealing of leaks in the compressed air lines, etc.

Details of the activities undertaken by the system attendants must be entered on maintenance cards and signed.

The weekly routine checks which must be undertaken by the system attendants are detailed in a checking plan. These include, for example:

- Checking all mechanical, movable parts for wear.
- Checking the pneumatic components for wear.
- Mechanical and control functional checking of the individual stations.
- Upon determining faults, checking the spare parts stock, immediately ordering any spare parts which are not available and determining the time of repair.

With careful implementation of the routine checks, 90%–95% of all necessary repairs can be undertaken at a time outside the system operating time as preventive maintenance activities. With the early identification of wear, operational breakdown and therefore also high costs can be largely avoided by preventive maintenance. The

380 The operation of automated assembly systems

maintenance departments normally undertake maintenance work outside the normal system working time. In this respect, it is necessary that the relevant system attendants be engaged in the maintenance operations so that their experience and expertise can be utilized fully.

13.4 Works safety

As yet, there are no specific regulations for automated assembly systems. Quite naturally, the safety regulations applicable to the operation of electrical and mechanical systems also apply to automated assembly systems.

The so-called crush points are a significant source of danger for the system operators and attendants. Since it is difficult to provide covers at these danger points, total encapsulation of all machines if preferable. Modern assembly machines are fully enclosed against unintentional or unauthorized access by transparent panels or foils.

Figure 13.10(b) shows a cover in the form of a fully enclosed guard with doors for access to the working area. In this respect, sliding doors require less space than hinged doors. A sliding door which can be slid upwards is clearly discernible in the example shown. When opening the door, the assembly system is switched off by a safety switch. Diagram (a) shows a totally enclosing cover which can be moved upwards as a complete unit by a centrally positioned cylinder for access into the machine.

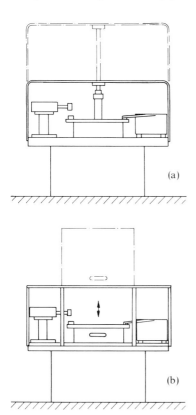

Figure 13.10 Safety cover arrangements for assembly machines: (a) totally enclosing cover with central cylinder; (b) sliding doors

Chapter 14
Outlook

Assembly is increasingly gaining in importance in the production of industrial products. In the past, the principal points of rationalization were to be found in the one-off parts production area and the related production technology such as metal cutting and basic forming, etc. The pressures to increase productivity and quality are also applicable to assembly. Assembly is therefore the rationalization potential of the future.

Industry must adapt itself to the fact that the life of products is becoming shorter and the number of variants greater. An added difficulty in this respect is that the administrative tendency in the form of shorter procurement time with the object of achieving zero stock production places ever increasing demands on production. This results in ever decreasing batch sizes in production. The necessity for economy is predominant and, in this respect, must not be regarded as an insular solution but as one link in the whole production process. A reduction in processing time and stocks must be possible with a productivity increase in assemblies. If the total industrial assembly is considered, in the German Federal Republic, for instance, it can be shown that the level of automation still hovers below the 10% mark. The reason for this is not the lack of assembly technology but rather more the design of products which are not suitable for automation.

The starting point for all rationalization efforts must therefore be the product design. Not every product or assembly can be designed so that automatic assembly is possible. The principal responsibility of product development is to ensure that the new product fulfils the customer's requirements. This means that achieving the prime objective of the product takes preference over an assembly-oriented product design. Compromises are always necessary between product development and production technology. This means that for the time being and also in the future the manual assembly of industrial products will account for a larger proportion of assembly operations than automated assemblies.

On account of the natural limits with regard to product design, in the future the following forms of assembly technology will have to be considered as viable alternatives:

- Manual single-point assembly or on-site assembly
- Manual flow line assembly
- Combined manual and automated flow line assembly
- Single-purpose automatic assembly machines for large-scale production
- Inline linked single-purpose automatic machines for large-scale production

- Flexible semi- or fully automated assembly cells
- Flexible semi- or fully automated assembly lines.

The principal objective of manual single-point or flow line assembly must include the following objectives:

- The greatest possible avoidance of secondary assembly activities.
- Work content improvement without overstressing the individual operatives by a limited allotment of differing operations.
- In spite of the flexible interlinking of individual work points with each other, decoupling of the work points from each other by the formation of intermediate buffers between work points in order to permit the development of individual work rates.
- Limited arrangement of operations for a more optimum and individual work point arrangement.

With combined flow line assembly, considerably more attention must be given to the manual activities in the future than has been the case in the past. The objective should not be restricted to a monotonous single-function activity but that a specific level of work content interest be achieved. In addition, with combined flow line assembly decoupling from the fixed machine cycle is, however, a system problem.

The problem of fully automated assembly firstly lies in the fact that the machines are dedicated machines which can rarely or only with great difficulty be re-equipped for other products, and secondly in the automated feed technique of the parts. Approximately 80% of all the faults which occur originate in parts feeding. In this respect, uniform quality of the parts and their cleanliness, etc. play a large role. There has been no significant advance in this field since the development of the vibratory spiral conveyor. The question therefore arises as to whether or not parts should be produced as bulk material and then be arranged during assembly instead of retaining an initially arranged condition achieved during production. Magazining techniques must be developed further. As far as possible, magazining operations should be related to manufacturing processes in which the magazining operation can be integrated as a secondary activity.

As a general rule, automatic assembly machines work on short cycle times. The desire to use this type of capital intensive equipment for a so-called ghost shift will remain an unfulfilled dream for just as long as a large personnel deployment is required for the rectification of faults in the feed area. In this respect, for example, magazine feed presents a possibility for improvement. The integration of parts manufacturing processes in assembly systems or else the integration of assembly processes in parts production plants are methods for a restricted assembly technique with low personnel requirement.

In the field of medium series production, automation is still largely limited to single operations. This area is the future field of application for so-called flexible-assembly equipment. Every flexible-assembly cell is in direct competition with manual assembly. The reason why, up to the the present, flexible-assembly cells have proved to be unsatisfactory in economic terms are to be found in the unfavourable price/performance relationship of flexible items of equipment such as robots and the related controls and also in the level of the necessary peripheral costs. Figure 7.1 shows the percentage distribution of the most frequently occurring single functions in assembly in the fields of precision and electrical engineering. The proportion of approximately 70% for handling and assembly can be split into half for handling and the other half for the provision and arranging of the single part. In this respect, assembly robots can only undertake the functions of handling and assembly which, depending upon the product complexity, account for 35% of the total involvement.

This means that only this proportion can exhibit the required level of flexibility by the application of a programmable item of equipment. All the other operations cannot be regarded as being flexible, but rigid, type-related operations. The end result is that considerable effort is necessary in the pheripheral arrangement in order to achieve economic results. The focal point of assembly robot development must be directed towards a favourable price/performance ratio, i.e. the items of equipment must either be quicker or cheaper. A further development in sensor technology is necessary to increase the intelligence. The so-called 'dip in the box' will be technically feasible within the foreseeable future; however, it does not represent an economic solution. Developments for techniques to retain an initially ordered parts condition are necessary for the development of the 'dip in the box' concept.

New technology, such as the application of CAD/CAM or CIM, may be suitable for bridging and reducing the conflicting aspects between product development and production technology. In the future, with shorter working hours, the capacity to increase productivity will have a decisive effect on the competitiveness of our economy. A key function in this respect lies in assembly technology. In addition to the technological requirements, it should not, however, be forgotten that the most important capital resource of a firm is the know-how and expertise of its employees. Highly qualified employees are required for the implementation of modern technologies and the optimization of whole processes. One of the key aspects for increasing productivity lies in the training and further education of our present and future technicians. In Section 1.2 (status of assembly in the production operation) the significance of assembly costs as a proportion of the production cost of a product is discussed in considerable detail. In spite of the significance of the cost of assembly in production, assembly technology is either not at all or inadequately taught at our professional colleges and universities in production or manufacturing technology courses. In the face of the proven importance of the rationalization potential of assembly technology, the earliest possible elimination of this deficiency is therefore a matter of paramount importance.

Chapter 15
References

1. *Warnecke, Löhr, Kiener* „Montagetechnik" Mainz Krausskopf-Verlag, 1975
2. *Lotter* „Arbeitsbuch der Montagetechnik" Mainz Vereinigte Fachverlage Krausskopf-Verlag Ingenieur-Digest, 1982
3. *Abele u. a.* „Einsatzmöglichkeiten von flexiblen automatischen Montagesystemen in der industriellen Produktion" Düsseldorf, VDI-Verlag, 1985
4. *Gairola* „Montagegerechtes Konstruieren" (Dissertation), Darmstadt, TU, 1981
5. *Witte* „Entwicklung montagegerecht gestalteter Produkte" Nürnberg, VDI-Tagung, November 1984
6. *Treer* „Automated Assembly" Dearborn-Michigan, USA, Society of Manufacturing Engineers, 1979
7. *Boothroyd* Design for Producibility – The road to higher productivity, Assembly Engineering, Hitchcock Publishing Company, 1982
8. *Lotter* „Montageerweiterte ABC-Analyse", Vortragsmanuskript, München, 5. Deutscher Montagekongreß, MIC, 1983
9. *Lotter* „Using ABC-Analysis for assembly", Bedford, UK, Assembly Automation, pp. 80–86, IFS, 1984
10. *Masing* „Handbuch der Qualitätssicherung" München, Carl-Hanser-Verlag, 1980
11. Autorenkollektiv Handbuch der Arbeitsgestaltung und Arbeitsorganisation Düsseldorf, VDI-Verlag, 1980
12. *Warnecke* „Der Produktionsbetrieb" Berlin-Heidelberg-New York, Springer-Verlag, 1984
13. *Lotter* „Primär-Sekundär-Montageanalyse als Planungsmethode für Montagelinien", Berlin, Wt-Zeitschrift für industrielle Fertigung, pp. 9–14, Springer-Verlag 1/1985
14. *Lotter* „Produktionssteigerung im Montagebereich der Feinwerktechnik" F+M, München, Hanser-Verlag, April 1985
15. *Lotter* The economic design of assembly processes, Bedford/England, IFS-Verlag, Assembly Automation, pp. 153–157, August 1982
16. *Lotter* Automated assembly in the electrical industry, Bedford/England, IFS-Verlag, Assembly Automation, pp. 29–35, 1984
17. VDI-Richtlinie 2860, VDI-Handbuch „Betriebstechnik", VDI-Handbuch „Materialfluß und Fördertechnik", Düsseldorf, VDI-Verlag, Oktober 1982
18. *Warnecke, Schraft* „Industrieroboter" Mainz Krausskopf-Verlag, 1979
19. *Schraft* „Industrierobotertechnik", Band 115, Kontakt & Studium, Grafenau, Expert-Verlag, 1984
20. *Warnecke, Schraft* Einlegegeräte zur automatischen Werkstückhandhabung, Mainz, Krausskopf-Verlag, 1973
21. *Auer u. a.* „Industrieroboter und ihr praktischer Einsatz", Band 36 – Kontakt & Studium, Grafenau, Expert-Verlag, 1981
22. *Wiendahl* Unveröffentlichtes Vorlesungsmanuskript, Hannover, Institut für Fabrikanlagen, Universität Hannover, 1984
23. *Grohe* „Wirtschaftliches Widerstandsschweißen in der Kleinteilefertigung", Mainz, Vereinigte Fachverlage Krausskopf-Verlag Ingenieur Digest „Verbindungstechnik", Heft 9, 1981

24. *Grohe* „Mechanisierung und Automatisierung an Widerstandschweißeinrichtungen", DVS-Berichte Band 40, Düsseldorf, DVS-Verlag, 1976
25. *Fa. Haas* Laser-Komponentensystem, Bedienungsanleitung, Schramberg, 1983
26. *Brunst* „Fügeverfahren", Mainz, Krausskopf-Verlag, 1979
27. *Borowski* „Das Baukastensystem in der Technik", Berlin, Springer-Verlag, 1981
28. *Lotter* „Integrierte Montagevorgänge in der Herstellung von Mikroschaltern", München, „Feinwerktechnik – Meßtechnik", Hanser-Verlag, 1984/2
29. *Lotter* „Just-in-time-Production durch Montageorganisation und Automatisierung", Vortragsmanuskript, Böblingen, JIT-Tagung, September 1985
30. *Wiendahl, Ziersch* „Verfügbarkeitsverhalten automatisierter Montageanlagen", VDI-Bericht Nr. 479, pp. 79–88, Düsseldorf, VDI-Verlag, 1983
31. *Dreger* „Vereinbarung zur Verfügbarkeit als Teil der Leistungsangaben eines Systems, QZ 20, Heft 2, pp. 35–39, 1975
32. *Tribus* „Planung und Entscheidungstheorie ingenieurwissenschaftlicher Probleme" Braunschweig, F.-Vieweg-Verlag, 1973
33. *Reisch, O.* „Die Berücksichtigung der Zuverlässigkeits- und Verfügbarkeitsanforderung bei der Planung von Maschinensystemen" (Dissertation), Hannover, Universität, 1978
34. *Lotter* „Aufbau und Einsatz Montagelinien am Biespiel der Feinwerktechnik, München, ibw-Kolloquium, TU, 1985
35. *Lotter* „Aufbau und Einsatz flexibler Montagesysteme" München, 6. Deutscher Montagekongreß, 1985
36. *Wiendahl* „Stand und Tendenzen der automatischen Montage", Hannover, Fachseminare IFA Universität, 1984
37. *Herzlieb* „Anlagenoptimierung durch stördatenbezogene Pufferdimensionierung und Störtaktabgleich", Hannover, Fachseminar IFA Universität, 1984
38. *Gehler* „Steigerung der Anlagennutzung automatisierter Montagesysteme in der Anlaufphase", Hannover, Fachseminar IFA Universität, 1984
39. *Feldmann* CAD-CAM Systeme für die Montage, München, 6. Deutscher Montagekongreß, 1985
40. *NN REFA* „Methodenlehrgang der Planung und Steuerung", München, Carl-Hanser-Verlag
41. *Ziersch* „Nutzungsverbesserung automatisierter Montageanlagen, Hannover, Fachseminar Institut für Fabrikanlagen Prof. Wiendahl, TU, 1984
42. *Lotter* VDI-Bericht Nr. 479, pp. 31–37, Düsseldorf, VDI-Verlag, 1983
43. *Lotter* MHI Loseblattsammlung, Abschnitt „Betrieb von automatisierten Montagesystemen", Landsberg, MI-Verlag, 1985
44. *NN* MTM-Handbuch, 3. Auflage, Hamburg, Deutsche MTM-Vereinigung e. V., 1981
45. *Schlaich* CADLAS – eine rechnergestützte Methode für Layouterstellung von automatischen Montagesystemen (IPA Stuttgart), VDI-Zeitung Band 127 No. 20, Düsseldorf, VDI-Verlag, 1975
46. *Klinger, H.* Speicherprogrammierbare Steuerungen (SPS), Anwendungsbeispiele – Aufbau und Programmierung – Unveröffentlichtes Schriftstück, Festo Didactic, Esslingen, 1986
47. VDI-Richtline 3258, Düsseldorf, VDI-Verlag, 1962

Further reading

Literature to which reference has not been made in the book, but which is, however, recommended for further study.

Andreasen, Kähler, Lund
    Montagegerechtes Konstruieren
    Springer-Verlag Berlin, Heidelberg, New York, Tokyo 1985
    Praxisnahe Anleitung zur montagegerechten Konstruktion

Riley, F. J.
    Assembly Automation
    IFS (Publications) Ltd., Bedford 1983
    Praktisch orientiertes Buch über Wahl und Anwendung von verkäuflicher Montageausrüstung

Heginbotham, W. B. (editor)
  Programmable Assembly
  IFS (Publications) Ltd., Bedford 1984
  Ausgewählte Beiträge über flexible Automatisierung. Mehrere Beispiele aus der Praxis

Rathmill (editor)
  Robotic Assembly
  IFS (Publications) Ltd., Bedford 1984
  Praxisbeispiele Robotereinsatz

Autorenkollektiv – Tabellenbuch Montage
  VEB Verlag Technik, Berlin 1984
  Datensammlung der Montagetechnik

Index

Note: illustrations are indicated by *italic page numbers.*

ABC analysis, assembly-extended, 9–35, 306, 338–339, *340*
Aids (for assembly), 21, *22*
Amortization calculation, 305–306
AMP plug pin, 28–29
Angle coder, 142–143
Annular parts, design of, 16, *18*
AQL (Acceptable quality level), 31
Arrangement difficulty, 15–17
　calculations for, 16, *17*
Arrangement features, feeder units affected by, 102–103
Articulated-arm robots, 144–147
　see also SCARA . . . robots
Artificial intelligence, 126
ASEA assembly robot system, 147–149, *150*, 289, *290*
Assembly, 1
　activities during, 1
　definition of, 26
　obstacles to rationalization, 2–3
　place in production operation, 2–3
Assembly aids, 21, *22*
Assembly costs, 33
Assembly direction, 20–21
Assembly lines,
　assembly machines combined to form, 210–214
　flexible-assembly cells interconnected by manual work points, 235–238
　operational characteristics of, 219
　uncycled, 218
Assembly machine systems, 205–210
Assembly machines,
　combining to form assembly lines, 210–214
　design of, 183–224
　horizontal drive shaft arrangement, 198–200, *201*, *202*
　multi-station,
　　electric-motor-driven, 195–205
　　oscillatory drives for, 200–205
　　pneumatically operated, 192–195
　oscillatory drive arrangement, 200–205
　practical example, 248–275

Assembly machines (*continued*)
　single-station, 184–186
　vertical drive shaft arrangement, 195–198
Assembly methods,
　selection of, 27–28
　subdivision of, 26, *27*
Assembly operations,
　efficiency of, 57–58
　PAP–SAP definition of, 62, 64–65
　robot, 228–229
Assembly robots, *see* Robots
Assembly sequence analysis, 306–310, 341–345
Assembly systems,
　availability of, 218–224, 323–325
　　factors influencing, 220–223
　　improvement of, 224
　flexible, 225–238
Assembly working spaces, 22–24
Assembly-extended ABC analysis, 9–35, 306
　checklist for, *35*, 35
　cost considerations, 33
　ease-of-assembly considerations, 20–26
　factors affecting, 9, *10*
　handling considerations, 13–20
　methods used, 26–29
　organizational implementation of, 33–34
　price considerations, 9–11
　quality considerations, 29–33
　stage-by-stage application of, *34*
　supply considerations, 11–13
　thermoswitch, 338–339, *340*
Assembly-oriented product design, 5
Audio cassettes, assembly of, 269–272
Automated assembly, quality level for parts, 31
Automated assembly lines, integration of manual work points in, 214–217
Automated assembly systems,
　assembly sequence analysis for, 306–310, 341–345
　availability of, 323–325, 348–349
　computer-aided planning of, 333–337
　cycle time of, 317, 326–327, 345
　evaluation of, 329–333

388 Index

Automated assembly systems (continued)
  financial investment aspects, 305–306, 327–329
  function analysis for, 316–317, 345
  future developments necessary, 382
  initial operation characteristics of, 324–325, 349
  layout planning for, 317–320, 345–347
  maintenance of, 377–380
  maintenance requirements of, 322
  modules for, 96–182
  operation of, 370–380
  optimized overall solution for, 333
  personnel requirements for, 321–322, 347–348
  planning of, 303–337
    practical example, 338–369
  product analysis for, 306, 338–341
  repair costs for, 379
  requirement list for, 303–306, 338
  safety precautions for, 380
  simulation of, 336–337
  spare parts for, 378–379
  system structure for, 324, 348–349
  workpiece carrier design for, 310–315, 345
Automatic flexible-assembly cells, 232–234
Automatic screw-inserting units, 161–168
  drop-tube-fed, 164–166
  feed-rail-fed, 166–168
  functions of, 163
  quality considerations for, 31–33, 163
  space for, 22–23, *24*
Automatic soft-soldering station, *180*
Automation, proportion in Germany, 381
Auxiliary assembly parts, 27
Auxiliary contact block, assembly of, 288–289
Availability, 218–224
  automated assembly system, 323–325, 348–349
  effect of SPC on, 247
  factors affecting, 218–224, 323–325
  individual stations, 323–324, 348
  wage premium related to, 376, *378*

Base parts, 6–7
  design of, 6
Basic analysis (PAP–SAP), 58–61
Belted material, 13, *14*
Bihler automatic punching-bending machine, 298
Blade wheel removal system, 104–105
Bonding,
  advantages of, 181, 182
  metering unit for, 181–182
Breakdowns, availability affected by, 221–222
Buffer capacities,
  machines, 213, 237, 238
  manual assembly systems, 47–48
Bulk material, 11, *12*

CAD layout planning, 333–336
CADLAS system, 334–336
  hardware for, 334, 335
  software for, 334–335

Cam controlled positioning units, 134, *138*
Cam drives, 131, *132*, *139*
  circular cyclic units, *151*, 155–157, *158*
Car fan motors, assembly of, 272–275
Car headlights,
  assembly of, 93–95
  assembly of clips on, 279–281
Cardan mounting, 147, 149
Centring of base parts, 6–7
Chamfers, 16–17, *18*
Checking stations, 190–192
  actions by, 191
  lift-movement, 192
  rigid-arrangement, 191
Chute magazines, 12, *13*, 294
Circular assembly tables, 50–53, *84*, *85*
Circular cyclic assembly machines,
  electric-motor-driven, 195–205
    horizontal shaft arrangement, 198–200, *201*, *202*
    vertical shaft arrangement, 195–198
  modules for, 206, *208*
  oscillatory drives for, 200–204
  pneumatically operated, 192–195
  practical example, *249*, *251*, *253*, *256*, *258*, *263*, *265*
Circular cyclic flexible-assembly cell, 234
Circular cyclic transfer equipment, 151–157
  cam-driven, 155–157
  feed systems used, 187–188
  Maltese-cross-driven, 153–155
  pneumatically driven, 152–153
Clamping springs, switch assembly, 82, *83*
Clamping workpiece carriers, 7
Coding discs, 142–143
Coding (of workpiece carriers), *47*, *94*, *258*, *273*, *279*, *311*, *315*, *349*
Coiled parts,
  design of, 17, *18*, *127*
  disentanglement test for, *126*
  feeding of, 126–129
Collection movement,
  human operative, 64, *65*
  robot, 228
Computer-aided planning, automated assembly systems, 333–337
Connection tags, manufacture of, 294
Contact breakers, assembly of, 292–302
Contact faces, 23, *25*
Conveyor-belt systems,
  combining of assembly machines by, 212
  workpiece carriers for, 315
  see also Double-belt systems
Cost considerations, 33
  PAP–SAP analysis, 75–77
  parts quality, effect of, 31–33
Cost responsibility, 4
Costs, factors affecting, 9–11
CP (continuous-path) control, 141
Cross head screws, 163
Cycle equalization allowance, 73

Cycle time,
    automated assembly system, 317, 326–327, 345
    availability affected by, 222–223
    robots, 227
Cycled transfer equipment, 150–159
    circular, 151–157
    longitudinal, 157–159
Cycling, line manual assembly, 54–56
Cylindrical parts,
    design of, 16, *18*
    feeding of, *104*, 107, *112*
Cylindrical workpieces, preferred positions of, *16*

Data exchange (by SPC), 247
Data reduction, 125
DEA assembly robot, 147, *149*
Decoupling, 211, 217, 232
Delivery movement,
    human operative, 64, *65*
    robot, 228
Depreciation, 330
Design,
    assembly-oriented, 5
    feed-oriented, 16–17, *18*
'Dip-in-the box' concept, 383
Direction (of assembly), 20–21
Discharge rails,
    design factors for, 114–117
    electromagnetically driven, 114, *115*
    gravity-fed, 113–114, *115*
    power-driven, 114, *115*
Disentanglement equipment, 126–127, *128*, 129
Disentanglement property, 126
Dish cams, 197
Dispensing containers, 37, *38*
Displacement-step-time diagrams, 192–194
Domestic appliance drives, assembly of, 282–287
Double-belt systems, 47, *49*, 73, 160, *217*, 218
Drives,
    industrial robots, 141–142
    positioning units, 130–131
    transfer equipment, 151–157
Drop-tube-type automatic screw-inserting machines, 164–166
Duplex design assembly machines, 257–258, 317

Ease of arrangement, 14–17
Ease of assembly, 21–22
Ease of disentanglement, 126
Ease of handling, 13–20
Easily deformable parts, feedability of, 102
Electric drives, 131, *132*, 142
Electric-motor-driven multi-station assembly machines, 195–205
Electronic position identification, 125–126
Enclosed contact faces, 23
Energy control, 133
Energy conversion, 133
Energy costs, 330

Ergonomics, manual work point, *37*
Examples (of assembly), *see* Practical example
Extended analysis (PAP–SAP), 79–81
External centring, 6, *7*

Fault buffers, 324
Fault description, 376, *377*
Fault occurrence time patterns, 321–322
Fault operation characteristic,
    detailed analysis of, 372–375
    rough analysis of, 371–372
Fault recording, 191
Faults, machine switch-off triggered by, *191*
Feed section,
    arrangement of, *190*
    reliability affected by, *190*
Feed-rail-fed automatic screw-inserters, 166–168
Feeder units, 102–129
    directional arrangements of, *188*
    parts with one arrangement feature, 103–108
    parts with several arrangement criteria, 108–124
    scoop segment used, 103–104
    secondary sorting equipment added, 122–123
Fels-type pneumatically operated swing positioning unit, 134, *137*
Festo FPC 404 controller, *241*, 242
Fine analysis (PAP–SAP), 61–65
Flexible-assembly cells,
    automatic, 232–234
    design of, 230–235
    integration with manual work points, 235–238
    semi-automatic, 231–232
Flexible-assembly systems,
    characteristics of, 226
    design of, 225–238
    future developments necessary, 382–383
    practical example, 275–291
Flow diagram, program controllers, *245*
Flow material, 12, *13*
FMS system, 161
Foot switch, 42
Foreign bodies, effect of, 31, 102
Form-locking, 28, 29
Formatted packaging, 11
Free space (for assembly), 22–24
Function analysis, 316–317, 345
Functional reliability, initial running affected by, 371–375

Galvanized parts, feedability of, 102
Grasping movement,
    human operative, 14, 64
    robot, *227*, 228
Grey image processing, 125
Grippers, 133–134, 139, 230
    multi-purpose, *230*, 277
    working space for, *25*

390   Index

Gripping movement,
　human operative, 64
　robot, *227*, 228

Hammering, characteristics of, *169*
Handling,
　classification of, 97–98
　definition of, 97
　symbols for, *100*
Handling considerations, 13–20
Handling equipment, 129–150
　axis designation for, 132
Headlight assembly,
　initial basis for assembly, 93–94
　new line assembly proposal, 94–95
Height-adjustable table, 85, *86*
Helical springs,
　design of, 17, *127*
　disentangling of, 126–127, *128*
　heat treatment of, 127, *128*, 129
High-pressure nozzle,
　assembly of, 264–269
　assembly sequence for, 307
Hoppers,
　blade wheel used, 104–105
　magnetic plate discharging system used, 105–107
　scoop segment used, 103–104
Horizontal articulated-arm robots, 144–147
　see also SCARA . . . robots
Human work output capacity, *36*
Hydraulic drives, 130, 141

IBP–Pietzsch feed unit, 125
Idle cycles, 250, 254
Immediate switch-off principle, 191
Inclined conveyors, 107–108
　part forms suitable for, *106*
Indexing tables, 119, *121*, 151, 187, *188*, *234*
　drives for, 152, 153
Indexing units, 42–43, 83, 90
Industrial robots *see* Robots
Initial running, prerequisites for, 370–376
Insulating-plate punched strip, manufacture of, 293
Integration,
　assembly operations into parts production processes, 297–298
　limits for, 302
　manual work points, 214–217
　parts manufacturing processes, 292–293
　parts production into assembly equipment, 299–302
Interest costs, 330
Interlocking parts, feed of, 126–129
Intermediate buffers, 59, 189, 238
Investment calculations, 364
　automated assembly system, 305–306, 327–329

Investment pattern, *305*
Investment risks, 365–368

Joggling operation, 27, 28
Joining methods, 26, *27*
Just-in-time production concept, 292–302
　advantage of, 301

Kinematics, robots, 136, 139
Knee switch, 42

Labour costs, *2*
Lap belt arrangement, 45, *46*, 47
Laser welding equipment, 174–175, 177–179
Lasers,
　fixed-head, 177, *178*, 179
　gas, 177, *178*, 179
　meaning of term, 175
　principle of, 175, 177
Layout planning,
　automated assembly system, 317–320, 345–347
　practical example, 318–320
　principles of, 317–318
Life curve (of product), 303, *305*
Line assembly,
　future developments necessary, 382
　manual assembly, 44–56
　manual transfer of assembled parts, 44–45
　　advantages of, 45
　　disadvantages of, 45
　　PAP–SAP analysis of, 59–60, 70–72
　　switch assembly, 81–82, *84*
　mechanical transfer of assembled part in arranged form, 46–56
　　advantages of, 56
　　cycling times for, 54–56
　　disadvantages of, 56
　　modular system, 50–53
　　PAP–SAP analysis of, 72–74
　　workpiece carriers used, 46–47, 53–54
　mechanical transfer of assembled part in unarranged form, 45–46
　　disadvantages of, 46
　　PAP–SAP analysis of, 60–61
　　transfer belts used, 45, *46*
　rectangular arrangements, *44*, 47, *48*, *73*, *90*, 91, 218
Longitudinal transfer equipment, 157–159
　horizontal drive shafts used, 199–200, *202*
　oscillatory drives used, 204–205, *207*
　overhead/underfloor systems, 159
　plate systems, 159, *160*
　practical examples, *267*, *274*
　rotary systems, 159, *160*

Machine hourly rate, 330–331
Machine occupation times, 300–301

Index 391

Magazine feed, 12, *13*, 268
Magnetic plate discharged hopper, 105–107
Maintenance,
　automated assembly system, 322, 377–380
　stored-program controllers, 246–247
Maintenance costs, 330
Maltese-cross drives, *151*, 153–155
MANALYS fault time determination system, 373–375
　outline of, *374*
　practical example of, 375–376
　report structure for, *374*
Manual assembly, 36–56
　definition of, 36
　future developments necessary, 382
　line assembly, 44–56
　organizational forms of, 39–56
　quality level for parts, 31
　single-point assembly, 40–43
　work points for, 37–39
Manual parts dispensers, 37, *38*, *40*, *67*
Manual work points,
　assembly tasks undertaken by, 216–217
　flexible-assembly cells interconnected by, 235–238
　integration into automated assembly lines, 214–217
　parts provision by, 214–216
　uncycled assembly lines including, 218
Material transfer, factors affecting, 18–20
Micro-jump movement, 109
Modular assembly system, 50–53
　possible arrangements for, *51*
Modules,
　automated assembly system, 96–182
　bonding, 181–182
　feeder, 102–129
　handling, 129–150
　purpose of, 96
　riveting, 168–171
　screw-insertion, 161–168
　selection of, 98
　soldering, 179–181
　transfer, 150–161
　welding, 171–179
MP (multi-point) control, 141
MTM (Methods Time Measurement),
　basic movements, 61–62, 226
　time measurement units, 62
Multi-purpose gripper, *230*, 277
Multi-spindle rotating-mandrel riveting unit, *174*
Multi-station assembly machines, 186–210
　electric-motor-driven, 195–205
　pneumatically operated, 192–195
Multi-tier buffer, *214*
Mushroom-head parts, feeding of, *19*, *104*, 105, *111*

Non-cycled transfer equipment, 159–161
　advantage of, 161

Notching, 169, *170*
Notching operation, 28–29
Number of parts, 7–9
Number of stations, availability affected by, 220–221, 323, 348

Operating system (of SPC), 242
Optical sensors, 123
Organizational forms, manual assembly of, 39–56
Organizational implementation, assembly-extended ABC analysis, 33–34
Oscillatory conveyors, *see* Vibratory spiral conveyors
Oscillatory drives, 200–205
Outlook (for assembly systems), 381–383
Overhead/underfloor longitudinal transfer systems, 159

PAP (primary assembly processes), definition of, 57
PAP–SAP analysis,
　application examples, 65–79
　basic analysis, 58–61
　cost considerations, 75–77
　extended analysis, 79–81
　fields of application, 58–65
　fine analysis, 61–65, 77–79
　robots, 226–229
　headlight assembly, 94
　line assembly, mechanical transfer of assembled part in arranged form, 72–74
　practical examples, 81–95
　single-point assembly, 66–70, *74*
　　parts provision by manual dispensers, 66–68
　　parts provision by manual dispensers and by vibratory spiral conveyors, 69–70
　　parts provision by paternoster, 68–69
　switch assembly, 82, 87–88
　switch element assembly, 89, 91–92
Parallel grippers, 133, *134*
Parts cleanliness, 102
Parts feed stations,
　design of, 189–190
　direction of feed in, *188*
　reliability of, 325
　specification for, *326*
Parts function symbols, handling processes, 98, *100*
Parts machining in assembly equipment, 293–294
Parts paternoster, *41*, 41–42, *68*, 69
Parts production processes,
　integration of assembly processes into, 297–299
　integration into assembly equipment, 299–302
Parts quality, 30–31
　availability affected by, 223, 323, 348
　cost implications of, 31–33
　initial running affected by, 370–372
Parts reduction, 7–9
Paste solders, 181

392 Index

Payment systems, 376, *378*
Personnel costs, 331
Personnel requirements, automated assembly system, 321–322, 347–348
Personnel training, availability affected by, 325, 349
Pick-up point, *188*
Piece-work time allowance, 75, 78
Planning (of automated assembly systems), 303–337
   practical example, 338–369
Plastic formed parts, feedability of, 102
Plate chain conveyors, 107
   see also Inclined conveyors
Plate longitudinal transfer systems, 159, *160*
Plate-type workpiece carriers, 83, *84*, 85, 89–90, 236
Plug parts, assembly of, 28–29
   free space for, *25*
Pneumatic drives, 131, 142
Pneumatically driven circular cyclic transfer units, 152–153
Pneumatically driven positioning units, *135*, *137*
Pneumatically operated multi-station assembly machines, 192–195
Pneumatically operated resistance welding machine, *176*
Portal-type flexible-assembly cell, 234, *235*
Portal-type positioning unit, 134, *137*
Portal-type robots, 147, *148*, *149*
Position identification, 125–126
Positioning point, *188*
Positioning units, 130–135
   control of, 132–133, *138*
   design of, 134–135
   drives for, 130–131
   grippers for, 133–134
   kinematic structure of, 131–132
Positioning of workpiece, 15, *16*
Practical examples, 248–291
   assembly machines, 248–275
   assembly sequence analysis, 306–309
   assembly system with integrated parts production, 294–297
   fault time determination by MANALYS, 375–376
   flexible-assembly systems, 275–291
   layout planning, 318–320
   planning of automated assembly systems, 338–369
Pre-assembly operations, 55–56, 86
Press-riveting, 168–169
Pressing, characteristics of, *169*
Primary–secondary analysis, 57–95
   see also PAP–SAP analysis
Printed circuit boards, fitting out of, 287–288
Process symbols, *100*, *101*
Product analysis, automated assembly system, 306, 338–341
Product design, 5
   and assembly situation, 306–309, 341–343

Product design (*continued*)
   factors affecting, 381
Program, 240, *244*
Programming,
   robots, 141
   stored-program controllers, 243–246
Programming equipment, 240–241
PTP (point-to-point) control, 140–141
Pumping (of lasers), 177
Punch-bending machines, 292–293, 298
Punch-rolling, characteristics of, *169*

Quality considerations, 29–33
   automated assembly system, 323, 348
   availability affected by, 223, 323
   screw-insertion, 163

Raised boss, 27
RASED principle, 372
Ratchet drive, *151*
Reach (and visual) area, 37, *38*
Reaching movement,
   human operative, 62–64
   robot, 227–228
Reaction bonding agents, 181
Reading list, 385
Rectangular arrangements, line assembly, *44*, 47, *48*, *73*, *90*, 91, 218, 236
References listed, 384–385
Relay switches, assembly of, 294–297
Release movement,
   human operative, 65
   robot, *227*, 229
Reliability, definition of, 219
Repeat effect, 43, 88
Reprocurement value, 75, 76
Requirement list, 303–306, 338
Resistance welding, 28, 173–174
Rigidity, assembly affected by, 24, 26
Riveting, *27*, 28
   comparison of processes, *169*
Riveting units, 168–171
Robots, 129–130, 135–150
   ASEA assembly robot system, 147–149, *150*, 289, *290*
   control of, 139–141
   cycle time of, 227
   DEA assembly robot, 147, *149*
   drives for, 141–142
   grippers for, *230*, 277
   horizontal articulated-arm robots, 144–147
   industrial, 143–144
   kinematics of, 136, 139
   measurement systems used, 142–143
   part systems of, 135, *140*
   primary–secondary fine analysis for, 226–229
   programming of, 141
   reaching movement, *227*, *229*
   SCARA, 144–147

Robots (*continued*)
  sensors for, 143–144
  swivel, 145
  teach-in programming, 141
  working space for, 229
Rocker, assembly of, 248–250
Rolling, characteristics of, *169*
Rosin-core solder, 179
Rotary longitudinal transfer systems, 159, *160*
Rotary tables, 42–43, 83
  assembly examples, 43
  line-assembly use, *50*, *84*, *85*, *90*, *91*
Rotating-mandrel riveting, 170–171
  design of tool, *173*
  machine for, *172*, *174*
  principle of, 170, *171*
Rotation locking, 29, *30*

Safety covers, 380
SAP (secondary assembly processes), definition of, 57
SCARA horizontal articulated-arm robots, 144–147
  practical examples, 231, 277, 289, *281*, 284, 287, *288*
  working area of, *146*, *147*
Scoop segment hopper, 103–104
Screw connections, 26, *27*
Screw-in operation, 28–29
  torque during, *162*
Screw-insertion units,
  description, 161–163
  drop-tube-fed, 164–166
  feed-rail-fed, 166–168
  space for, 22–23, *24*
  swing-action drop tube for, 165
  swing-action feed head for, 165–166
  see also Automatic screw-inserting units
Secondary sorting equipment, 122–123
Semi-automatic flexible-assembly cells, 231–232
Sensors,
  feed section, 190
  flexible-assembly system, 277
  future developments necessary, 382
  industrial robots, 143–144
  optical, 123
  positioning of, 371–372
Separating devices, 118–121
  distribution of parts by, 119–122
  indexing table discharge, 119, *121*
  segment device, 119, *120*
  transverse action slider, 118–119
  two-pin slider, 119, *120*
Separation slider, *188*
Service life, 303–304, *305*, 338
Sheet-metal formed part,
  discharge of, *117*
  feeding of, *19*, *112*
Short-duration stoppages, 324
Signal conversion, 133

Signal detection, 132
Signal processing, 132
Silver solders, 181
Simulation technique, 336–337
Single part, handling and assembly of, PAP–SAP analysis for, 77–79
Single-point assembly, 40–43
  arrangement examples for, 40–43
  PAP–SAP analysis of, 63–64, 66–70, *74*
  parts provision by dispensers, 40–41
    PAP–SAP analysis of, 66–68
  parts provision by manual dispensers and by vibratory spiral conveyors, PAP–SAP analysis of, 69–70
  parts provision by paternoster, 41–42
    PAP–SAP analysis of, 68–69
Single-station assembly machines, 184–186
Skeleton belt arrangement, 45, *46*, 47
Soldering units, 179–181
Space costs, 330
Spare parts stocking, 378–379
Spherical parts,
  feeding of, *104*
  scoop design for, *104*
Spot-welding, 172, 179
Spray nozzle-spray head, assembly of, 255–260
Stability, effects of, 24, 26
Stamped part, feedability of, 102
Stored-program controllers (SPC), 195, 239–247
  contact plan for, *246*
  data exchange by, 247
  design of, 239–240
  flow diagram for, 240, *244*
  maintenance of, 246–247
  modules of, 239–240, 241
  operating system of, 242
  programming of, 243–246
  programs used, 240, *244*
Strip feed, 12
Supply condition, assembly affected by, 11–13
Surface buffer, 95
Swing-action drop tube (for screw-inserter), 165
Swing-action feed head (for screw-inserter), 165–166
Switches,
  assembly of, 311–313
    initial basis for, 81–82
    layout of system for, 320
  assembly sequence for, 307, *308*, *309*
  components of, *82*
  new line assembly proposal for, 83
  cycle of, 85–87
  efficiency of, 88
  layout of, *84*
    PAP–SAP analysis of, 87–88
  relay switches, assembly of, 294–297
Switch block, assembly of, 275–279
Switch element,
  assembly of, 297–299
  initial basis for assembly, 88–89
  new line assembly proposal, 89–91

Switch element (*continued*)
  new line assembly proposal (*continued*)
    efficiency of, 91, *92*
    layout of, *90*
    PAP–SAP analysis of, 91–92
Swivel robots, 145
  see also SCARA . . . robots
Symbols,
  handling, *100*
  process, *100*
System structure, automated assembly systems, 324, 348–349

'Teach-in' programming (of robots), 141
Television cameras, position identification by, 125–126
Terminal block, assembly of, 260–264, *265*
Thermoswitch,
  assembly sequence for, 308, *310*, 341–345
  automated assembly system for, 349–364
    availability of, 348–349
    cycle time of, 345
    evaluation of, 364–365
    investment aspects, 364, 365–369
    layout planning for, 345–347
    machine I, 350–352
    machine II, 352–354
    machine III, 354–355
    machine IV, 355–357
    machine V, 357
    machine VI, 357–359
    machine VII, 359–361
    machine VIII, 361–362
    machine IX, 362–364
    personnel requirements for, 347–348
    workpiece carrier for, 313–315, 345
  components of, 339–341
  function analysis for, 345
  product analysis of, 338–341
  production requirements for, 338
Tiered buffers, 213–214, 215, 237–238
Time-displacement diagrams, *186*, *194*, *206*
TMU (Time Measurement Unit), definition of, 62
Tolerances, 31
Tong grippers, 133, *134*
Torque, screwing-in, 162
Touch-release flap, 37, *38*
Tower-type rotary magazine, 41
Transfer belts, 45, 46
Transfer equipment, 150–161
  cycled, 150–159
  non-cycled, 159–161
Transverse action slider, 118–119
Turned parts, feedability of, 102
Two-hand technique, 42
  PAP–SAP analysis of, 67
  switch assembly, 87

Uncycled assembly lines, 218

Utilization, 219–220

Valve plates, assembly of, 250–255
VDI,
  definition of handling, 97
  symbols for handling processes, *100*
  symbols for production processes, *100*, *101*
Vibration-plate disentangling unit, 129
Vibratory spiral conveyors,
  alignment elements in, *110*
  arrangement of parts in, 109–110
  baffle plates, 108
  blockage prevention elements in, *110*, *111*
  capacity of, 123–124
  combined with auxiliary parts hopper, 124, *125*
  design of, *108*
  discharge rails for, 113–117
  distribution of parts from, 121–122
  faults due to, 189
  functions of, 98
  height deflectors in, *110*, *111*
  manual assembly, 69–70, *71*, 76, *78*, 83
  movement in, 109
  optimum fill height for, 124
  oscillation frequency of, 109
  output rate of, 124
  removal of parts from, 119–121
  secondary sorting equipment used with, 122–123
  separation of parts from, 117–119
  standard arrangement elements for, 109, *110*, *111*, *112*
  symbolic representation of, *101*
  tilt edges in, *112*
  types of, 110–113
Visual (and reach) area, 37, *38*

Welding, 28
  advantage of, 172
  laser, 174–175, 177–179
  resistance, 173–174
  spot welding, 172, 179
  units, 171–179
Welding current control types, *175*
Work order sequence, 309
Work point, manual assembly,
  arrangement of, 37–39
  features of, 39
  line assembly, *49*, *50*
  rotary table used, 42–43, *50*
  single-point assembly, *40*, *41*, *42*
  standard dimensions, 37
Work-point cost calculations, 331–333, 365, *366–367*, 369
  automatic screw-insertion, 32–33
Working space, 22–24
  robots, 229

Workpiece,
  behaviour, *98, 99*
  carriers,
    assembly transport, 212
    coding of, *47*, 94, 258, 273, 311, 315, 349
    combining of assembly machines by, 212–213
    contamination of, 372
    counter buffers for, 50, *52*
    design of, 310–315, 345
    location points on, 54
    manual assembly, 46–47, 53–54
    multi-tier buffer for, *214*

Workpiece (*continued*)
  carriers (*continued*)
    plate-form, 83, *84*, 85, 89–90
    pre-assembly operations carried out on, 55
    standard sizes of, 161
    surface buffers for, *95*
    switch assembly, 83, *84*, 85
    transverse movement of, 161, *162*
  characteristics, *98*
  parameters, 98–99
Works safety, 380

Zig-zag magazine, 12, *13*